Principles of Food
Chemistry

Principles of Food Chemistry

Contributors

Suzymeire Baroni, Izabel Aparecida Soares et al.

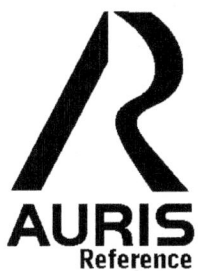

www.aurisreference.com

Principles of Food Chemistry

Contributors: Suzymeire Baroni, Izabel Aparecida Soares et al.

Published by Auris Reference Limited

www.aurisreference.com

United Kingdom

Principles of Food Chemistry

ISBN: 978-1-78154-871-4

British Library Cataloguing in Publication Data
A CIP record for this book is available from the British Library

Printed in the United Kingdom

Exclusively distributed by CBS Publishers & Distributors Pvt. Ltd.

Sales & Distribution Rights only for India, Pakistan, Bangladesh, Sri Lanka, Nepal and Bhutan.This book is not to be sold outside these territories.

Contents

List of Abbreviations

AXOS	Arabinoxylo-oligosaccharides
COS	chitooligosaccharides,
CBE	Cocoa butter equivalents
CPA	Cyclopiazonic acid
DP	Degree of polymerization,
DON	Deoxynivalenol
DNA	Deoxyribonucleic acid
EU	European Union
EVOO	Extra virgin olive oil
FA	Fatty acid
FID	Flame ionization detector
FDA	Food and Drug Administration
FOS	Fructooligosaccharides
GLC	Gas liquid chromatography
GDL	Glucono-δ-lactone,
GDP	Gross domestic product
HBV	Hepatitis B virus,
HCC	Hepato-cellular carcinoma
IMO	Isomalto-oligosaccharides
LDPE	Low-density polyethylene
NDOs	Non-digestible oligosaccharides
NMR	Nuclear magnetic resonance
PLS	Partial least square
POS	Pectin-derived oligosaccharides
PET	Polyethylene terephthalate
PUFAs	Polyunsaturated fatty acids
PVC	Polyvinyl chloride
PCA	Principle component analysis
RP	Reversed-phase

List of Contributors

Suzymeire Baroni
Veterinary Medicine -Federal University of Paraná-Palotina, Brazil

Izabel Aparecida Soares
Federal University South Border- Realeza- Paraná, Brazil

Rodrigo Patera Barcelos
Veterinary Medicine -Federal University of Paraná-Palotina, Brazil

Alexandre Carvalho de Moura
Federal University South Border- Realeza- Paraná, Brazil

Fabiana Gisele da Silva Pinto
State University of Western Parana, Centre for Science and Health, Brazil

Carmem Lucia de Mello Sartori Cardoso da Rocha
University State of Maringá, Departament of Cell Biology and Genetics, Brazil

Ana Rodríguez-Bernaldo de Quirós
Department of Analytical Chemistry, Nutrition and Food Science, Faculty of Pharmacy, University of Santiago de Compostela, A Coruña E-15782, Spain

Noelia Viqueira Varela
Department of Analytical Chemistry, Nutrition and Food Science, Faculty of Pharmacy, University of Santiago de Compostela, A Coruña E-15782, Spain

Raquel Sendón
Department of Analytical Chemistry, Nutrition and Food Science, Faculty of Pharmacy, University of Santiago de Compostela, A Coruña E-15782, Spain

Makun Hussaini Anthony
Department of Biochemistry, Federal University of Technology, Minna, Nigeria

Dutton Michael Francis
Food, Environment and Health Research Group, University of Johannesburg, South Africa

Njobeh Patrick Berka
Food, Environment and Health Research Group, University of Johannesburg, South Africa

Gbodi Timothy Ayinla
Ibrahim Badamasi Babangida University, Lapai, Nigeria

Ogbadu Godwin Haruna
Sheda Science and Technology Complex,Federal Ministry of Science and Technology, Abuja, Nigeria

Makoto Kanauchi
Department of Food Management, Miyagi University, Japan

Sakiko Hatanaka
Industrial Technology Institute, Miyagi Prefectural Government, Japan

Makoto Shimoyamada
School of Food and Nutritional Sciences, University of Shizuoka, Japan

Seyed Fazel Nabavi
Applied Biotechnology Research Center, Baqiyatallah University of Medical Sciences, P.O. Box 19395-5487, Tehran 14359-16471, Iran

Arianna Di Lorenzo
Department of Drug Sciences, Medicinal Chemistry and Pharmaceutical Technology Section, University of Pavia, Pavia 27100, Italy

Morteza Izadi
Health Research Center, Baqiyatallah University of Medical Sciences, Tehran 14359-16471, Iran

Eduardo Sobarzo-Sánchez
Laboratorio de Química Farmacéutica, Facultad de Farmacia, Universidad de Santiago de Compostela, Santiago de Compostela 15782, Spain

Maria Daglia
Department of Drug Sciences, Medicinal Chemistry and Pharmaceutical Technology Section, University of Pavia, Pavia 27100, Italy

Seyed Mohammad Nabavi
Applied Biotechnology Research Center, Baqiyatallah University of Medical Sciences, P.O. Box 19395-5487, Tehran 14359-16471, Iran

Tathiana Souza Martins Meyer
School of Pharmacy, Federal University of Rio de Janeiro, Rio de Janeiro, Brazil

Ângelo Samir Melim Miguel
School of Pharmacy, Federal University of Rio de Janeiro, Rio de Janeiro, Brazil

Daniel Ernesto Rodríguez Fernández
School of Pharmacy, Federal University of Rio de Janeiro, Rio de Janeiro, Brazil

Gisela Maria Dellamora Ortiz
School of Pharmacy, Federal University of Rio de Janeiro, Rio de Janeiro, Brazil

J. M. N. Marikkar
International Institute for Halal Research and Training, International Islamic University Malaysia, Malaysia

M.E.S Mirghani
International Institute for Halal Research and Training, International Islamic University Malaysia, Malaysia

I. Jaswir
International Institute for Halal Research and Training, International Islamic University Malaysia, Malaysia

Nicola Caporaso
University of Naples Federico II, Department of Agriculture, Via Università 100, Portici 80055 (NA), Italy
The University of Nottingham, Division of Food Sciences, Sutton Bonington, Loughborough, LE12 5RD, United Kingdom

Samiirah Chummun
Department of Agriculture and Food Science, Faculty of Agriculture, University of Mauritius, Mauritius

Hudaa Neetoo
Department of Agriculture and Food Science, Faculty of Agriculture, University of Mauritius, Mauritius

Preface

Food chemistry is the study of chemical processes and interactions of all biological and non-biological components of foods. The biological substances include such items as meat, poultry, lettuce, beer, and milk as examples. It is similar to biochemistry in its main components such as carbohydrates, lipids, and protein, but it also includes areas such as water, vitamins, minerals, enzymes, food additives, flavors, and colors. The text *Principles of Food Chemistry* contains a complete overview of the chemical and physical properties of the major and minor food components and their changes during processing, handling and storage. First chapter discusses about microbiological contamination of homemade food. In second chapter, the migration of three chemicals, benzophenone, 1,4-diphenylbutadiene and Uvitex OB from low-density polyethylene samples into the food simulant, 50% ethanol (v/v), is studied. Different levels of aflatoxin contamination in foods and feeds have been described in third chapter. Fourth chapter focuses on new cheese-like food production from soy milk. The aim of fifth chapter is to analyze the available scientific data, published over the last five years, regarding the antibacterial effects of cinnamon and its active constituents such as cinnamaldehyde and eugenol. Sixth chapter focuses on the biotechnological production, health benefits and applications of non-natural oligosaccharides in the food industry. A review on application of chromatographic and infra-red spectroscopic techniques for detection of adulteration in food lipids has been presented in seventh chapter. Last chapter describes the factors affecting the virgin olive oils composition, with a focus on the aroma compounds, biophenols and the resulting sensory profile.

Chapter 1

MICROBIOLOGICAL CONTAMINATION OF HOMEMADE FOOD

Suzymeire Baroni[1], Izabel Aparecida Soares[2], Rodrigo Patera Barcelos[1], Alexandre Carvalho de Moura[2], Fabiana Gisele da Silva Pinto[3] and Carmem Lucia de Mello Sartori Cardoso da Rocha[4]

[1] Veterinary Medicine -Federal University of Paraná-Palotina, Brazil

[2] Federal University South Border- Realeza- Paraná, Brazil

[3] State University of Western Parana, Centre for Science and Health, Brazil

[4] University State of Maringá, Departament of Cell Biology and Genetics, Brazil

INTRODUCTION

The consumption of healthy food is a consumer's right and the duty of the manufacturing industry. Health authorities are duty bound to prepare and enforce laws to protect the population's health. The supply of food free from health risks to the population is actually a challenge. In fact, contaminated food may cause serious infections and jeopardize the health of the population.

Owing to their frequency, food-caused infections are a very grave issue to public health. They may cause hazards ranging from a simple intestine discomfort to cases that are more serious, such as neurological disorders and death, because of the high number of microorganisms involved in a simple epidemic event.

Fresh or processed animal-derived food may harbor several pathogenic microorganisms that cause physiological disorders in people who consume them. When food eventually contaminated by disease-causing microorganisms is consumed, pathogens or their metabolites invade the host's fluids or tissues and trigger serious types of diseases, such as tuberculosis. They are conveyed by non pasteurized milk or by cheese contaminated by bacterial populations of *Mycobacterium bovis* and *M. tubercolosis* or by*Brucella abortus*, gram negative bacteria, intracellular pathogen that cause undulant fever and arthritis in human beings.

Bacteria, fungi, protozoa and viruses are the main microorganism groups that cause food disorders. Due to their diversity and pathogenesis, bacteria are by far the most important microbial group commonly associated with food-transmitted diseases. High rated agents in food infections are *Salmonella* sp.,*Campylobacter* sp and *Listeria monocytogenes* due to their importance in eventual sequelae. The microbiological health risks in fowl consumption and its raw products include contamination by the above food pathogens.

Besides being one of the principal causes of food-derived diseases since its attack generally involves a great number of people, the genus *Salmonella* is associated with economic liabilities, commercial damage and decrease in production due to its frequency and extension. These facts occur because of the great number of food products that may be contaminated by this bacterium, namely, food with high humidity, protein and carbohydrate rates, such as beef, pork, chicken, eggs, milk and their derived products, highly liable to deteriorate. The contamination process by pathogenic bacteria in humans may be caused by poor hygiene conditions during processing involving sick people and animals or involving feces from infected agents. Bacteria-contaminated food may also be hazardous to public health due to the excessive growth in bacteria populations at food surface or within the food. These bacteria may come from the environment and cause toxins that develop into serious health problems on intake.

Hand-manipulated meat, sausages, salamis and cheese are among the most consumed products worldwide. They are also liable to high microbiological contamination due to their manufacturing process.

The World Health Organization and the Food and Agriculture Organization of the United Nations have published reports and studies developed in several regions of the planet highlighting the pathogen risks to populations and suggested the protection of food consumers through special industrial, operational, commercial and residence care. The need for great attention in food safety is a self-evident topic. In fact, improvements in food processing methods and conscience-awareness with regard to food safety by all involved in the food production chain will surely reduce the incidence of food-originated diseases.

MICROBIOLOGICAL CONTAMINANTS OF MILK AND HOMEMADE FRESH CHEESE

Milk is one of the most complete food featuring high levels of protein and mineral salts. However, due to the availability of nutrients and almost neutral pH, milk is highly perishable. It is highly liable to microbial growth and requires thermal treatment for its conservation. Pasteurization prolongs milk conservation time, conserves its natural characteristics and preserves it safe for human consumption. High temperatures are involved so that the product's pathogenic microbiota are eliminated with no changes in its physical and chemical constitution. However, people in rural regions still drink milk in natura and use it thus as prime matter for the manufacture of derived products.

The hygienic obtaining of milk is the first critical factor within the manufacturing process of cheese and other products. In fact, the animal, equipments and environment at milking may be an important contamination source by microorganisms Faults during milking and processing coupled to inadequate conservation temperatures at the selling outlets are factors that contribute towards the commercialization of milk products with microbiological characteristics that go against health norms and legislation. The quality of milk and that of its products is a highly relevant factor for positive industrialization success since both the dairy and the consumer are interested in the outcome. In some case, however, a significant increase in the price of milk ensues. Milk is a product that should come from healthy herds, with good meals and managements, and from farms with proper technical installations that guarantee conservation during transport up to the dairy factory.

Since the number of milk contaminants increases at a slow rate from the moment of their introduction, the importance of adequate conservation of recently obtained milk should be underpinned as a basic practice for the maintenance of its quality. Milk should be submitted at low temperatures immediately after the milking process, with the consequent avoidance of the proliferation of unwanted microorganisms.

As a milk-derived product, cheese is frequently a food-originating pathogen vector. This is especially true for handmade fresh cheese manufactured from raw milk, lacking any maturation process. The product's microbial contamination is relevant for the industry because of financial liabilities, and for public health because of the risks in food-transmitted diseases.

Several studies have shown that a product's quality and durability largely depend on the prime matter used in manufacturing. It is practically impossible to improve the qualities of a derived product, such as cheese, with a high number of microorganisms present in raw milk.

FRESH MINAS CHEESE

Fresh Minas cheese (traditionally manufactured in the state of Minas Gerais, Brazil, whence its name) is defined by the Brazilian Ministry of Health (Decree 146) as fresh cheese obtained by enzyme coagulation of milk with curds and other appropriate coagulant enzymes, supplemented or not by the activity of specific lactic bacteria. According to the Technical Rules for the Identification and Quality of Milk Products [7], fresh Minas cheese may be classified as cheese with low moisture or semi-hard cheese with moisture ranging between 36 and 45.9%; cheese with high moisture or moderate mass cheese with 46 to 54.9% moisture; and very high moisture cheese or soft mass cheese, with not less than 55% moisture.

The processing of fresh Minas cheese comprises the following stages: milk pasteurization, coagulation, cutting, draining, milling, salting, packing and cooling. Since the manufacturing of this type of cheese is highly simple, many small, medium-sized and large dairies are interested in its fabrication. In fact, it is the most common type of cheese found in fairs, bars and grocers. The cheese is normally placed in a common non-vacuum plastic bag and closed by a metal seal [9].

According to the Brazilian Association of Cheese Industry (ABIQ), Brazil produces 400,000 tons of cheese per year, of which 240,000 tons are produced under federal, state and municipal inspection. Most production (95%) is consumed by common people.

The intake of fresh cheese may be risky for the consumer's health. However, Decree 861/1984 basically prohibits the sale of fresh cheese manufactured from the raw milk of cows, goats or sheep, pure or mixed. Milk should undergo pasteurization or other equivalent thermal treatment. Current legislation was published after several registers of human brucellosis caused by fresh cheese. In defiance of the law, the homemade manufacture of cheese in certain regions of Brazil is not done with pasteurized milk. Consequently, the consumption of homemade cheese brings to the fore old dangers such as brucellosis (Maltese fever) and other infectious diseases.

In spite of the legal prohibition against the commercialization of fresh and tender cheese manufactured from raw milk, the sale of homemade fresh Minas cheese occurs openly and everywhere in Brazil [. This is partially due to a greater yield, simpler processing and lack of product's maturation in the fabrication of this type of cheese, with low costs for the consumer and a fast return of expenditure to the manufacturer

Food protection authorities classify microbial biological contamination as a main danger to public health. Who has constantly raised its voice on the need to restrict food contamination by health-impairing biological agents. Although microbial quality of food is of paramount importance, registration at the Federal Inspection Service does not guarantee lack of pathogens in food [13].

Food-derived diseases may be caused by several microorganism groups that include bacteria, fungi, yeasts, protozoa and viruses. Due to their diversity and pathogenesis, bacteria are by far the most important microbial group and commonly associated with food-transmitting diseases

Bacteria are microorganisms largely spread throughout the natural world and may be found in every type of environment [14]. They cause diseases in humans, animals and plants and deteriorate food and other materials. On the other hand, they may be useful too when they compose the human being's normal microbiota and are used in the production of food as symbiotic in agriculture and medicine.

In spite of certain unreliable Brazilian statistics, it is believed that food-derived diseases in Brazil are high [15]. In fact, several studies estimate that 12% of hospitalization cases in Brazil occur because of infectious intestinal diseases [16].

Occurrences of food-derived diseases are normally associated with certain risk factors, or rather, procedures that benefit toxin infections. The following may be highlighted: faults in food refrigeration; conservation of warm food at room temperature; food prepared many hours earlier for later consumption with inadequate conditioning during the interval; faults in the cooking process; handling of food by people with inadequate personal hygiene practices, or with lesions or with contaminating diseases; usage of contaminated prime matter; faults in the hygiene of utensils and other equipments in food preparation; favorable environmental conditions for the growth of etiological agents; food obtained from unreliable sources; inadequate storage; use of utensils which release toxic residues; intentional or accidental addition of toxic chemicals to the food; usage of water with uncontrolled drinkability features; water contamination from damages in the supply system [17].

Problems in the manufacture of cheese in Brazil are related to precarious conditions of milk, bad conditions during the manhandling of cheese and the lack or deficiency of refrigeration throughout the production chain. These factors worsen the situation and establish contamination conditions which favor the development of microorganisms at several places [18].

Whereas some microorganisms contribute beneficently towards the processing, safety and quality of certain food products, other organisms are involved in processes with unwanted effects in food and for the consumers' health. There are two categories of food-transmitted microbial diseases: food intoxication and infection by food. In food intoxication, the person ingests toxins that are pre-formed by microorganisms in the food. The toxin causes damage to the organism. Examples comprise botulinum toxin that binds itself to the nerve terminals at the muscle level and impedes the release of acetylcholine neurotransmitter, and staphylococcus toxin that acts on the brain's vomiting-center [19]. Infection by food occurs when the pathogen, such as by *Salmonella typhy* and other serotypes, is ingested and multiplies itself, causing diseases in the intestine tract and often in other organs [20].

The sale of animal-derived food in fair stalls without any refrigeration and without any protection against dust and insects may alter their quality. In the case of cheese, it is sold in portions or slices and thus the external incorporation of biological or non-biological foreign matter is dangerous due to faults in the handling of the product during commercialization, poor hygiene of the stalls and utensils used, and crossed contamination between exposed products [21].

Food microbial contamination is unwanted and dangerous within food microbiology. This aspect should be faced with great strictness. The acknowledgement of possible hygiene deficiency implying in food contamination brings to the fore microorganism groups, comprising indicators, and pathogenic microorganisms that find an excellent environment in food for their development and even for the release of toxic substances [22]. Total and thermotolerant coliforms, such as *Staphylococcus aureus,* fungi, yeasts and even *Salmonella* spp., should be highlighted among the microorganisms whose presence and numbers indicate the quality of the product.

The above mentioned microorganisms, causes of several types of pathogenesis, are transmitted to humans because of lack of hygiene, bad habits of handlers, inefficient production processes, maintenance or re-heating of food at inadequate temperatures and also by non-adequate conditions in industries where the food is produced [23].

Most microorganisms, whose pathogenicity in humans depends on their variegated presence in food, are relatively sensitive to high temperatures. In fact, they are destroyed by the adequate cooking of eventually contaminated food or by pasteurization processes.

The Brazilian Agency for Health Vigilance (ANVISA) established, by Decree RDC 12 of the 2nd January 2001[24], the microbiological Standards for several types of food, described in Table 1.

So that food-caused disease cases and events could be characterized, the populations should be informed on the symptoms of each, such as mild diarrheas and vomiting since these are considered as a "passing illness" and not necessarily associated with food consumption [25].

Table 1: Microbiological Standards for Food: cheese with high moisture (55%)

Microorganism	Quantity
Coliforms at 45°C	5x10² MPN/g
Staphylococcus aureus	5x10² CFU/g
Salmonella sp.	Absence in 25g

[i] - * MPN (most probable number), CFU (colony forming unit). Source: ANVISA/2001[24]

According with registers, more than a billion cases of acute diarrhea are detected in less-than-5-year-old children in developing countries yearly, with 5 million deaths. Between 1999 and 2001, in the state of Paraná, Brazil, 67.1% of food epidemics were caused by bacteria. Moreover, out of 1389 notified epidemics, 38.6 were confirmed in the laboratory; 29/7% were confirmed clinically or epidemiologically suspect and 31.6% were of unknown etiology [25].

World cheese production is slightly above 19 million tons. Cheese production increased more than 76.3% during the last thirty years, or rather, from approximately 10.8 million tons in 1978 to more than 19 millions in 2008. The expansion of milk-producing regions and production increase throughout recent years provided a highly relevant presence of Brazilian production within the world market of milk-derived exports. Concern is therefore high with regard to the quality of commercialized goods for internal and external consumption.

Family-run agriculture in Brazil has an important share in the milk production chain, with approximately 86% of milk producers. However, the production and management of these milk producers are foregrounded on a homemade basis with scanty technical assistance and high influence of cultural factors that may put to risk consumers' health. Technical and educational orientation through the introduction of healthy manufacturing practices are deemed necessary to minimize contamination risks and food intoxication by the product.

Research in all Brazilian regions, where the production and commercialization of cheese is undertaken mainly by small producers, has demonstrated the risk of toxin infections in the consumption of these products by the population.

The curd-cheese is the most produced and consumed milk-derived product in the northeastern region of Brazil. Several investigations [26] have shown that the handling and carelessness in hygiene within the production system have made it foremost as a contamination source. The manufacturers are transmission vectors of the pathogen *Staphylococcus aureus* and others that may cause food intoxication. The presence of positive coagulase staphylococcus witnesses the lack of hygiene and sanitary conditions during the production, processing, distribution, storing and commercialization stages of samples of curd-cheese. Sanitary education of the producers and the spreading of processing techniques based on good manufacturing practices are mandatory.

Researches in the state of Mato Grosso, in the Mid-Western region of Brazil, (Loguercio & Aleixo 2001) [27] have shown the poor hygiene and sanitary conditions that characterize the production of fresh Minas cheese. *Staphylococcus aureus* bacteria rates higher than those permitted by current legislation are rife. The need for more sanitary surveillance and orientation by government authorities is urgent.

Research work in the southeastern region of Brazil [28] (Salotti et al 2006) evaluated the microbiological quality of fresh Minas cheese samples. Results from the hinterlands of the state of São Paulo, Brazil, showed non-compliance to rules established by the Brazilian Agency for Sanitary Vigilance (ANVISA) for 83.4% of homemade products and 66.7% for industrial samples with regard to thermotolerant coliforms. In the case of positive coagulase *Staphylococcus*, 20% of homemade samples and 10% of industrial products failed to comply with the ANVISA regulations. Microbiological results revealed the potential risk of the product for consumers.

After analyzing samples of fresh Minas cheese in Minas Gerais for coliforms and *E. coli*, a recent study [29] showed the presence of microorganisms, above the rates allowed by current legislation, in 30% of cheese with certificate; 70% of cheese without certificate and 61.4% of mild cheese. Since *E. coli, Proteus, Providencia, Serratia, Klebsiella* and *Enterobacter* were identified within the Enterobacteriaceae isolated in fresh Minas cheese, the risk to public health when the products are consumed is amply demonstrated.

Was reported [30] on the risk in the consumption of fresh Minas cheese by the population of the state of Paraná, southern Brazil. Samples inspected by the Federal Inspection Service of Santa Helena PR Brazil revealed that only 15% were in accord to ANVISA standards. All homemade cheese samples and 70% of inspected ones were not according to legislation. Studies [31] confirmed the above results and reported that 50% of samples of analyzed cheese had thermotolerant coliforms, 100% had positive coagulase *Staphyloccocus* and 12.5% had *Salmonella* sp. These samples were inadequate for human consumption since they were not consonant to cheese microbiological standards.

MICROBIOLOGICAL CONTAMINANTS OF JERKED BEEF

One of the most traditional products of the northeastern region of Brazil is jerked beef which may be characterized as a nutrition food with high calorie rates and widely accepted by consumers for its peculiar sensorial features. Jerked beef is produced from cuts derived from all parts of cattle carcass, salted and dried, with longer durability when compared to that of fresh meat [32].

Due to different nomenclature in Brazil, such as 'carne-de-sertão', 'carne serenada', 'carne de- viagem', 'carne-mole', 'carne-do-vento', 'cacina' or more simple still, dehydrated meat, jerk beef is often confused with another type of salted beef, albeit industrialized, called 'charque' or dried salted meat [33].

Jerked beef was first used in the northeastern region of Brazil as an alternative to preserve beef surplus which could not be consumed immediately and so that the meat would not deteriorate quickly due to difficulties in its preservation especially among the poor population with no refrigeration equipments. Favorable climate conditions and availability of seawater salt, fresh meat could be preserved by being dehydrated and salted.

Currently the above-mentioned preservation process is less relevant due to the introduction of refrigeration. However, many people from different regions of Brazil, especially from the northeast, became accustomed to the produce's characteristic taste and continued to produce jerked beef will less amounts of salt and frequently without exposure to the sun.

Each Brazilian state developed its own technology and thus produced jerked beef with different characteristics with regard to aspect, taste, color, amount of salt and shelf life. The states of Rio Grande do Norte and Ceará are the greatest producers of jerked beef mainly due to climatic conditions that favor the food's dehydration. In fact, jerked beef passed from a locally consumed product and used in certain food receipts to wider conditions. In fact, it is appreciated throughout Brazil and in several meal preparations. Jerked beef may be found in big city centers such as São Paulo and Rio de Janeiro, in homes and restaurants, outside the restricted circle of northeastern cuisine [34], and in the menu of the poorest worker [35,36].

Owing to the popularization of homemade salting technique, jerked beef production follows typically regional norms. Consequently, it is produced in a highly rudimentary way under inadequate sanitary conditions [37,38]. Analysis of the hygiene conditions in the production and commercialization of jerked beef in the region of Itapetinga BA Brazil may be brought forward as an example of the popularization of the technique. In fact, 73.3% of the shopkeepers interviewed admitted that they themselves produced the jerked beef on sale in their shops. Whereas 63.6% used non-inspected meat, 27.3% used meat inspected by municipal health officers and only 0.1% was inspected by federal health officers. Jerked beef was stored and commercialized in 71% of the shops at room temperature, which favored the multiplication of contaminant microorganisms and flies. These facts bring health risks to consumers and jeopardize the product's physical aspects [39].

Salting technique consists in the removal of water from the meat tissues; decrease in water activity ensues, inhibits microbial development and the speed of unwanted reactions of the final product. When salted beef is conserved without any type of refrigeration, its shelf life is higher than that of fresh meat [40]. However, jerked beef has low sodium chloride (NaCl) rates, between 5 and 6%, high moisture, between 65 and 70% [35,41,42] and water activity of 0.92. It may be characterized as partially dehydrated meat in which water activity is not sufficient decreased to avoid microbial development (and consequently degradation) or the production of microbial toxins [43,44].

Although the literal translation of the jerked beef in Portuguese is 'meat exposed to the sun', it is actually only rarely exposed to the sunrays during the dehydration process. The end product is a semi-dehydrated homemade product with four-day shelf-life at room temperature and up to eight days under refrigeration [43,45,41].

Data on the physical and chemical qualities of jerked beef sold in butcheries and supermarkets in João Pessoa PB Brazil showed that water activity in all samples was relatively high, between 0.898 and 0.967, and that the rates of sodium chloride (NaCl) ranged between 3.73% and 9.79%. Consequently, NaCl employed in the process was insufficient to decrease water activity in the product and thus it did not have a significant inhibitory action in the development of most microorganisms in the beef [46]. Lack of standardization in the quality of jerked beef was also assessed in samples collected at inspected shops. Mean rates of water activity were 0.94±0.02. The same average was obtained for samples collected in shops without any health inspection [47]. Variations in sodium chloride rates were also registered in the samples. Techniques for more efficient conservation are required to decrease such risks since it is a type of food with contamination possibilities throughout the manufacturing process.

With regard to the microbiological contamination of jerked beef, the transformation by which meat in natura is processed into jerked beef requires that technological alterations modify the initial microbiota by which the salting and dehydration process selects more tolerant microorganisms for such conditions [48]. Pathogens that may contaminate jerked beef comprise *Clostridium perfringens*, *Staphilococus aureus*, *Salmonella*, verotoxin-producing *Escherichia coli*, *Campylobacter*, *Yersinia enterocolítica,Listeria monocytogenes, Aeromonas hydrofila*, and other deteriorating bacteria [49]. However, low NaCl rates used in jerked beef is one of the factors that trigger microbiological development since decrease in water activity is insufficient to hinder the development of deterioration-producing bacteria of the genus *Pseudomona*. It also provides proper conditions for the growth of gram-positive bacteria as those of the genus *Staphylococcus* [38].

Samples of jerked beef from the north of the state of Minas Gerais, Brazil, showed that the amount of mesophile aerobic bacteria, an index of food hygiene quality, was between $2.0x104$ UFC/g and $8.9x108$ UFC/g. Psichrotrophic bacteria were found in 93.33% of samples, between $5.4x103$ UFC/g and $2.9x106$ UFC/g. Results show poor hygiene in the manufacture of jerked beef [50]. Similar results were reported in samples of jerked beef commercialized in João Pessoa where the number of mesophile bacteria ranged between $1.8x105$ and $7.5x107$ UFC/g, with a clear correlationship between mesophile contamination and hygiene and sanitary standards [42].

High thermotolerant coliform rates, which also demonstrate unsatisfactory hygiene and sanitary conditions during the processing stages in the manufacture of jerked beef, were also registered in most jerked beef samples sold in butcheries and supermarkets in João Pessoa PB Brazil [46]. However, total coliforms in food did not report recent fecal contamination or the occurrence of enteropathogens [51,52]. However, Brazilian sanitary laws did not regulate the presence of this microorganism group in meat.

The commercialization of jerked beef in health inspected or not in the region of João Pessoa PB Brazil has been evaluated and results showed high rates in both groups. Ninety-six samples were analyzed and high contamination by feces-derived microorganisms was reported. *Staphylococcus ssp.* rates were high in both groups, with a low frequency for *S. aureus* [47]. *Staphylococcus aureus* rates were higher than 5logUFC/cm2 in 50% of jerked beef samples commercialized in butcheries and supermarkets in João Pessoa PB Brazil. The above amounts demonstrate high contamination causing gastrointestinal disorders in consumers [53].

Mesophile microorganisms *Salmonella sp.* and *Staphylococcus aureus* in jerked beef commercialized at room temperature and under refrigeration in Campina Grande PB Brazil showed no significant difference in *S. aureus* counts for samples commercialized at room temperature and under refrigeration. *Salmonella* ssp. was detected in 40% of jerked beef samples commercialized at room temperature and in 30% of samples under refrigeration.

Another source of contamination in the commercialization of jerked beef may be found in supermarkets, open market stalls and butcheries. Data reveal that the utensils used in 75% of these outlets were not exclusively for meat cutting and that the handling of money and food was common practice in 25% of the businesses. Aprons, disposable caps and clean closed shoes were only found in 25% of the shops.

The inadequate washing of hands and other habits such as talking during the handling and commercialization of food were also reported in all commercial enterprises [54]. It has been verified that in João Pessoa, supermarkets had the best hygiene and sanitary profile in jerked beef quality, whereas open markets and stalls in fairs had the worst [42]. In the latter case, meat is exposed without any type of protection and any passerby may handle it at will.

Investigations were carried out with regard to alien matter, such as flies, acarids, larvae, insects, feathers and others, found in jerked beef sold in 20 (90.9%) shops in Diadema SP Brazil, specialized in typical products from the northeastern region of Brazil. Exposure of products without any wrappings is an excellent condition for attacks by insects, especially flies, and rodents, making it improper for human consumption in the wake of health-hazard matter [55]. Almost all jerked beef is manufactured and sold in small shops and specifically prepared for people who appreciate the product. Consequently, lack of sanitary rules for its production, precarious conditions in its commercialization, storage without refrigeration and its exposure without any protection characterize jerked beef in such conditions as haphazard to public health.

MICROBIOLOGICAL CONTAMINANTS IN MEAT FILLINGS (SAUSAGES MADE FROM BEEF AND FOWL MEAT, SALAMI)

Animal-derived food conveys a host of microorganisms dangerous to human health. The incidence of toxin infections in Brazil is high, although statistics are rather lacking on the matter. Bacteria causing toxin infections are widely distributed although their main natural habitat is the human or animal intestine tract [14]. The most common bacteria in food contamination are of the genera *Escherichia, Salmonella, Shigella, Yersinia, Vibrio, Brucella, Clostridium, Listeria, Campylobacter, Bacillus cereus*and *Staphylococcus aureus* [56]. Sausages, widely used in Europe, is a type of food stuffed with meat from swine, fowls, goats, cattle and fish, seasoned with several types of spicy ingredients. Sausages are a highly popular food in Brazil, easily accessible to all classes of people and consumed throughout the country. Sausages have great acceptance in the southern and southeastern regions due to a more Europeanized culture.

Brazilian swine breeding has a very important role in several sectors of Brazilian economy. It produces jobs and intensifies demand of agricultural products in the industrialization and commercialization of animal-derived products. Besides providing excellent animal protein to the population, the meat industry exports meat and important economical assets are aggregated [57].

Data by the Brazilian Association of Production and Exportation Industry of Pork (ABIPECS) showed that approximately 65% of the Brazilian pork production is directed towards the internal market through industrialized products. Among the processed products, the fresh Tuscan-type sausage, made exclusively from pork, uses the less important animal parts as food, with great acceptance among the population.

Pork and its derived products undergo bacterial alterations owing to several factors such as animal health and fecal contamination by *Escherichia coli* highly relevant worldwide as a microorganism hazardous to animal and population health involving hygiene and sanitary issues [57]. The same author evaluated the occurrence of *E.coli* in swine in the abattoirs of Rio de Janeiro, Brazil, from which the Tuscun-type sausages were made. Different parts of the animal used in the stuffing process were examined and concluded that, depending on the meat and the manufacturing process, sausages were not fit for consumption.

Toxoplasma gondhii in fresh pork sausages commercialized in Botucatu SP Brazil was evaluated by researches [58]. Pork represents one of the main sources of infection by *T. gondii* in humans. Swine were the most important animals in the process of toxoplasmosis transmission [59,60,58,61]. Mendonça's data did not show any evidence of *T. gondii* in the samples, perhaps due to salt, used in the manufacturing process, which eliminated the microorganism.

The occurrence of food infection by pork sausages contaminated with *Salmonella sp*. has been suggested [62]. Brazilian sanitary laws [63] make it mandatory that the microorganism should be lacking in 25% so that human intoxication may occur. However, such possibility may vary since it depends on serotype and the person's health conditions and tolerance. Mürmann's results [62] showed that 24% of pork sausages samples were contaminated by *Salmonella enterica*.

Contamination by *Salmonella sp* in pork may occur in pens through contact with feces, lack of hygiene and sanitation in the installations and by other animals during the transport, waiting or pre-finishing period. A high increase of *S. enteriditis* in food toxin infections in humans and in aviary products has been reported in Brazil since the 1990 [64].

Fecal coliforms, positive coagulase staphylococcus, *Salmonella* spp and *Campylobacter* spp in fresh sausages were evaluated [65]. When the hygiene and sanitary quality among the different types of fresh sausages was compared, pork sausages had the worst scores with regard to risks in public health, as ruled by the RDC n.12 of Anvisa [63].

The authors also registered that most samples were not in accordance to microbiological standards and thus hazardous to consumer's health. Another datum refers to the absence of *Campylobacter* spp in the samples, perhaps due to sodium chloride concentrations over 1.5% that may have inhibited these microorganisms.

Was analyzed [66] the presence of *Listeria* spp, principally *L. monocytogenes*, during the manufacture of fresh mixed-meat sausages in three abattoirs, supervised by state health authorities, in Pelotas RS Brazil. Results showed that all samples from the three abattoirs were contaminated by *Listeria* spp, of which the most frequent species was *L. innocua* (97.6), followed by *L. monocytogenes* (29.3%) and *L. welshimeri* (24.4%).

When the hygiene and sanitary conditions in the manufacture of fresh sausages in the northwestern region of the state of Paraná, Brazil, were analyzed [67] data failed to show any microbiological contamination that would jeopardize the health of the consumer. The manufacture of these samples followed strict handling and processing procedures.

On the other hand, another authors [68] studied the prevalence of antimicrobial resistance by serotypes of *Salmonella* isolated from fresh pork sausages and found significant quantities of the above in samples collected in the southern state of Santa Catarina, Brazil. These serotypes resisted the antimicrobial products sulfonamide and tetracycline (81%); ampicillin (50%) and chloramphenicol (31.25%). Was evaluated the microbiological quality of fresh sausages in two towns of the state of Minas Gerais, Brazil [69]. Results confirmed positive coagulase *Staphylococcus* in 35% of samples which made them improper for human consumption. The same author also demonstrated that 35% of samples were contaminated by thermotolerant fecal coliforms above the maximum limits.

The consumption of chicken meat and its derivates has recently increased considerably in Brazil due to price decrease, good quality and practical cuttings provided [70]. Per capita consumption increased from 10 kg to 35.4 kg, only slightly lower than beef consumption (União Brasileira de Avicultura) [71]. The products' quality is highly important and a great concern to health authorities, food industry and consumers. Chickens bred for human consumption may host several pathogenic microorganisms such as *Campylobacter jejuni, Salmonella* sp and *E. coli* [72,73].

Rall investigated [70] the hygiene and sanitary conditions of chicken meat and several types of sausages commercialized in the interior of the state of São Paulo, Brazil, by determining the Most Probable Number of coliforms at 45°C. The same authors also analyzed the presence of *Samonella sp* by the traditional method and by PCR. Data showed that 40% of the 75 sausage samples analyzed were improper for human consumption due to excess in coliforms and 7 samples (9.3%) were positive for*Salmonella* sp. (9.3%). Research by PCR increased to 56% *Salmonella*-positive samples. When the frequency rate of *Salmonella* was added to the microbiological limits for coliforms, it might be concluded that 86.7% of sausages were improper for human consumption.

In their research in the northwestern region of the state of São Paulo, Brazil, others authors [74] found contamination by *Salmonella* in 16% of chicken sausages samples. The most relevant item in the above result may be the handling of the product during processing, coupled to the exposure of the meat to several contamination sources or to already contaminated chickens that provided the contamination of the final product.

The above authors researched the microbiological quality of industrialized avian products and their derivates in another region of the state of São Paulo. Research determined the presence of*Campylobacter jejuni* and *Salmonella sp.* Sausages samples analyzed were 42.8% positive for *C. jejuni*and 28.5% for *Salmonella sp.*

The presence of microorganisms in the above research works suggests the need for greater care during the handling and preparation of sausages that may be eaten in natura, without any heating treatment that would reduce the number of microorganisms causing toxin infections [75].

Vienna sausage may be defined as an industrialized meat-stuffed product obtained from the emulsion of animal meat to which are added a variety of ingredients and condiments, filling a natural or artificial casing, and submitted to proper thermal process [76]. Vienna sausages are highly popular in Brazil due to their low costs and for the manufacturing of the ubiquitous hot dog.

The physical and chemical characteristics of Vienna sausages should contain a maximum of 65% moisture, 30% fat, 2% starch, 7% total carbohydrates, 12% protein. Fresh sausages should be under permanent refrigeration (0°C to 5°C) from manufacture until consumption, with expiry period after 48 hours [77].

Vienna sausages samples of the hot-dog type were analyzed in Niterói and Rio de Janeiro RJ Brazil to detect thermotolerant coliforms, positive coagulase *Staphylococcus*, *Clostridium* spp and *Salmonella*spp by conventional methods with the necessary modifications [78]. When compared to health norms, results showed that 33% of samples were inadequate for consumption due to the presence of their isolated microorganisms.

Salami is another highly appreciated product in southern Brazil. Its homemade manufacture started in the early 20th century with an enormous variety of industrialized types that differed in composition, casing, size of meat and fats, spices, smoking process and maturation period prior to commercialization. Researchers revaluated the various characteristics [79] of salamis produced by small- and medium-sized agro-industries in the southern state of Santa Catarina, Brazil. Bacteria *Staphylococcus aureus*,*Salmonella* spp, *Listeria monocystogenes* and *E.coli* were researched in the products. Although results did not identify contamination by *Salmonella* spp, the *E. coli* and *S. aureus* counts were significant, but within the reliability parameters.

Was analyzed the quality [80] of salami in the interior of the state of São Paulo, Brazil, and verified that, despite samples with *E. coli* and fecal coliforms, all samples were within health standards. Nevertheless, 60% of samples were contaminated by *Staphylococcus aureus* and 22% were unhealthy for consumption.

FINAL CONSIDERATIONS

Owing to their importance for public health, the correct handling of meat and milk products required greater attention, care and supervision from the competent health authorities. Since there is great cultural diversity in food manufactured in Brazil, the direct intervention of all the sectors involved within the food production chain is mandatory to warrant healthy and reliable products and thus a decrease in diseases caused by food contamination.

REFERENCES

1.   Arcuri EF, Brito MAVP, Brito JRF, Pinto SM, Angelo FF, Souza GN, Qualidade Microbiológica do leite refrigerado nas fazendas. Arquivo Brasileiro Medicina Veterinária Zootecnia, Belo Horizonte, 2006; 58(3): 440-446.

2.   Lange CC, Brito, JRF. Inluência da qualidade do leite na manufatura e vida de prateleira dos produtos lácteos: papel das altas contagens microbianas. In: Brito, J.R.F; Portugal, JA (Eds.) Diagnóstico da qualidade do leite, impacto para a indústria e a questão dos resíduos de antibióticos. Empresa Brasileira de Pesquisa Agropecuária (Embrapa), Juiz de Fora, p. 117-138, 2003.

3.   Gomes HÁ, Gallo CR. Ocorrência de *Staphylococcus aureus* e

produção de enterotoxinas por linhagens isoladas a partir de leite cru, leite pasteurizado tipo C e queijo Minas frescal comercializados em Piracicaba,SP. Revista Ciência e Tecnologia de Alimentos, 1995; 15(2): 158-161.

4.   Gonçalves CA, Vieira LC. Obtenção e higienização do leite in natura. Empresa Brasileira de Pesquisa Agropecuária (EMBRAPA). Amazônia Oriental, Belém: Documento 141. 2002. 28p.

5.   Olivieri D. de A. Avaliação da qualidade microbiológica de amostras de mercado de queijo mussarela, elaborado a partir do leite de búfala (Bubalus bubalis). 61 p. Dissertação (Mestrado) - Escola Superior de Agricultura Luiz de Queiroz, Piracicaba, SP, 2004.

6.   Huhn S, Hajdenwurcel JR, Moraes JM de, Vargas OL. Qualidade microbiológica do leite cru obtido por meio de ordenha manual e mecânica e ao chegar à plataforma.Revista do Instituto de Laticínios Cândido Tostes, 1980; 35(209): 3-8.

7.   Brasil. Ministério da Agricultura, do Abastecimento e da reforma Agrária. Portaria n. 146, de 07 de março de 1996. Diário Oficial da União, Seção I, Brasília, DF, p 3977-3886, 1996.

8.   Vieira DAS, Neto JPM. Elaboração de queijos frescais em pequena escala. Informe Agropecuário, 1982; 8(88): 28-29.

9.   Loguercio AP, Aleixo JAG. Microbiologia de queijo tipo Minas Frescal produzido artesanalmente. Ciencia Rural, 2001;31(6): 1063-1067.

10. Hoffman FL, Cruz CHG, Vinturim TM. Qualidade microbiológica de queijos comercializados na região de São José do Rio Preto, SP. Revista do Instituto de Laticínios. Cândido Tostes,1995; 50: 42-47.

11. Furtado MM, Mosquim MCA, Fernandes AR, Silva CAB da. Laticínios diversificados. In: Silva CAB da, Fernandes AR. Projetos de empreendimentos agroindustriais: produtos de origem animal. Viçosa: UFV, 2003. 308p.

12. Almeida Filho ES. Características microbiológicas do queijo Minas frescal, produzido artesanalmente e comercializado no Município de Poços de Caldas, MG. Dissertação (Mestrado) - Faculdade de Ciências Agrárias e Veterinárias, Campus de Jaboticabal da UNESP; 1999.

13. Pinto PSA, Germano MIS, Germano PML. Queijos minas: problemas emergentes de vigilância sanitária. Revista Higiene Alimentar,1996; 10(44): 22-27.

14. Cunha CP, Nascimento MGF, Jesus VLT, Nascimento ER, Corbia ACG. Queijo tipo minas frescal com e sem serviço de Inspeção Federal -

Contaminação por coliformes fecais e *Escherichia coli*. V Congresso Brasileiro de Higienistas de Alimentos. Foz de Iguaçu, PR, 1999. Revista Higiene Alimentar,1999; 13(61): 34-35.

15. Franco, GMB, Landgraf, M. Microbiologia dos alimentos. São Paulo: Atheneu; 1996.

16. Santos, TBA, Balioni, GA, Soares, MMSR, Ribeiro, MC. Condições higiênico-sanitárias de alfaces antes e após tratamento com agentes antibacteriano. Revista Higiene Alimentar, 2004; 18(121): 85- 88.

17. Nascimento, MS, Silva, N, Catanozi, MPLM. Avaliação microbiológica de frutas e hortaliças frescas, comercializadas no município de Campinas, SP. Revista Higiene Alimentar, 2003;17(114): 73-76.

18. CENEPI/FUNASA/MS. Manual integrado de prevenção e controle de doenças transmitidas por alimentos, 2001.

19. Lisita, MO. Evolução da população bacteriana na linha de produção do queijo minas frescal em uma indústria de laticínios. 61 p. Dissertação (Mestrado) - Escola Superior de Agricultura Luiz de Queiroz, Piracicaba, SP, 2005

20. Silva, JEAE. Manual de controle higiênico sanitário de alimentos. São Paulo: Varela; 1999

21. Trabulsi, LR. Microbiologia. São Paulo: Atheneu; 1999.

22. Correia, M, Roncada, M.J. Características microscópicas de queijo prato, mussarela e mineiro comercializados em feiras livres da Cidade de São Paulo. Revista de Saúde Pública, junho 1997; 3 (31).

23. Franco, RM, Almaida, LEF. Avaliação microbiológica de queijo ralado, tipo parmesão, comercializado em Niterói, RJ. Revista Higiene Alimentar, 1992; 6(21): 33-36.

24. Brasil, Agência Nacional de Vigilância Sanitária (ANVISA). Resolução RDC nº 12 de 02 de Janeiro de 2001.

25. Silva, LF. Procedimento operacional padronizado de higienização como requisito para segurança alimentar em unidade de alimentação. Dissertação (Mestrado em Ciência e Tecnologia de Alimentos) - Universidade Federal de Santa Maria, RS, 2006.

26. Tigrel, DM, Borelly, MAN. Pesquisa de Estafilococos coagulase-positiva em amostras de "queijo coalho" comercializadas por ambulantes na praia de Itapuã (Salvador-BA). Revista Ciências medica e biologia, 2011;.10(2): 162-166.

27. Loguercio, AP, Aleixo, JAG. Microbiologia de queijo tipo minas frescal produzido artesanalmente. Ciência Rural, 2001; 31(6).

28. Salotti, BM, Carvalho, ACFB, Amaral, LA, Vidal-Martins, AMC, Cortez, AL. Qualidade Microbiológica do queijo minas frescal comercializado no município de Jaboticabal, SP, Brasil. Arquivos do Instituto Biologico, 2006;73 (2): 171-175.

29. Okura, M.H. Avaliação microbiológica de queijo tipos minas frescal comercializados na região do triângulo mineiro Tese (Doutorado em Microbiologia) - Tese (doutorado) - Universidade Estadual Paulista, Faculdade de Ciências Agrárias e Veterinárias, 2010.

30. Pinto, FGS, Souza, M, Saling, S, Moura, AC. Qualidade microbiológica de queijo Minas Frescal comercializado no Município de Santa Helena, PR, Brasil. Arquivos do Instituto Biológico, 2011; 78 (2): 191-198.

31. Oliveira, DF, Tonial, CEC. Sazonalidade como fator interferente na composição físico-quimica e avaliação microbiológica de queijos coloniais. Arquivo Brasileiro de medicina veterinária e zootecnologia, 2012; 64(2): 521-523.

32. Vasconcelos O. Por cima da carne seca. Revista Globo Rural, 1986; 1 (5): 15-20.

33. Lira GM, Shimokomaki M. Parâmetros de qualidade da carne de sol e dos charques.Higiene Alimentar, São Paulo,1998; 44 (13): 66-69.

34. Serviço de Informação da Carne [SIC]. Charque, carne de sol, carne seca. Desenvolvido pelo Comitê Técnico do SIC. São Paulo. Disponível em: http://www.sic.org.br/charque.asp (Acesso 14 de agost 2012).

35. Nóbrega DM, Schneider I S. Contribuição ao estudo da carne de sol visando melhorar sua conservação. Revista Higiene Alimentar. 1983; 2(3): 150-4.

36. Gouvêa JAG, Gouvêa AAL. Tecnologia de fabricação da carne de sol. Bahia: Rede de Tecnologia da Bahia– RETEC/BA, 2007. 23 p. Dossiê Técnico.

37. Menucci TA. Avaliação das condições higiênico-sanitárias da carne de sol comercializada em "casas do norte" no município de Diadema- SP. 2009. 121 f. Dissertação (Mestrado em Saúde Pública) – Faculdade de Saúde Pública, Universidade de São Paulo, São Paulo, 2009.

38. Azevedo PRA, Morais MVT. A tecnologia da produção da carne de sol e suas simplificações nos aspectos higiênico-sanitários. Revista Nacional da Carne, São Paulo, 2005; 29(98): 12-13.

39. Ramos ALS, Ramos EM, VIANA EJ. Avaliação das condições higiênicas na produção e comercialização da carne de sol na região de Itapetinga, BA. Revista Higiene Alimentar, 2007; 21(150): 371-374.

40. Picchi, V, Cia G. Fabricação do charque. Boletim do Centro de Tecnologia de Carnes, 1980; 5: 11-30.

41. Shimokomaki M, Olivo R. Suplementação de vitamina e melhora a qualidade de carnes e derivados. In: Shimokomaki, M. et al. (Ed.). Atualidades em ciência e tecnologia de carnes. São Paulo: Varela, 2006. cap. 11, p. 115-121

42. Farias SMOC. Qualidade da carne de sol comercializada na cidade de João Pessoa-PB, Dissertação de Mestrado- UFPB/CT, 142f. 2010.

43. Lira GM, Shimokomaki M. Parâmetros de qualidade da carne de sol e dos charques. Revista Higiene Alimentar. 1998; 12(58): 33-5.

44. Felicio PE. Carne de sol – Produto artesanal, de consumo regional, tem potencial para ser fabricado e comercializado no país todo. ABCZ, 2002.

45. Costa EL, Silva JA. Avaliação Microbiológica da carne de sol elaborada com baixos teores de cloreto de sódio. Revista Higiene Alimentar 2001; 21(2): 149-53.

46. Costa EL, Silva JA. Qualidade sanitária da carne de sol comercializada em açougues e supermercados de João Pessoa – PB. Bol. CEPPA. Curitiba. 1999; 17(2): 137-44

47. Costa EL, Silva JA. Avaliação Microbiológica da carne de sol elaborada com baixos teores de cloreto de sódio. Revista Higiene Alimentar. 2001; 21(2): 149-53.

48. Maca JV, Miller RK, Acuff, GR. Microbiological, sensory and chemical characteristics of vacuum-packaged ground beef patties treated withs salts of organic acids. Journal of Food Science, Chicago,1997; 62(3): 591-596.

49. Gill CO. Microbiological contamination of meat during slaughter and butchering of cattle, sheep and pigs. In: The microbiology of meat and poultry. Londres: Blackie Academic & Professional, 1998. cap. 4, p. 119-157

50. Cruz ALM. Produção, comercialização, consumo, qualidade microbiológica e características físico-químicas da carne de sol do Norte de Minas Gerais, Dissertação de Mestrado,Montes Claros, MG: ICA/ UFMG, 2010.

51. Franco BDGM, Landgraf, M. Microrganismos patogênicos de importância em alimentos. In: Microbiologia dos alimentos. São Paulo: Atheneu, 2008. cap. 4, p. 33-82.

52. Silva N, Junqueira VCA, Silveira NFA, Tanawaki MH, Dos Santos, R S, Gomes R A R. Manual de Métodos de Análise Microbiológica de Alimentos. 4° edição. São Paulo. Ed. Livraria Varela, 2010.

53. Costa EL, Silva JA. Qualidade sanitária da carne de sol comercializada em açougues e supermercados de João Pessoa – PB. Bol. CEPPA Curitiba 1999; 17(2): 137-44.

54. Miranda PC, Barreto NSE. Avaliação Higiênico-Sanitária de diferentes Estabelecimentos de Comercialização da Carne-de sol no Município de Cruz Das Almas-Ba. Revista Caatinga, Mossoró, 2012; 25(2): 166-172.

55. Mennucci TA, Marciano MAM, ATUI, MB, Polineto A, Germano PML. Study on contaminant materials within "sun dried meat (jerked beef)" at the "Northern Houses. Revista Instituto Adolfo Lutz 2010; 69(1): 47-54.

56. Pinto A. Doenças de origem microbiana transmitidas pelos alimentos. Millenium 1996; (4): 91-100.

57. Franco R. *E.coli*: ocorrência em suínos abatidos na grande Rio e sua viabilidade experimental em linguiça frescal tipo toscana.Tese doutorado. Universidade Federal Fluminense; 2002.

58. Mendonça A O. Detecção de *Toxoplasma gondii* em linguiças comercializadas no município de Botucatu- SP. Tese doutorado. Universidade Estadual Paulista Julio Mesquita Filho; 2003.

59. Durbey J P. Refinemente of pepsin digestion method for isolation of *Toxoplasma gondii* from infected tissues. Veterinary Parasitology 1998; 74: 75-77.

60. Gamble H R, Murrel K D. Detection of parasites in food. Parasitology 1998; (117): 97-111.

61. Tenter A M. Current knowledge on the epidemiology of infections with *Toxoplasma*. The Tokai Journal of Experimental and Clinical Medicine 1998; 23 (6): 391.

62. Mürmann L. Avaliação do risco de infecção por *Salmonella* sp em consumidores de linguiça frescal de carne suína em Porto Alegre –RS. Tese doutorado. Universidade Federal do Rio Grande do Sul; 2008.

63. Brasil. Agência Nacional de Vigilância Veterinária. Resolução n°12 de 12 de janeiro 2001. Regulamento técnico sobre os padrões microbiológicos para alimentos. 2001. Disponível: http:/WWW.anvisa.gov.br/Regis/resol/12_oirac.num. (Acesso: 10 agosto 2012).

64. Fuzihara T O, Fernandes S A, Franco B D. Prevalence e dissemination of *Salmonella* serotypes along the slaughtering process in Brazilian small poultry slaugtherhouses. Journal Food Protection 2000; 63: 1749-1753.

65. Cortez A L L, Carvalho A C F B; Amaral L A, Salotti B M, Vidal-Martins A M C. Coliformes fecais, Estafilococos coagulase positiva (ECP) e*Campylobacter* ssp em linguiça frescal. Alimentos Nutrição. 2004 15 (3): 215-220.

66. Silva W P, Lima A S, Gandra E A, Araújo M R, Macedo M R P, Duval E H. *Listeria* spp. no processamento de linguiça frescal em frigoríficos de Pelotas, RS, Brasil. Ciência Rural 2004; 34 (3): 911-916.

67. Corazza M L, Mantovani D, Filho L C, Costa S C. Avaliação higiênico-sanitária de linguiças tipo frescal após inspeção sanitária realizada por órgãos federal, estadual e municipal na região noroeste do Paraná. Revista Saúde e Pesquisa 2011; 4 (3): 357-362.

68. Spricigo D A, Matsumoto S R, Espindola M L, Vaz E K, Ferraz S M. Prevalência e perfil de resistência antimicrobianos de sorovares de *Samonella*isolados de linguiças suínas tipo frescal em Lages, SC. Arquivo Brasileiro Medicina Veterinária e Zootecnia 2008; 60 (2): 517-520.

69. Marques S C, Boari C A, Brcko C C, Nascimento A R, Picolli R H. Evaluation of hygienical-sanitary type frescal commercialized in the cities of Três Corações and Lavras- MG. Ciência Agrotecnologia 2006; 30 (6): 1120-1123.

70. Rall V L M, Prado J G, Candeias J M G, Cardoso K F G, Rall R, Araujo Junior J P. Pesquisa de Salmonella e das condições sanitárias em frangos e linguiças comercializadas na cidade de Botucatu. Brazilian Journal Veterinary Research Animal Science Brazilian. Journal. 2009; 46(3): 167-174.

71. União Brasileira de Avicultura. Relatório anual 2005-2006. Disponível em ☐ http://www.uba.org.br/ubanews_files/rel_uba_2005_06.pdf☐ (Acesso em: 23 jul.2012).

72. Lucey B, Feurer C. Greer P, Moloney P, Cryan B, Fanning S. Antimicrobial resistence profiling and DNA amplification Fingerprint (DAF) of thermophlic *Campylobacter* spp in human, poutry and porcine samples from Cork region of Ireland. Journal of Applied Microbiology 2000; 89 (5): 727-734.

73. Natrajan N, Sheldon B W. Inhibition of *Salmonella* on poultry skin protein and polysaccharide-based films containing a nisin formulation. Journal of Food Protection 2000; 63 (9): 1268-1272.

74. Carvalho A C F B, Cortez A L L. *Samonella* spp in carcasses, mechanically deboned meat sausages and chiken meat. Ciência Rural 2005; 35 (6): 1465-1468.

75. Carvalho A C F B, Cortes A L L. Contaminação de produtos avícolas industrialização e seus derivados por *Campylobacter jejuni* e *Salmonella* sp. Arquivo Veterinária 2003; 19 (1): 057-062.

76. Brasil, Ministério da Agricultura, Pecuária e Abastecimento. Secretaria de Defesa Agropecuária – MAPA/SDA. Instrução Normativa N° 4 de 31 de março de 2000.-Aprova os Regulamentos Técnicos de Identidade e Qualidade de Carne Mecanicamente Separada, de Mortadela, de Lingüiça e de Salsicha -Diário Oficial da União, Brasília, DF, p.6, de 05 de abril de 2000. Seção 1.

77. Ferreira M C, Fraqueza M J, Barreto A S.Avaliação do prazo de vida útil da salsicha fresca. Revista Portuguesa de Ciências Veterinárias 2007; 102 (561): 141-143.

78. Martins L L, Santos J F, Franco R M, Oliveira L A T, Bezz J. Bacteriological study in bovine and chikem hot dog type- sausages sold in vacuumed packing-case and retail commercialized in Rio de Janeiro city and Niterói, RJ/ Brazil supermarkets. Revista. Instituto. Adolfo Lutz 2008; 67 (3): 215-220.

79. D'Agostini F P, Campana P, Degenhart R. Qualidade e identidade de embutidos produzidos no baixo Vale do Rio do peixe, Santa Catarina-Brasil. E. Tech Tecnologias para Competitividade Industrial 2009; 2 (2): 1-13.

80. Hoffmann F L, Garcia-Cruz C H, Vinturim T M, Carmello M T. Qualidade microbiológica do salame. B. Ceppa 1997; 15 (1): 57-64.

Chapter 2

STUDY OF THE MIGRATION OF THREE MODEL SUBSTANCES FROM LOW DENSITY POLYETHYLENE INTO FOOD SIMULANT AND FRUIT JUICES

Ana Rodríguez-Bernaldo de Quirós , Noelia Viqueira Varela and Raquel Sendón

Department of Analytical Chemistry, Nutrition and Food Science, Faculty of Pharmacy, University of Santiago de Compostela, A Coruña E-15782, Spain

ABSTRACT

In the present work, the migration of three chemicals, benzophenone, 1,4-diphenylbutadiene and Uvitex® OB from low-density polyethylene samples into the food simulant, 50% ethanol (v/v), was studied. The key parameters of the diffusion process, the partition and diffusion coefficients, were calculated by using a mathematical model based on Fick's Second Law. As expected, the diffusion coefficients increased with temperature and the values obtained ranged between 3.87×10^{-11} and 1.00×10^{-8} cm^2/s. Furthermore, the migration in different fruit juices was also evaluated and the results indicated that benzophenone migrated to a greater extent in comparison with the other two migrants in all beverages analyzed. To quantify the migrants, a high- $\times$ 3 mm, 3 μm particle size) and using a gradient elution system consisting of Milli-Q water and acetonitrile. The total analysis time did not exceed 8 min.

INTRODUCTION

Polymeric materials including, polyethylene, polypropylene, polyvinyl chloride (PVC), polystyrene, polyethylene terephthalate (PET), and so on, have been widely used in the food packaging industry in many applications.

The major concern related to their use is the migration of low molecular substances, such as additives, residual monomers and oligomers from the material to the food. Different factors, such as the time and temperature of contact, the surface area of the material in contact with the food, the nature of the food and the migrant, *etc.*, control the migration process [1]. The migration from a plastic material into the food is a predictable physical process that, in most cases, follows Fick's Laws [2].

It is generally accepted, that substances with low molecular weight (<1000 Da) can be absorbed in the gastrointestinal tract, and therefore may cause potentially hazardous effects on consumers' health [3]. The Regulation EU No 10/2011 [4] issues the Union list of monomers, other starting substances, macromolecules obtained from microbial fermentation, additives and polymer production aids authorized in the manufacture of plastic materials and articles, as well as the restrictions which they are subject to.

Due to the complexity of the food matrices and the analytical difficulties to determine the migrating substances in foodstuffs, the legislation allows the use of food simulants to conduct the migration assays. For cloudy drinks like juices and nectars, the simulants that must be used are 3% acetic acid (w/v) (B) and 50% ethanol (v/v) (D1) [4].

In the present study, three model migrants with different physico-chemical properties and uses, 1,4-diphenylbutadiene (DPBD), benzophenone and Uvitex® OB, were selected. DPBD belongs to fluorescent whitening agents; Uvitex OB is an optical brightener, both compounds act protecting the colour of plastic materials from adverse environmental conditions and aid to prevent yellowing and discoloration that polymers can suffer from under unfavorable conditions. Benzophenone, is a photoinitiator for UV-cured inks, commonly used for printing the external face of packaging [5].

The aim of the present work was to study the migration kinetics of these three selected migrants from LDPE (low-density polyethylene) samples into the food simulant, 50% ethanol (v/v); thus, the key parameters of the diffusion process, partition and diffusion coefficients, were calculated by using a mathematical model based on Fick's Second Law. Additionally, the migration in fruit juices was also evaluated.

MATERIALS AND METHODS

Chemicals and Standard Solutions

Standards of benzophenone (BZP) (CAS No. 119-61-9; M.W. 182.221; log P (octanol/water) 3.18) (purity 99%) and 1,4-diphenylbutadiene (DPBD) (CAS

No. 886-65-7; M.W. 206.287; log P (octanol/water) 5.290) (purity 98%) were supplied from Sigma Aldrich (Steinheim, Germany) and Uvitex® OB (CAS No. 7128-64-5; M.W. 430.5694; log P (octanol/water) 7.22) (purity 99%) was obtained from Fluka. Ethanol absolute, acetonitrile and tetrahydrofuran were provided by Merck (Darmstadt, Germany). Water used for all solutions was obtained from a Milli-Q water purification system (Millipore) (Bedford, MA, USA).

A primary stock solution of the three model migrants, of known concentration (1000 μg/mL), was prepared in tetrahydrofuran. Working solutions within the range of 0.05–5 μg/mL were prepared by dilution with acetonitrile. Solutions were stored at 4 °C in amber flasks to protect from light.

Plastic Films

Low-density polyethylene (LDPE) films additivated with the three model migrants (BZP, DPBD and Uvitex® OB) were used to conduct the migration assays. The films were prepared by an extrusion process.

Food Samples

The eight commercial juices used in the study were purchased in local supermarkets. Their fruit composition and characteristics are summarized in Table 1.

Table 1: Composition and characteristics of the commercial fruit juices

Sample	Fruit Composition	Suspended Solids	pH
Juice 1	Grape, apple, strawberry, raspberry, Currant, purple carrot, cranberry and ginseng extract		2.83
Juice 2	Orange, mango and guarana extract		3.67
Juice 3	Tomato	x	4.21
Juice 4	Orange, carrot puree, carrot, lemon and orange pulp	x	3.39
Juice 5	Orange, carrot and lemon		2.94
Juice 6	Peach and soy seed		3.92
Juice 7	Orange	x	3.44
Juice 8	Orange		3.46

x, juice with suspended solids.

Migration Tests

Migration Kinetics into 50% Ethanol (v/v)

Films were cut into 2 cm × 5 cm (10 cm²) pieces and weighed. Then, they were put in tubes with screw caps containing 50 mL of the food simulant, 50%

ethanol (v/v). The time-temperature conditions are presented in Table 2. At selected time intervals, aliquots of the food stimulant were removed, filtered through a 0.45 μm PTFE membrane filter (Advanted, Toyo Roshi Kaisha, Ltd., Utsunomiya-shi, Japan) and analyzed by high-performance liquid chromatography.

Table 2: Time-temperature conditions used in the migration assays

Temperature (°C)	Time (h)
10	2; 4; 12; 24; 48; 96; 168; 219; 675
20	2; 4; 8; 12; 24; 48; 96; 168; 219; 675
40	1; 2; 4; 8; 12; 24; 48; 96; 168; 219
60	0.5; 1; 2; 4; 8; 12; 24; 48; 96; 168

Migration in Fruit Juices

To perform the migration assays in fruit juices, 10 cm² of the additivated films were put in contact with 50 mL of the beverages in tubes with screw caps. The time-temperature conditions were 10 days at 40 °C. Moreover, for samples 7 and 8, the tests were also carried out at 20 °C. After that, the film was removed and cleaned with paper tissue. In order to simplify the analysis instead of determining the quantity of migrants in the foodstuff, the migrant that remained in the polymer was determined and it was assumed that all the missing migrant was in the food.

The migrants were extracted as follows: the plastic films were placed in flasks with 50 mL pure ethanol and stored in an oven at 70 °C for 6 h. Aliquots were removed from the flasks, filtered and analyzed by HPLC. In order to check the initial concentration of the model migrants, the films were extracted as indicated. The determined concentrations were, 124.30, 43.75 and 568.56 mg/kg for BZP, DPBD and Uvitex® OB, respectively. A second extraction under the same conditions was performed to assure a complete extraction.

Chromatography

Analyses were performed in a chromatographic system consisting of an HP1100 quaternary pump, a degassing device, an autosampler, a column thermostatting system, a diode array UV detector and Agilent Chem-Station for LC and LC/MS systems software. Chromatographic separation was carried out on an Ace 3 C18-HL column (30 × 3 mm, 3 μm particle size) thermostated at 30 °C. Milli-Q water (A) and acetonitrile (B) were used as a mobile phase. The gradient elution program was as follows: 0–2 min, 50% B isocratic; 2–5 min, linear gradient 50%–100% B; 5–15 min, 100% B isocratic and then a post-time of 5 min. The flow rate was 0.7 mL/min and the injection volume was 20 μL. BZP was detected at 256 nm, DPBD at 330 nm and Uvitex® OB at 372 nm.

Identification and Quantification

The identification of the migrants was made by comparison of their retention times and spectra with those of pure standards. Quantitation was performed on the basis of linear calibration plots of peak area against concentration. Calibration lines were constructed within the 0.05–5 μg/mL range.

RESULTS AND DISCUSSION

Chromatographic Method

A suitable chromatographic separation and peak-resolution of the three model migrants were achieved by using a reversed stationary phase and a binary solvent gradient composed of Milli-Q water and acetonitrile. Furthermore, the proposed method was rapid, under these conditions the analysis was completed within 8 min.

The linearity of the method was tested by using a series of BZP, DPBD and Uvitex® OB standards of known concentrations. Calibration curves were constructed using seven concentration levels of standard solutions, and they were fitted to a linear equation. Each point of the calibration curve is the average of two peak-area measurements. Parameters of linearity, the linear equation, and the determination coefficients are shown in Table 3. All compounds showed a good linearity, correlation coefficients were, in all cases, greater than 0.9988.

Table 3. Parameters of linearity of the model migrants

Migrant	Retention Time	Equation	r^2
BZP	1.383 min	$y = 183.39x + 6.78$	0.9994
DPBD	5.208 min	$y = 485.65x + 11.57$	0.9998
Uvitex® OB	7.180 min	$y = 213.17x + 6.88$	0.9988

The limits of detection, calculated according to ACS guidelines [6] (defined as signal three times the height of the noise level), were 0.025 mg/L for BZP, 0.01 mg/L for DPBD and 0.005 mg/L for Uvitex® OB. Similar LOQ were obtained by Gandhimathi et al. [7].

Migration Kinetics in 50% Ethanol (v/v)

Mathematical Model

The key parameters of the migration process, namely partition and diffusion coefficients were determined. A mathematical model based on Fick's Second Law (Equation (1)) was used to assess the migration of the model contaminants into the food simulant, 50% ethanol (v/v):

$$\frac{\partial C_p}{\partial t} = D \frac{\partial^2 C_p}{\partial x^2}$$

(1)

where Cp is the concentration of the migrant in the polymer at time t (s) and position x and D is the diffusion coefficient in $p(cm^2/s)$.

An analytical solution of this differential equation that describes the diffusion kinetics was proposed by Crank (1975) [8]; after a slight modification, this can be expressed by the following equation (Simoneau, 2010) [9]:

$$\frac{m_{F,t}}{A} = c_{P,0}\rho_P d_P \left(\frac{\alpha}{1+\alpha}\right) \times \left[1 - \sum_{n=1}^{\alpha} \frac{2\alpha(1+\alpha)}{1+\alpha+\alpha^2 q_n^2} \exp\left(-D_p t \frac{q_n^2}{d_P^2}\right)\right]$$

$$\alpha = \frac{1}{K_{P/F}} \frac{V_F}{V_P}$$

where, $m_{F,t}$ is the mass of the migrant transferred from P (LDPE films) into F (food simulant) after time t, (mg); A is the area of P in contact with F (cm^2); $C_{P,0}$ is the initial concentration of the migrant in P (mg/kg); ρ_P is the density of P (g/cm^3); t is the migration time (s); d_P is the thickness of P (cm); V_P is the volume of P (cm^3); V_F is the volume of F (cm^3); q_n is the positive roots of the equation $\tan q_n = -\alpha \cdot q_n$; D_P is the diffusion coefficient of the migrant in the polymer (cm^2/s); $K_{P/F}$ is the partition coefficient of the migrant between P and F.

The partition coefficient ($K_{P/S}$) indicates the relative solubility of the model migrant between the polymer and the food simulant when equilibrium is reached [10,11,12].

Partition coefficients between the film and food simulants were calculated according to the following equation:

$$K_{P/S} = \frac{C_P}{C_S}$$

where: $K_{P/S}$ is the partition coefficient between the polymer and the food simulant. C_P is the concentration of substance in the polymer at equilibrium, in µg/g. C_s is the concentration of substance in the food simulant at equilibrium, in µg/g.

The migration kinetics of the selected model migrants were carried out at four different temperatures and in the food simulant, 50% ethanol (v/v). This is the food simulant of choice for "juices and nectars and soft drinks containing fruit pulp, musts containing fruit pulp, liquid chocolate" [4].

Table 4 summarized the key parameters—partition and diffusion coefficients—of the migration process. The migrant Uvitex® OB presented the highest $K_{P/S}$ values (Table 4) in comparison with the other migrants studied, showing that this compound has a higher affinity to the polyethylene; on the contrary, BZP presented the lowest values.

Table 4: Partition ($K_{P/S}$) and Diffusion (D) coefficients for the food simulant studied

Migrant	Temp. (°C)	$K_{P/S}$	D (cm²/s)
BZP	10	2.84	1.78×10^{-9}
	20	2.84	5.74×10^{-9}
	40		
	60		
DPBD	10	64.01	6.45×10^{-10}
	20	5.08	1.48×10^{-9}
	40	7.68	7.81×10^{-9}
	60	34.72	1.00×10^{-8}
Uvitex® OB	10	>1000	3.87×10^{-11}
	20	606.62	6.79×10^{-11}
	40	168.27	4.74×10^{-10}
	60	414.24	2.08×10^{-9}

When evaluating the diffusion coefficients, the highest values correspond to BZP, which could be explained because of its lower molecular weight; this effect has been reported in a study with different photoinitiators [13]. Naturally, for all migrants, D augmented with the temperature. The results obtained for DPBD at 20 °C and 40 °C were three orders of magnitude higher than the values at 25 °C (3.7×10^{-12} cm²/s) and 40 °C (7.5×10^{-12} cm²/s) obtained by Sanches-Silva *et al.*(2008) [14] for an LDPE-orange juice system. It could be attributed to the insoluble solids presented in the orange juice. In the case of BZP, Sanches-Silva *et al.* (2009) [12] obtained a similar D value at 25 °C (3.1×10^{-9} cm²/s) in 30% ethanol (v/v). The migration kinetics of selected migrants into 50% ethanol (v/v) at 10 and 20 °C are illustrated in Figure 1.

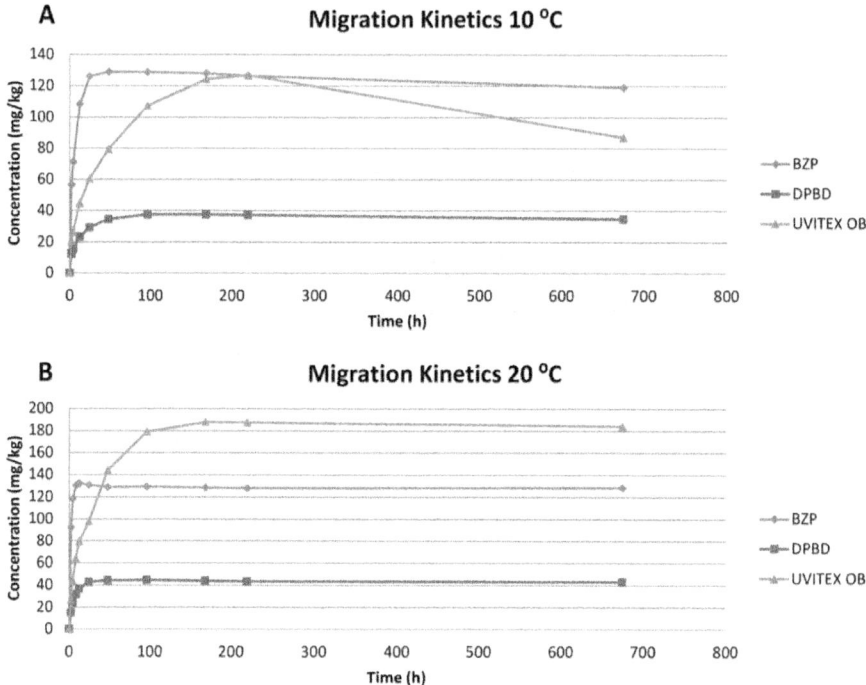

Figure 1: Migration kinetics of BZP, DPBD and Uvitex® OB obtained from LDPE films into 50% ethanol (v/v) at 10 °C (**A**) and 20 °C (**B**).

To study the effect of the temperature on the diffusion of the model migrants, DPBD and Uvitex® OB into 50% ethanol the Arrhenius Equation was used.

$$D = D_0 e^{-\frac{E_A}{RT}}$$

where D is the diffusion coefficient (cm²/s), D_0 is the preexponential factor (cm²/s), E_A is the activation energy (kJ/mol), R is the gas constant (8.314 × 10^{-3} kJ/molK), and T is the temperature (K). E_A values were determined from the slope by representing the 1/T *versus* logarithm of D. The E_A is the energy necessary for a given compound to move through the polymeric matrix [15,16]. The D_0, E_A and correlation coefficients (r^2) obtained were 0.16 cm²/s; 45.1 kJ/mol; 0.931 and 26.4 cm²/s; 64.51 kJ/mol; 0.993 for DPBD and Uvitex® OB, respectively. The Arrhenius Equation allows the prediction of the diffusion coefficient for any temperature between 10 °C and 60 °C.

Migration in Fruit Juices

The amount of BZP, DPBD and Uvitex® OB that had migrated into the fruit juice after 10 days at 40 °C (and at 20 °C for samples 7 and 8) expressed as the percent with respect the initial concentration in the plastic films is presented in Figure 2. BZP migrated to a greater extent in comparison with the other two migrants in all beverages investigated, whereas Uvitex®OB migrated to a much lesser extent. The highest concentrations of BZP and DPBD were found in beverage 6—This juice has soy seed in its composition—on the contrary, the highest concentration of Uvitex® OB was found in beverage 1, which has the lowest pH (2.83).

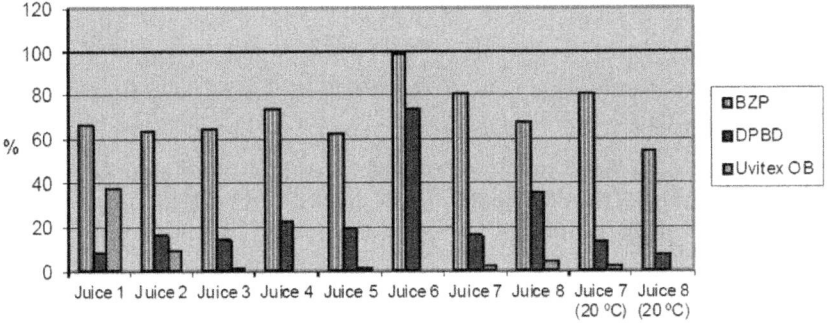

Figure 2: Percentage of BZP, DPBD and Uvitex® OB migrated in fruit juices after 10 days at 40 °C and 10 days at 20 °C.

To evaluate the effect of the pulp on the migration process, two identical orange juices with and without pulp were analyzed. In general, higher concentrations of migrants were detected in the juice that contains pulp; this behavior was also observed in other studies [14,17]. As it has been reported in the study conducted by Sanches-Silva *et al.* [14], the fraction of fat associated to the insoluble solids seems to be a key factor that affect the migration.

CONCLUDING REMARKS

Briefly, in the present work, the migration kinetics of three model migrants, BZP, DPBD and Uvitex® OB from LDPE films into the food simulant, 50% ethanol (v/v), were determined. A mathematical model based on Fick's Second Law was used to calculate the diffusion and partition coefficients, the D values obtained ranged between 3.87×10^{-11} and 1.00×10^{-8} cm^2/s. The migration in fruit juices was also studied and results show that the presence of pulp in the beverage affects the migration process.

ACKNOWLEDGMENTS

The study was financially supported by the "Consellería de Cultura Educación e Ordenación Universitaria" from the Xunta de Galicia (Galicia, Spain), Ref. No. GRC 2014/012.

R. Sendon is grateful to the "Parga Pondal" Program financed by "Consellería de Innovación e Industria, Xunta de Galicia" for her postdoctoral contract. The authors are also grateful to Gonzalo Hermelo Vidal, Cristina Casal and Patricia Blanco Carro for their excellent technical assistance.

REFERENCES

1. Barnes, K.A.; Sinclair, R.C.; Watson, D.H. *Chemical Migration and Food Contact Materials*; Woodhead Publishing Limited: Abington Hall, Abington, Cambridge, UK; CRC Press LLC: Boca Raton, FL, USA, 2007; pp. 1–12.

2. Brandsch, J.; Mercea, P.; Rüter, M.; Tosa, V.; Piringer, O. Migration modelling as a tool for quality assurance of food packaging. *Food Addit. Contam.* 2002, *19*, 29–41.

3. European Food Safety Authority. Note for Guidance for Petitioners Presenting an Application for the Safety Assessment of a Substance to be Used in Food Contact Materials Prior to Its Authorization (Updated on 30/07/08). Available online: http://www.efsa.europa.eu/de/search/doc/21r.pdf (accessed on 30 July 2015).

4. Union Guidelines on Regulation (EU) No 10/2011 on Plastic Materials and Articles Intended to Come into Contact with Food. Available online: http://ec.europa.eu/food/food/chemicalsafety/foodcontact/docs/10-2011_plastic_guidance_en.pdf (accessed on 30 July 2015).

5. Paseiro-Cerrato, R.; Rodríguez-Bernaldo de Quirós, A.; Sendón, R.; Bustos, J.; Santillana, M.I.; Cruz, J.M.; Paseiro-Losada, P. Chromatographic methods for the determination of polyfunctional amines and related compounds used as monomers and additives in food packaging materials: A state-of-the-art review. *Compr. Rev. Food Sci. Food Saf.* 2010, *9*, 676–694.

6. American Chemical Society (ACS). Subcommittee of environmental analytical chemistry. *Anal. Chem.* 1980, *52*, 2242–2280.

7. Gandhimathi, M.; Murugavel, K.; Ravi, T.K. Migration study of optical brighteners from polymerpacking materials to jam squeeze and fruit drink by spectrofluorimetry and RP-HPLC methods. *J Food Sci. Technol.* 2014, *51*, 1133–1139.

8. Crank, J. *The Mathematics of Diffusion*, 2nd ed.; Clarendon: Oxford, UK, 1975; pp. 44–68.

9. Simoneau, C. Applicability of Generally Recognised Diffusion Models for the Estimation of Specific Migration in Support of EU Directive 2002/72/EC. Available online: http://www.ibebvi.be/src/Frontend/Files/Labo/5/files/guideline%20modelling_70a.pdf (accessed on 30 July 2015).

10. Tehrany, E.A.; Desobry, S. Partition coefficients in food/packaging systems: A review. *Food Addit. Contam.* 2004, *21*, 1186–1202.

11. Sanches-Silva, A.; Cruz, J.M.; Sendón, R.; Franz, R.; Paseiro-Losada, P. Migration and diffusion of diphenylbutadiene from packages into foods. *J. Agric. Food Chem.* 2009, *57*, 10225–10230.

12. Sanches-Silva, A.; Andre, C.; Castanheira, I.; Cruz, J.M.; Pastorelli, S.; Simoneau, C.; Paseiro-Losada, P. Study of the migration of photoinitiators used in printed food-packaging materials into food simulants. *J. Agric. Food Chem.* 2009,*57*, 9516–9523.

13. Sanches-Silva, A.; Pastorelli, S.; Cruz, J.M.; Simoneau, C.; Castanheira, I.; Paseiro-Losada, P. Development of a method to study the migration of six photoinitiators into powdered milk. *J. Agric. Food Chem.* 2008, *56*, 2722–2726.

14. Sanches-Silva, A.; Cruz Freire, J.M.; Franz, R.; Paseiro-Losada, P. Time-temperature study of the kinetics of migration of diphenylbutadiene from polyethylene films into aqueous foodstuffs. *Food Res. Int.* 2008, *41*, 138–144.

15. Limm, W.; Hollifield, H.C. Modelling of additive diffusion in polyolefins. *Food Addit. Contam.* 1996, *13*, 949–967.

16. Graciano-Verdugo, A.Z.; Soto-Valdez, H.; Peralta, E.; Cruz-Zárate, P.; Islas-Rubio, A.R.; Sánchez-Valdes, S.; Sánchez-Escalante, A.; González-Méndez, N.; González-Ríos, H. Migration of a-tocopherol from LDPE films to corn oil and its effect on the oxidative stability. *Food Res. Int.* 2010, *43*, 1073–1078.

17. Franz, R.; Welle, F. Migration measurement and modelling from poly (ethylene terephthalate) (PET) into soft drinks and fruit juices in comparison with food simulants. *Food Addit. Contam.* 2008, *25*, 1033–1046.

Chapter 3

AFLATOXIN CONTAMINATION IN FOODS AND FEEDS: A SPECIAL FOCUS ON AFRICA

Makun Hussaini Anthony[1], Dutton Michael Francis[2], Njobeh Patrick Berka[2], Gbodi Timothy Ayinla[3] and Ogbadu Godwin Haruna[4]

[1]Department of Biochemistry, Federal University of Technology, Minna, Nigeria

[2]Food, Environment and Health Research Group, University of Johannesburg, South Africa

[3]Ibrahim Badamasi Babangida University, Lapai, Nigeria

[4]Sheda Science and Technology Complex, Federal Ministry of Science and Technology, Abuja, Nigeria

INTRODUCTION

A groan can almost be audible from scientists who work or have an interest in the field of mycotoxicology (fungi and mycotoxins) to the effect of "not another review publication on AF". It is true that since the discovery of AF in the early 1960s, a huge literature on AF has developed. In one way, this literature can be considered appropriate, as these mycotoxins set the scene for a new burst of activity in the contamination of feeds and foods with mycotoxins, those produced by filamentous fungi, which has shown that these substances are not merely academic novelties, but have important effects in questions of food quality and human and animal health. Further, the principal member of this mycotoxin group, AFB_1, one of the most carcinogenic natural products formed in nature (D'Mello, 2003), a major cause of hepatocellular carcinoma (liver cancer) in both animals and humans is rated as a Class IB carcinogen by the International Agency for Research on Cancer (IARC) (IARC, 1993) meaning that it is a proven cancer-inducing agent. It also occurs ubiquitously in the environment contaminating many different food and feed commodities. Rather interestingly, critics of the excessive focus on AF may perhaps, point out that one of the reasons it is intensely investigated is the ease with which it can be detected and measured due to its strong fluorescence and ultra violet (UV) light absorbing properties (Bhatnagar and Ehrlich, 2002) which skews attention away from other important mycotoxins not containing a chromophore such as the trichothecenes (TH) and fumonisins (FB). Another irony of the kudos given to AF is because of it discovery stemming mainly from the condition observed in Britain called Turkey X disease, where it was associated with deaths of tens of thousands of turkey poults (Blount, 1961) is that many of the symptoms of the disease fit those of cyclopiazonic acid (CPA) and later analysis of the groundnut used to produce feed for these birds was found to contain (CPA) in addition to AFB_1.

Whatever the merits and demerits of again reviewing literature on AF, this review has a specific purpose which has not been thoroughly explored in the past. This is to look at the ravages that this toxin has caused in Africa, not only from an animal and human health point of view, but also from the economic consequence of having agricultural commodities contaminated with these toxins. Although most countries of the world can be affected by AF, it is sub-Saharan Africa (SSA) that has suffered the most. It is difficult to defend this statement because of the lack of detailed information on the occurrence of AF in African crops. This is because, much of SSA agriculture occurs in impoverished rural areas and the lack of technical infrastructure in many African countries does not allow for routine quality control of even commercially produced commodities, never mind those produced by rural populations for their own consumption. Much of the data on the incidence of mycotoxins in SSA countries is generated by non-African agencies (e.g., IARC) (IARC 1993) or by academic institutions. This review attempts to bring together information from studies conducted over the years and attempts to discuss it in a way that will give some idea on the mycotoxin problem in the continent both from an agricultural, economic and health points of view. For reasons already highlighted, occurrence and effects of AF, principally AFB1, is the best representative in doing this.

FACTORS ENCHANCING THE PREVALENCE OF AFLATOXIGENIC FUNGI AND AFLATOXINS IN AFRICA

Prior to AF contamination, the food material must be infected with fungi that have the genetic capacity to synthesize and deposit the toxins on the foods and feeds before or after harvest. Only species of the genus Aspergillus are endowed with the 23 genes responsible for synthesis of AF. Members belonging to this genus are most abundant in the tropics and as such, are major food spoilage agents in warmer climates. The genus is metabolically versatile producing over twenty mycotoxins. Of the over 180 species of Aspergillus, only a few are aflatoxigenic. After the discovery of AF in the 1960s, A. flavus and A.

parasiticus of the section Flavi were the only known AF producers producing the B and B/G types of AF, respectively (Blount, 1961). Other aflatoxigenic species that subsequently emerged are A. nomius (B and G types), A. bombycis (B and G AF), A. ochraceoroseus, and A. pseudotamarii (B type), but they occur less frequently (Peterson et al., 2001, Ito et al. 2001)). A. tamarii, A. parvisclerotigenus (B types), A. rambellii and certain members of Aspergillus subgenus Nidulantes namely: Emericella venezualensis (Frisvad et al., 2005) and E. astellata (Frisvad et al., 2004), have now been included in the growing list of aflatoxigenic species. A. arachidicola sp. Nov. and A. minisclerotigenes sp. Nov that produce both forms of the toxin, are the latest emerging aflatoxigenic species (Pildain et al. 2008). The unexpected new comer is A. niger, an ochratoxin (OT) producer which was discovered over four decades ago but was never associated with AF synthesis. However, in a search for aflatoxigenic fungi in Romanian medicinal herbs, Mircea et al. (2008) showed the capacity of some strains of A. niger to produce AFB1. All these known AF producing fungi particularly A. flavus are common and widespread in nature, and have been shown as fungal contaminants of African foods and feeds according to Atehnkeng et al. (2008), Essono et al. (2009), Njobeh et al. (2009) and Makun et al. (2011).

Despite the fact that a strain of mould has the genetic potential to produce a particular mycotoxin, the level of production would in part be influenced by the nutrients available.

Typically, moulds require a source of energy in the form of carbohydrates or vegetable oils in addition to a source of nitrogen either organic or inorganic, trace elements and available moisture for growth and toxin production. Cereals particularly oilseeds adequately provide all these nutrients and so are considered ideal substrates for growth of fungi and consequently, toxin synthesis (FAO, 1983). However, even amongst the cereals, mycotoxins contamination varies with size and integrity of seed the coat with the small, compact grains (wheat, rice, oat, sorghum) and those encapsulated in hard seed coats (beans and soybeans) being less susceptible to fungal infection and mycotoxin formation than larger grains such as maize (Stössel, 1986). In Africa, grains contribute about 46% of the total energy intake (FAOSTAT, 2010), this figure may even be higher in rural SSA, where cereals and tubers and roots are virtually the sources of nutrition. This over dependence on grains, an ideal substrate for AF synthesis in rural Africa is a major reason for the high AF load in the continent.

Probably the two most important environmental components favouring mould growth and AF production are hot and humid conditions. Although the optimum temperature and moisture content for growth and toxin production for the various aflatoxigenic fungi varies, many of them achieve best growth and toxin synthesis between 24OC and 28OC (Schindler et al., 1977) and seed moisture content of at least 17.5% (Trenk and Hartman, 1970; Ominski et al., 1994). These conditions approximate the ambient climatic conditions in most parts of Africa and hence also account for the high prevalence of the toxins in the continent. As to how climate change will alter the AF situation in the continent, is still a matter of public debate. However, in an attempt to predict how extreme climatic conditions associated with climate change can affect AF contamination, Paterson and Lima (2010) adduced that while anticipated warmer (33°C) and more humid conditions might increase AF prevalence in Europe, perhaps the reverse might be the case in tropical Africa as most aflatoxigenic fungi will not survive the expected 40°C. Even though this prediction is plausible, it is pertinent to state herein that if there will be a rise in temperature and a rainfall reduction in all parts of Africa except for East Africa where an increase in mean annual rainfall is expected (IPCC, 2007), it is only reasonable to assume that many regions of the continent will experience droughts while East Africa would be hotter and more humid and given that these conditions favour AF production (Mutegi et al. 2009)), climate change might exacerbate the AF crisis in Africa. In agreement with this deduction, Paterson and Lima (2010) also speculated that if temperature does not increase as envisaged, then droughts might be more frequent. Drought conditions actually constitute stress factors to plants rendering them vulnerable to Aspergillus infection (Robertson, 2005; Holbrook et al., 2004) with ensuing increase in AF pollution. An indelible sign that droughts prop up AF contamination is the fact that these conditions preceded the fatal outbreak of acute human aflatoxicosis that occurred in Kenya in 2004 (Afla-guard, 2005; CDC, 2004).

Soil is another natural factor that exerts a powerful influence on the incidence of fungi. Crops grown in different soil types may have significantly different levels of AF contamination. For example, peanuts grown in light sandy soils support rapid growth of the fungi, particularly under dry conditions, while heavier soils result in less contamination of peanuts due to their high water holding capacity which helps the plant to prevent drought stress (Codex Alimentarius Commission, 2004). Produce harvested from land on which groundnut has been planted the previous year were more highly infested by A. flavus and contained more AF than crops grown on land previously planted with rye, oats, melon or potatoes (Martin and Gilman, 1976). Likewise, previously fungicide treated soil has been shown to reduce incidence of A. flavus in groundnuts to very low levels. Accordingly, it might also be useful to add recommendations on appropriate crop rotation programmes and soil treatments in order to reduce the hazards from mycotoxins.

The presence of other microorganisms either bacteria or other fungi may alter AF elaboration on food materials. When A. parasiticus was grown in the presence of some bacteria; Streptococcus lactis and Lactobacillius casei, AF production was reduced (Ominski et al., 1994). Meanwhile, fungal metabolites such as rubratoxins from Penicillium purpurogenum; cerulenin from Cephalosporium caerulens; and Acrocylindrium oryzae enhance AF production even though they repress growth of AF producing fungi (Smith and Moss, 1985). This type of positive interaction between fungi in the same food matrix with regards to AF synthesis couple with multi-occurrence of mycotoxins from the different fungi which could at the least have additive if not synergistic health impact on the host (Speijer and Speijer, 2004) worsen the AF plight in Africa because such simultaneous co-occurrence of fungi and mycotoxins in African agricultural commodities is very common phenomenon as indicated by many workers including Makun et al. (2007), Makun et al. (2009) and Njobeh et al. (2009). Much more than in other parts of the world, insects, termites, rodents and birds constitute a major problem to food safety and availability in Africa. The accessibility of these pests and predators to crops is made possible by the available deplorable storage facilities (bags, 'rhumbu') and traditional postharvest preservation techniques (drying grains on rocks and bare floor) applied in the region. Besides eating up parts of the human rations, they boost the susceptibility of crops to microbial infestation and infection with a correlating swell in AF occurrence (Hell et al. 2000). In the course of feeding on crops in the field or during storage, these animals physically wound kernels and tubers. Mechanical damage resulting from such actions of pests disrupts the seed coat and facilitates the penetration of fungal inoculum into the interior of the grains. In addition to the physical damage, these animals transmit spores from other plants and environmental surfaces to inoculate the already defective kernels and tubers and as such, help to distribute moulds widely throughout a bulk mass of grain or feed. The metabolic activities of pests especially insect larvae also produce metabolic water and heat (Sinha, 1984) that are beneficial for mould growth. The activities of these insects and birds occasioned by poor storage amenities (Hell et al., 2000; Udoh et al., 2000) provoke fungal growth and toxin production (Agboola, 1992; Wagacha and Muthomi, 2008). According to the review of Bankole and Adebanjo (2003) on the mycotoxin situation in West Africa, the commonest insects that spread A. flavus in preharvest maize in the region are lepidopteran ear borer Mussidia nigrivenella, Sitophilus zeamais and Carpophilus dimidiatus.

Agricultural practices have profound impact on AF contamination of foods. Population drift from agrarian rural areas to urban settlements in search for employment has led to reduced workers on the farm and so 'off season' harvesting during early rains is increasingly common. Also because farmers have improved varieties of grains particularly maize, crops are now grown and harvested twice during the planting season and so harvesting of the first set of these crops is done in the wet months of July and August. Such 'off season' harvesting promotes growth of aflatoxigenic fungi and toxin synthesis (Kaaya and Warren, 2005). Harvesting methods that enhance seed breakage would also increase the degree of mycotoxin formation. Aspergillius flavus was observed to be more abundant in kernels from pods gathered by combined harvesting than from hand harvested pods and the respective AFB1 content were 1780 and 140ppb (Martin and Gilman, 1976). This is in agreement with the suggestion that certain modern agricultural management practices may create unique ecological niches which select toxigenic fungi (Bilgrami et al., 1981). Meanwhile, as Africa is experiencing a boost in mechanized farming, there are no commiserate control measures put in place to reduce the negative impact of this agricultural revolution on mycotoxin pollution.

Fungi are generally aerobic organisms; therefore storage atmosphere deficient in oxygen would lead to reduced metabolism and consequently mycotoxin production. Javis (1971) and Agboola (1992) reported that a reduction in oxygen content of storage environment from 5% to 1% and increase in carbon dioxide content to above 20% dramatically reduced the growth of A. flavus and AF production. Commodities are better stored anaerobically with the addition of organic acids such as propionic acid as preservatives in storage systems which do not absorb moisture or enhance moisture migration (Ominski et al. 1994). Unfortunately traditional storage facilities in Africa are devoid of such standard storage environments (Hell et al., 2000; Udoh et al., 2000). Unwholesome trade practices have become problematic in guaranteeing food safety in the developing world. Since fungal proliferation and mycotoxin formation increase with duration of storage (Hell et al. 2000) when favourable conditions prevail, the hoarding of commodities under poor storage circumstances from October to July/August for sale during period of scarcity in order to maximize profits which has become a common practice in rural Africa exaggerates AF contamination problem. Similarly, the tradition of mixing grains of different grades in order to improve the quality of contaminated grains especially when one contains a large number of fungi spores will provide inoculum for the good grade and probably contaminate the toxin-free grain with AF. Other compelling factors that worsen the AF burden in Africa are public ignorance of the existence of the toxins; complete absence or lack of enforcement of regulatory limits; and introduction of contaminated food into the food chain which has become inevitable due to shortage of food supply caused by drought, wars and other socioeconomic and political insecurity (Wagacha and Muthomi, 2008).

OCCURRENCE OF AF IN FOODS AND FEEDS

AF producing fungi particularly A. flavus are common and widespread in nature and most often found when certain grains are grown under stressful conditions such as drought. The moulds occurs in soil, decaying vegetation, hay and grains undergoing microbiological deterioration and invades all types of organic substrates when the conditions are favourable for growth particularly in hot and humid situations (Ominski et al.,1994). Under such a suitable environment, aflatoxigenic fungi contaminate foods and feeds directly or indirectly. In direct contamination, the product is infected with aflatoxigenic fungi with subsequent toxin production. Indirect contamination occurs when food or feed was previously contaminated with AF producing fungi and although the fungi has been removed or killed during processing, AF still remains in the final product. Such contamination of cereals and oilseeds is the main point of entry of many mycotoxins in the human and animal dietary systems particularly in Africa (Smith and Moss, 1985). Fungal infection of agricultural produce is inevitable but while in developed countries, they get removed from the food chain, in most parts of Africa, moulded foods are part of the daily diets. All foods and feedstuffs are vulnerable to fungal contamination but the nature and degree of aflatoxigenic fungal contamination will depend on the presence or absence of AF in the product. Whereas the identification of toxigenic fungal contaminants is an undoubted pointer to a potential risk, positive conclusions can only be made with certainty by quantifying the suspected toxins which is why this critical review of levels of AF in food commodities is an empirical assessment of Africa's health condition with regards to mycotoxicoses. The crops that frequently support growth of AF producing fungi and subsequent toxin production are thus principal sources of exposure to AF that include but not limited to cereals (maize, sorghum, pearl millet, rice, wheat), oilseeds (peanut, soybean, sunflower, cotton), spices (chile peppers, black pepper, coriander, turmeric, ginger), and tree nuts (almond, pistachio, walnut, coconut, brazil nuts) (FAO, 1983). The other possible sources of entry of mycotoxin into animal and human systems include fruits, vegetables, animal tissues and animal products and fermented products (Jarvis, 1976).

There are at least 14 natural occurring AF known to exist, however, only six have public health and agricultural significance and they include AFB1, B2, G1 and G2. The other two are AFM1 and M2 which are the hydroxylated forms of AFB1 and B2, respectively that are secreted in animal (including humans) tissues and fluids. There is no region of the world that is free from AF problem but the strict food and feed quality control programmes put in place in the developed countries greatly reduce the AF burden in those countries. This however, is not the case in Africa and to some extent, in Asia. Besides the lack of regulation of mycotoxins in the continent, the prevailing hot, humid climate and subsistence on foods suitable for AF contamination and other factors discussed in Section 2.0, make Africa the most AF vulnerable region in the world. Hence, a current overview of its natural occurrence in different raw and processed food commodities that are major sources of the toxins in the continent which will not only reflect the disparity in human exposure to AF worldwide, but also contributes to a better understanding of Africa's health afflictions is the primary objective of this section.

Raw Agricultural Products

Nuts and Oilseeds

These products are the most investigated of all foods and feeds with regards to AF contamination. The reasons being that they are the most susceptible to AF in addition to their high protein content particularly in the case of groundnut which has made them priceless components of many animal and human diets. Data on AF incidence and concentrations in groundnut are consistently incriminating with several reports regularly accounting for extremely high levels of the toxins (Table 1) in all regions of the continent. Table 1 shows over 60% prevalence of AF across Africa at very critical levels. The incidence and levels reported vary with season. Highly contaminated samples were obtained during the rainy season (90% frequency at 12-937 µg/kg) than in the dry season (53.2% prevalence at 15-390 µg/kg) (Kamika and Takoy, 2011). The highest contaminations noted in Malawi (ICRISAT, 2010) were in districts that were prone to late session rains which enhance favourable conditions for post harvest contamination. Similarly, Mutegi et al. (2009)

demonstrated that wet and humid weathers are associated with severe AF contamination than drier locations. Serious contamination of Nigerian groundnut is a recurring problem since the 1960s when up to 8000 µg/kg were reported in 1976 (Opadokun, 1992) from the Northern region of the country. Even after heat treatment that should reduce AF levels in contaminated samples (Ogunsanwo et al. 2004), dry-roasted groundnuts from South Western part of the country contain as high as 165 µg/kg (Bankole et al.2005). Bankole and Adebanjo (2003) reviewed the AF contents in groundnut cake (range: 20 – 455 µg/kg) and snack (30 µg/kg) from Nigeria and concluded that the levels of contamination in these products were toxicologically unsafe. Homemade and unrefined groundnut oil contains AF at levels ranging between 20 and 2000 µg/ kg in Nigeria (Obidoa and Gugnani, 1992). It is pertinent to state herein that in

1988, the deaths of some primary school children in Nigeria were associated with incriminating levels of AF in groundnut cake 'kulikuli' (Fapohunda, 2011). The AF problem of peanuts persists even in South Africa where food safety standards are adhered to. This came to limelight in 2001 when the South African Primary School Nutrition Programme received substantial media coverage on unacceptably high AF content in sandwich containing peanut butter given to school children. The government then established a national monitoring programme to survey AF in groundnuts and peanut products. Between July, 2003 and March 2004, 1140 peanut and groundnut products samples were analyzed and accordingly, about 30% of analyzed samples did not comply with the legal limit of 10 μg/kg (5 μg/kg for AFB1) and concentrations as high as 560μg/kg were obtained (MRC, 2006).

AF contamination of cottonseed has been a major concern worldwide as extremely high contents ranging between 200,000 to 300,000μg/kg were reported much earlier in samples exported to the European markets (Smith and Moss, 1985). A survey of cottonseed for AF during such periods (between 1976 and 1979) when Nigeria was a major exporter of the crop, revealed over 60% (17/28) incidence rate with mean value of 105μg/kg recorded (Opadokun, 1992). Gbodi (1986) subsequently showed moderate contamination of the cash crop grown in the semi temperate climate of Plateau State, Nigeria (Table 1). Mycotoxin research on cottonseed in Nigeria has since ceased to exist as the country no longer depends on agriculture as its main source of revenue. When AF contaminated cottonseed is processed into oil, the toxins are concentrated in the residual cottonseed meal and cake which are often used as feed components of livestock feeds. The cottonseed oil also contains residual amount of the toxins at certain levels but this usually depends on the extent of contamination of the seeds from which it was obtained. Aside from cottonseed, its products are also very important sources of human and animal exposure to AF. Abalaka (1984) found 26.1μg/kg and 12.6 μg/kg of AF in groundnut and cottonseed oil samples from the Guinea Savannah region of Nigeria. All cottonseed meal samples investigated for AF in South Africa (Table 1) contained the toxins in which 42 of the 60 samples analyzed exceeded the maximum level for feeds. The AF menace in cottonseed is not limited to the sub-Saharan region of the continent as aflatoxigenic fungi and the two B forms of the toxin were found naturally contaminating cottonseed, cottonseed meal and cake from Egypt (Mazen et al. 1990). Melon seed another important oilseed in West Africa has been shown to be prone to fungal and AF contamination (Table 1) at largely unsafe levels (Bankole et al. 2006). Also Opadokun (1992) reported high incidence (73%) of AF in Nigerian melon seed at mean content of 19μg/kg.

Table 1: Some reports of aflatoxin levels in some foods commodities from Africa

Commodity	Country	Type of Aflatoxin	Incidence rate	Range (µg/kg)	Mean level ± SD (µg/kg)	Reference
Barley	Tunisia	AFB1	11/25	3.5-11.5	18.4 ±27.3	Ghali et al. (2008)
Cheese	Libya	AFM₁	15/20	0.11-0.52		Elgerbi et al. (2004)
Cow milk	Nigeria	AFM₁	3/22	-	≤ 2.04	Atanda et al. (2007)
	Sudan	AFM₁	42/44	0.22 – 6.90	2.07	Elzupir and Elhussein, (2010)
	South Africa	AFM₁	42/42	0.04 – 1.32	0.12	Dutton et al. (2010)
	Kenya	AFM₁	474/613	0.005-0.78	0.064	Kang'ethe and Lang'a (2009)
	Cameroon	AFM₁	10/63	0.006-0.527	-	Tchana et al. (2010)
Cowpea	Benin	AFB1	3/92		3.58	Houssou et al. (2009)
	Cameroon	AF	5/15	0.2-6.2	2.4	Njobeh et al. (2010)
Dried Beef	Nigeria	AFB	10/10	0.003		Oyero and Oyefolu (2010)
		AFG	10/10	0.004		
Dried Chilli	West Africa	AFB1	1/30	3.2	3.2	Hell et al. (2009)
Dried figs	Morocco	AFB1	1/20	0.28	0.28	Juan et al. (2008)
Dried okra	West Africa	AFB1	3/30		5.4	Hell et al. (2009)
		AFB2	1/30		0.6	
Dried raisins	Morocco	AFB1	4/20	3.2-13.9	10.7±2.3	Juan et al. (2008)
Egg	Cameroon	AF	28/62	0.002-7.68	0.82± 1.71	Tchana et al. (2010)
Fresh Beef	Nigeria	AFB	10/10	0.02		Oyero and Oyefolu (2010)
		AFG	10/10	0.03		
Groundnut	DR Congo	AFB1	43/60	1.5-937	229.07	Kamika and Takoy, (2011)
Groundnut	Malawi	AFB1	/1189	0-3871		ICRISAT (2010)
Groundnut	Kenya	AF	170/769	0-7525	< 4	Mutegi et al.(2009)
Local beer	Malawi	AF	5/5	8.8-34.5	22.3±4.93	Matumba et al. (2011)
Maize	Malawi	AFB		0.0-1335		ICRISAT (2010)
Maize	Nigeria	AFB1	55/55	0-1874	257.82	Atehnkeng et al. (2008)
		AFB2		0-608		
		AFG1		0-937		
		AFG2		0-286		
Maize	Ghana	AF	30/30	6.20-29.50	13.596	Akrobortu (2008)
Maize	Uganda	AF	22/49	1.00-1000		Kaaya and Warren, (2005)
Maize flour	Morocco	AFB1	16/20	0.23-11.2	1.57± 0.78	Zinedine et al. (2007) [b]
Melon seed	Nigeria	AFB1	37/137	2.3-47.7	14.2	Bankole et al.(2006)
Milk[c]	South Africa	AFM1	98/114	Max: 2.07	0.15	Lishia et al. (data unpublished)
Milk[d]	South Africa	AFM1	85/85	Max: 2.48	0.14	
Millet	Nigeria	AFB1	12/49	1370.28-3495	2587.47±78.23	Makun et al.(2007)
Mouldy Sorghum	Nigeria	AFB1	93/168	0-1164	199.51-259.90	Makun et al. (2009)
Pasteurized milk	Morocco	AFM₁	48/54	0.001-0.117	0.018	Zinedine et al. (2007)[a]
Powdered milk	Nigeria	AFM₁	19/100	0.02-0.41	0.136	Makun et al. (2010)
Powdered soymilk	Nigeria	AFB1	30/30	4.58-19.76	11.53	Adebayo-Tayo et al. (2009)
		AFB2		2.57-11.54	6.04	

Commodity	Country	Type of Aflatoxin	Incidence rate	Range (µg/kg)	Mean level ± SD (µg/kg)	Reference
Raw cow Milk	Egypt	AFM₁	19/50	0.023 – 0.073	0.049 ± 0.017	Amer and Ibrahim (2010).
Rice	Nigeria	AF	21/21	27.7–371.9	82.5±16.9	Makun *et al.* (2011)
Roasted Groundnut	Nigeria	AFB1	68/106	5-165	25.5	Bankole *et al.* (2005)
		AFB2	28/106	6-26	10.7	
		AFG1	12/106	5-20	7.2	
		AFG2	3/106	7-10	8.0	
Smoke-dried fish	Nigeria	AFB1	11/11	1.505-8.11	3.46	Adebayo-Tayo *et al.* (2008)
		AFG1	11/11	1.810-4.51	2.94	
Sorghum	Tunisia	AFB1	58/93	0.34-52.9	9.9±11.5	Ghali *et al.* (2009)
		AFB2	45/93	0.11-3.7		
		AFG1	3/93	0.45-0.7		
Sorghum	Malawi	AFB1	2/15	1.7-3.0	2.35±0.65	Matumba *et al.* (2011)
Sorghum	Ethiopia	AFB1		0-26		Ayalew *et al.* (2006)
Soybean	Cameroon	AF	2/5	0.2-3.9	2.1	Njobeh *et al.* (2010)
Wheat	Kenya	AFB1	23/50	0-7	1.93	Muthomi *et al.* (2008)
	Tunisia	AFs	15/51	4.0-12.9	6.7±2.4	Ghali *et al.* (2008)
	Nigeria	AFB1		17.10-20.53	19.00±1.67	Odoemelam and Osu, (2009)
Wheat & products	Algeria	AFB1	30/53	0.13-37.42	>5	Riba *et al.* (2010)
	South Africa	AFB1	13/238	0.5-2.0	>2	Mashinini and Dutton, (2006)

Low AF contamination of cowpea, soybean and their products is reported in many parts of the world. The data on these nuts from Cameroon and Benin (Table 1) showed only trace amount of AF in few samples. Similarly, only 3 positive samples were demonstrated in 268 cowpea samples analyzed between 1975 and 1983 from Nigeria (Opadokun, 1992). In the same review, it was shown that all two samples of soybean oil assayed did not contain AF while 41 of 55 palm kernels were contaminated with low amount of the toxin. High seed coat integrity, ensuring limited access and low moisture content are responsible for the low susceptibilities of these nuts to aflatoxigenic fungi (Stossel, 1986) with consequent rare occurrence of the toxins in them. In spite of the low presence of AF in soybean, intolerable levels of these carcinogens were found in branded and unbranded powdered soymilk, a processed product of soybean in Nigeria (Adebayo-Tayo et al. 2009). Soymilk has become a popular infant recipe in Africa because it is a cheap source of water soluble protein, carbohydrates and oil and so human exposure to AF can arise in countries using it for human consumption. Although producing fungi particularly Aspergillus flavus are considerably natural contaminants of cocoa bean (Sanchev-Hervas et al. 2008), AF are rarely detected in cocoa bean and at very low concentration. In a survey for mycotoxins in cocoa and cocoa products sponsored by the Foundation of German Cocoa and Chocolate Industry, Hamburg, Beucker et al. (2005) screened over 200 samples and found AFB1 at concentrations below 2µg/kg. Literature search for data on high AF in cocoa from Africa was unsuccessful and that could be a clear indication that these toxins are not problematic in this cash crop. However, ochratoxin A is the mycotoxin of interest to the cocoa (Arogeun and Jayeola, 2005) and soybean (Smith and Moss, 1985) industries.

Cereals

Cereals are rich sources of minerals, vitamins, carbohydrates, oils and proteins but when refined majority of the nutrients are lost leaving mostly carbohydrates and are therefore grown mainly for energy. Grains as they are sometimes called provide more food energy worldwide than any other crop and thus are staples. They are staple for two third of the earth's population, providing 85% of the world's food energy and protein intake (FAOSTAT, 2006). Cereal consumption is moderate in developed countries however in Africa and Asia, it is a daily sustenance. In Africa, cereals contribute 46% of the total energy intake; however, this figure could be as high as 78% in some African countries (FAOSTAT, 2010). According to figures made available by the afore mentioned statistics division of FAO, the six most cultivated and hence consumed grains worldwide in order of decreasing production are maize, rice, wheat, barley, Sorghum and millet, and of these major grains maize, wheat and rice together account for 87% of all cereal production worldwide and 43% of all food calories. Because of their rich nutrient composition, cereals support fungal growth and mycotoxin production excellently on the farm, during storage and after processing into foods and feeds with the small grains (sorghum, rice, wheat, millet and rice) being less susceptible to fungal and toxin contaminations than the larger grains like maize. And since these ideal substrates for mycotoxin contamination are highly consumed globally, they constitute the most remarkable sources of mycotoxins (especially the most prevalent of mycotoxins; AF) to animals and human beings. While there are other cereals (oats, rye, triticale, fonio, buckwheat and quinoa) of global significance, this review will focus on the six major ones which are pertinent to Africa. Maize is one of the important staple foods in Africa and is now widely grown for animal feeds. It is the third most cultivated food commodity in the continent after cassava and sugar cane but it ranks the first in term of food energy supply in Africa. AF are regularly detected in maize throughout the world and the recent serious contamination which was associated with drought led to fatal human aflatoxicosis in Kenya (Afla-guard, 2005; CDC, 2004) as previously stated in Section 2.0. The natural occurrence of AF in maize from some African countries has been reviewed (Table1). A regional effect on the incidence of the toxins has been demonstrated with lower contents observed in the drier North African countries (Morocco; 0.11-11.2 µg/kg) than amounts reported for such SSA countries as Uganda, Nigeria, Ghana and Malawi having up to 1874 4µg/kg. Monitoring for AF in maize samples from different regions of Africa showed disturbingly high levels of contamination above 1000 µg/ kg with many of the samples containing AF contents exceeding the CODEX regulatory limit of 20 µg/kg (five times more lenient than the EU guideline of 4µg/kg). Over 77% of the tested samples from Malawi were contaminated with AF with 90% of the positive samples not qualified for the EU export market (ICRISAT, 2010). Atehnkeng et al. (2008) previously analyzed 55 samples of maize from 11 districts across three agro-ecological zones of Nigeria for AF and mean values ranging between 30.9 -507.9 µg/kg were recovered from the 10 districts, which were far beyond all known acceptable levels. Maxwell et al. (2000) also found AF at alarming concentrations of between 3,000-138,000 µg/ kg in Nigerian preharvest maize samples. In a survey for AF in Ghana which spanned for a decade, the toxin had over 50% prevalence with more than half of the investigated samples having levels exceeding the EU maximum limit (Akrobortu, 2008). Similarly, stored maize from Uganda contain unsafe AF levels (Kaaya and Warren, 2005; Kaaya and Kyamuhangire, 2006). Very high AF level of 46,400µg/kg was found in samples from Kenyan local markets (CDC, 2004).

There is a general paucity of reports on mycotoxin in African indigenous rice. Opadokun (1992) reported AFB in only 13 of the 279 rice samples analyzed in Nigeria with only one having a level above 5µg/kg. One reason why mycotoxicologists are not attentive to local African rice relative to other crops such as maize and peanuts for example is due to the fact that it is entirely consumed within the continent and not available in the European markets. Makun et al. (2011) demonstrated a 100% prevalence of AF in Nigerian rice at unsafe levels (range: 28 - 372 µg/kg) (Table 1), and also showed critical contamination by ochratoxin A (OTA) and presence of deoxynivalenol (DON), fumonisins (FB) and zearalenone (ZEA) at trace levels. Although, AF were not found in Ugandan grown rice, however, they have been reported in rice from Cote D'Ivoire (1.5 – 10 µg/kg) and Kenya (294-1050 µg/kg) (Kaaya and Warren, 2005). Data from these few studies underscore the need for more mycotoxin surveys of indigenous rice in Africa.

Wheat does not thrive well in tropical climates so African wheat production is concentrated in the narrow strip along the Mediterranean coast from Morocco to Tunisia, in the Nile valley (Egypt), and in parts of South Africa, Kenya, Ethiopia, Zimbabwe and Sudan. Very little is grown in West Africa however, there is an increase in wheat production in Nigeria and the Democratic Republic of Congo in recent years. The grain is produced mainly for bread making but its bran is increasingly used as component of animal feeds. In spite of the limited production (18.6 million tonnes per year) of the cereal in Africa, 50.4 million tonnes is consumed yearly in the continent (FAOSTAT, 2010). Data on AF in local African wheat (Table 1) reveal moderate contamination across the continent with the highest level of contamination occurring in Nigeria. Over 57% incidence rate of AFB1 in 53 Algerian preharvest and stored wheat samples was established by Riba et al. (2010), with only 5 analyzed samples having contents above the EU legal limit. The Tunisian grown wheat had safe levels of AFB1 (Ghali et al. 2008). Of the 51 samples screened from the country, 37% were positive for AFB[1] at levels below 3.4μg/kg. Similarly, with regards to $A_{FB}1$, all South African wheat samples analyzed by Mashinini and Dutton (2006) qualified for the EU market, though it occurred simultaneously with DON, ZEA, OT and FB in some of the samples. Although AFB_1 contents in the Kenyan wheat were low (<7 μg/kg), most of the samples were also concurrently contaminated with low but significant levels of DON, T-2 toxin, ZEA and AFB1 (Muthomi et al. 2008). Such co-occurrence of mycotoxins could result in additive or synergistic effects in the host animal (Miller, 1995) as will be discussed in-depth subsequently. Considering the EU and other international and national legal limits, the Nigerian wheat with AFB1 levels above 17 μg/kg (Odoemelam and Osu, 2009) is of low quality and thus, unfit for human consumption. Makun et al. (2010) also found extremely high AFB_1 contaminations of wheat marketed in Minna, Nigeria at unacceptable levels (range: 40-275 μg/kg) in 27 of the 50 tested samples. The severe contamination levels of this crop in Nigeria making it a principal source of mycotoxins raise public health concerns and underscore the need for the regulation of mycotoxins in SSA.

Because it was replaced by maize as a staple food commodity in many rural settlements in Africa (Bandyopadhyay et al. 2007), sorghum (also known as guinea corn) is another cereal that has been neglected for some time. However, its rising industrial profile as a suitable raw material for beer brewing has seen to its re-emergence at the world market such that as of 2007, sorghum production in Africa increased significantly even to the detriment of rice and wheat production (FAOSTAT, 2010). The renewed focus on sorghum is also because it is one of the most drought tolerable crops and such high water-use efficient characteristics makes it the crop of choice to boast food security in drought stricken regions of Africa and for the future against the anticipated water scarcity in the world. Sorghum is a staple grain for over 750 million people in Africa, Asia and Latin America (CODEX, 2011) that is traditionally grown mainly in the semi-arid tropics for human consumption and production of local alcoholic drinks is now a component of animal feeds. Regardless of its inherent resistance to mould infestation due to its high composition of fungicidal principles; phenols and tannin (US Grain Council, 2008), fungal contamination constitutes a major biotic constraint to sorghum improvement and production worldwide. It is estimated that annual economic losses in Asia and Africa due to mould are in excess of US $130 million (Chandrashekar et al. 2000). AF is the most investigated mycotoxin in this crop being reported in nine countries at levels of up to 3,282 μg/kg in Brazil (CODEX, 2011). A few reports from some major sorghum producing countries (Table 1) generally reveal a moderate prevalence and low concentrations of AF in the northern part of the continent though many of the contaminated samples from Tunisia (62.50%) had higher contents than the EU limit (Ghali et al. 2009). Meanwhile all the samples from Malawi met the EU standard (Matumba et al. 2011). The samples analyzed by Makun et al. (2009) were mouldy which might explain the observed increased incidence and levels of contamination. The relevance of such biased data cannot be overemphasized as mouldy grains do normally enter the human and animal food chain in Africa. In another unbiased study using representative samples, Odoemelam and Osu (2009) supported the findings of Makun et al. (2009) by reporting unacceptable levels of AFB1 (between 27 to 36 μg/kg with a mean value of 31±3.4) in Nigerian grown sorghum. Twenty six of 69 sorghum samples from Uganda contained the toxins of which 10 of the positive ones had levels over 100μg/kg (Kaaya and Warren, 2005). Besides the entry of AF into human foods via contaminated sorghum, the carryover of substantial amount of these toxins from contaminated sorghum into traditional beer and beverages (Okoye, 1987; Matumba et al. 2011) poses additional risk of AF exposure. This should attract much attention to toxic contaminants of the grain but the reality is that there is limited information on mycotoxins in sorghum from Africa which is not commensurate to the escalating economic value of the cereal. Thus, the decision taken by the Codex Committee on Contaminants in Foods at her 5th session held in The Hague, The Netherlands in March, 2011, to prepare a discussion paper on 'mycotoxin in sorghum grain' is a very laudable initiative.

Millet (Pennisetum spp.) is the other traditional cereal replaced by maize in Africa in the last three decades. It is resistant to drought and so has been extensively cultivated in arid and semi arid regions. Millet ranks the sixth most important grain in the world, sustaining a third of the world's population and is the fifth most cultivated crop in Africa after maize, sorghum and wheat (Obilana et al. 2002). The occurrence of AF in this grain has largely been reported in West Africa particularly in Nigeria, its principal producer in the continent. In a review on chemical safety of traditional grains, Brimer (2011) reported that the average AF content of freshly harvested millet from West Africa was 4.6μg/kg. Odoemelam and Osu (2009) found higher values (range: 34 – 40 μg/kg; mean content: 37.5±2.5 μg/kg) in samples from the forest region of Nigeria. Makun et al. (2007) obtained the highest value of 3,495 μg/kg (Table 1) in wet season samples of millet that have been stored for over a year in Niger State, Nigeria. A 1971 report on AF contamination of Ugandan millet as reviewed by Kaaya and Warren (2005) revealed a low incidence (9/55) and AF level of 9 µg/kg. The limited reports on mycotoxins in millet in Africa are understandable as it is one of the 'lost crops of Africa' and that it is not an export crop. However, the anticipated 4 and 8 times reduction in risks of AF related problems if sorghum and millet, respectively, replace maize as primary staples (Bandyopadhyay et al. 2007), should bring these two African traditional crops to the front burner of mycotoxin research worldwide.

There is generally low natural incidence of AF in barley and derived products around the world. The levels reported in Tunisian grain (Table 1) is not an exception to this observation. Very low levels (<3.9 µg/kg) were also observed in barley from South Africa (Maenetje and Dutton, 2007). Because barley is grown in Africa for foods and lager beer production, good agricultural and manufacturing practices along with strict quality control measures are usually put in place in order to gain acceptability in competitive markets and this therefore accounts for the low prevalence of toxic substances in the grain as evident in these few reports. This sort of monitoring should be extended to all other foods and feedstuffs.

Animal Feeds

A majority of animals reared in Africa are free ranged. During the dry season, pastoralists and their flocks travel hundreds of kilometres in search of greener pastures. Invariably, most domesticated African animals are raised on low grade cereals, informal pasture and domestic wastes. Commercially produced animals on the other hand, are essentially fed compounded feeds that are composed primarily of cereals, oilseeds and their by-products including those of animal origin. Even though the main ingredients of the diets of both the roaming and farm animals have been shown in subsection 3.1.1 and 3.1.2 to be heavily loaded with AF in the continent, a review on the occurrence of AF in animals feeds is still appropriate as processing might alter AF levels in the final product. More so, commercial livestock farming is now a major industry in Africa and thus feeds have become a major source of exposure of humans to food borne toxins via consumption of food products obtained from animals fed AF contaminated feeds. In view of the anticipated disparity in toxin content between the feed ingredients used and the compound feed produced, Mngadi et al. (2008) analyzed South African animal feeds and the raw ingredients used in manufacturing them for AF, FB, ZEA and OTA. Accordingly, data revealed 17 of the 23 samples tested positive for AF (Table 2) and over 7 of these samples having levels equal or exceeded the legislated limit of 20µg/kg for animal feeds. The raw ingredients namely cottonseed cake, sunflower oil cakes, molasses meal and bagasse were between 4 to 8 times more contaminated than the animal feeds and this disparity was attributed to the heat and other physical and chemical treatments (Rahana and Basappa, 1990) employed during processing that might have eliminated some of the toxins. As shown in Table 1, there is a high prevalence (84.6%) of carcinogenic AF in animal feeds from Kenya with 455 of the tested 830 samples having levels exceeding 5µg/kg, the WHO/FAO limits for feeds destined for diary animals. Poor storage facilities, use of moulded maize for feed production and the absence of monitoring for AF during processing were the reasons for the high frequency and levels of AF in Kenyan feeds (Kang'ethe and Lang'a, 2009). Data on AF in animal feeds from Nigeria (Table 1) clearly points out that while retailing from open bags provokes AF contamination, manufacturing of feeds using materials less susceptible to fungi including wheat offal and palm kernel limits contamination (Adebayo-Tayo and Ettah, 2010). The 6 positive samples among the 13 feeds screened were mainly from retailed shops by the aforementioned workers had no AF presence in the wheat and palm kernel base feeds. In accordance with EU regulations, poultry feeds from Morocco were safe for animal consumption (Zinedine et al. 2007). In light of the significant prevalence of AF in African feeds and feeding stuffs, it is recommended that the use of rapid and sensitive mycotoxin test kits by farmers, manufacturers and consumers to monitor the quality of products will largely lessen the AF burden in the continent.

Table 2: Some reports of aflatoxin levels in some animal feeds and fruits from Africa

Commodity	Country	Type of Aflatoxin	Incidence	Range (µg/kg)	Mean level ± SD (µg/kg)	Reference
Animal feeds	South Africa	AF	17/23	0.8-156	38.9	Mngadi et al. (2008)
Animal feeds	Kenya	AFB1	703/830	0.9-595	8.9-46.0	Kang'ethe and Lang'a (2009)
Animal feeds	Sudan	AF	36/56	4.1-579.9	130.6	Elzupir et al. (2009)
Animal feeds[b]	South Africa	AF	99/108	3.2-950	112.54	Mwanza et al. (data unpublished)
Bush mango seeds	Nigeria	AFB1	20/20	0.2-4.0	1.5	Adebayo-Tayo et al.
		AFG1	20/20	0.3-4.2	1.5	(2006)
Cottonseed	Nigeria	AFB1	3/8	0.0-271		Gbodi (1986)
		AFB2	3/8	0.0-36.6		
		AFG1	2/8	0.0-183		
		AFG2	1/8	0.0-9.1		
Cottonseed meal[a]	South Africa	AFB1	60/60	13.4-75.7	24.90	Reiter et al. (2011)
		AFB2		1.0-5.4	1.84	
		AFG1		6.6-64.3	17.91	
		AFG2		0.3-3.3	0.90	
Pistachio	Morocco	AFB1	9/20	0.04-1430	158±6.3	Juan et al. (2008)
Pistachio	Tunisia	AF	21/40	0.24-122.4	21.8±38.0	Ghali et al. (2009)
Poultry feeds	Morocco	AFB1	14/21	0.05-5.38	1.26± 0.65	Zinedine et al.
		AFB2		0.03-0.58	0.18±0.18	(2007)
		AFG1		ND	ND	
		AFG2		ND	ND	
Poultry/livestock feeds	Nigeria	AFB1	6/13	0.0-67.9	15.5	Adebayo and Etta (2010)

Vegetables

Reports on fungal and mycotoxin contaminations of vegetables are exceptionally uncommon. One of the most recent and novel works in this area is that of Hell et al. (2009) who studied 180 dried vegetable samples across three countries in West Africa including Benin, Mali and Togo. AF concentrations were determined in dried okra, hot chilli, tomatoes, melon seed, onion and baobab leaves. The results as seen in Table 1 is one of the very first few reports on AF in dried okra and hot chilli after that of Obidoa and Gugnani (1992). AFs were not detected in the other dried commodities but however, the toxins have been detected in fresh vegetables elsewhere. For instance, Muhammad et al. (2004) found the toxins present in fresh tomatoes marketed in Sokoto, Nigeria. Similarly, Sahar et al. (2009) also showed the presence of AF in fresh tomatoes, pumpkin, powder chillies and coriander (dry) grown in Pakistan. Other fruits and vegetables in which the toxins were found in that study include cucumber, persimmon, peanut and peach. Interestingly, Obidoa and Gugnani (1992) had earlier found AF in dried okra, onions, dry pepper and table foods sold at restaurants. The ready-to-eat dishes mainly "gari" (cassava products) and beans served with vegetable soups had AFB1 at levels ranging between 8 and 61 µg/kg with total AF values ranging from 31.21 to 268µg/kg. The data suggest that AFs are common contaminants of most African table foods. Another vegetable that abhor aflatoxigenic fungi and consequently contain AF is oyster mushroom (Jonathan and Esho, 2010). The severe paucity of information on mycotoxins in vegetables necessitates increased research in the area.

Fruits

In a review of mycotoxins in fruits and fruit-processed products, Fernandez-Cruz et al. (2010) revealed that the commonest mycotoxin contaminants of fruits worldwide are patulin (PAT), OTA, AF and Alternaria toxins, and that natural AF contamination has been reported in oranges, apple and apple juices, dried apricots, dates, prunes, musts, dried figs and raisins. According to the review, AFs are most frequently reported in dried figs and raisins worldwide at significant levels of up to 550 and 63µg/kg, respectively. Furthermore, their occurrence was reported in apple juice from Egypt and dried raisins and figs from Morocco. Other reviewed literature of interest in this area is that provided by Trucksess and Scott (2008) and Barkai-Golan and Paster (2008). Reports on mycotoxins in fruits are not many in Africa with one reported in pawpaw and pineapple being contaminated by Aspergillus flavus in Maiduguri, Nigeria (Akinmusire, 2011). Baiyewu et al. (2007) showed the presence of the toxins in pawpaw from South Western Nigeria. Some data on AF incidence in Africa

(Table 2) presented low occurrence of the toxins in African fruits. Juan et al. (2008) found AFB1 at levels exceeding the EU limit of 2µg/kg in 5 and 20% of 20 samples each of pistachio and dried raisin, respectively, with the highest value of 1430 µg/kg detected in pistachio. While all dried fig samples were found fit for the EU market with regards to AFB_1 contents, 15% of them had AF levels above the 4 µg/kg, a maximum recommended limit of EU. The use of traditional processing and preservation methods for fruits in rural Morocco that provide optimal conditions for mould growth and mycotoxin formation was adduced as the cause for the increased prevalence of these AF in fruits. AF analysis in Tunisian pistachio performed by Ghali et al. (2009) found unacceptable (according to EU standards) concentrations of AF in 17% of the samples. Bush mango (Irvingia spp) is a sweet tasting fruit. The pulverized form of the dried seeds is used as condiment and thickeners for soups and stew in West Africa. Investigation into its AF content in Nigeria by Adebayo-Tayo et al. (2006) showed a 35% non compliance with the EU AFB1 standard for fruits. There is need for AF surveys in commonly eaten fruits in Africa namely, oranges, banana, plantain, guava, dates, etc. in order to properly ascertain the extent to which fruits expose the African population to the toxins.

Roots and Tubers

Roots and tubers are a major source of nutrition in the world after cereals. They are basic diets for about a billion people in the developing countries, providing 10% of world's food energy and protein intake (Shewry, 2007). They account for 40% of food eaten by half the population of SSA, contributing 20% of total energy intake in the continent (FAOSTAT, 2010). Cassava, potatoes, yam and taro form the bulk of roots and tubers consumed worldwide. Cassava and yam are not vulnerable to AF contamination (Bankole et al. 2006) and even the processed products such as cassava and yam chips and their flour have low contamination rates. Analysis of cassava and yam chips from Benin showed no contamination by AF (Gnonlonfin et al. 2008) neither was the toxins found in cassava products from Tanzania (Muzanila et al. (2000), Nigeria (Jimoh and Kolapo, 2008) and Ivory Coast (Kastner et al. 2010). However, data obtained from Cameroon (Table 1) show low levels of the toxins in stored cassava chips at levels greatly dependent on processing practices, storage facilities and duration of storage (Essono et al. 2009). Higher prevalence and contents of AF are observed in yam based products than in cassava products. Apart from the 100% frequency (Table 1) shown in Yam chips from Benin (Bassa et al. 2009), 23% of the 107 samples analyzed by Mestres et al. (2004) from the same country had AF levels over the 15 µg/kg CODEX standard value for total AF. AFB_1 and G_1 were detected in yam chips from Nigeria at levels ranging from 5-27µg/kg (Jimoh and Kolapo, 2008). AFB_1 was also found in 22% of samples from Nigeria (Bankole and Mabekoje, 2003) and a larger survey conducted later in same country found the toxin at a prevalence rate of 54.2% in the same products at toxicologically significant levels (range: 4–186 µg/kg; mean: 23 µg/kg), 32.3% with AFB_2 (range: 2-55 µg/kg), 5.2% were positive for AFG_1 (range: 4-18 µg/kg), while 2 samples contained AFG_2) (Bankole and Adebanjo, 2003). The present unwholesome practice of storing and marketing high moisture cassava and yam products is responsible for the contamination in Africa (Essono et al. 2009) and should be discouraged.

Animal Products

Milk and Milk Products

Animals fed AFB_1 and B_2 contaminated feeds excrete into their milk the less toxic AFM_1 and M_2, respectively. AFM1 is of particular interest being the hydroxylated metabolite of AFB1 and is known to have 2-10% of the carcinogenic potency of the parent compound (Zinedine et al. 2007). The carryover of this carcinogen in cow at a transfer ratio (consumed AFB1 to excreted AFM_1) of 200:1 (Smith and Moss, 1985) which could be as high as 40:0.05 (JECFA, 2001) into human and animal milk that are the main sources of nutrition for infants whose vulnerability due to undeveloped immune system is obvious, poses serious health concern. Its stability to heat, cold storage, freezing and drying (Yousef and Marth, 1985) during processing makes dairy products another important source of AFM_1 exposure. Milk and milk products are traditionally staple food commodities for the nomadic population of Africa. They are recognized by the elites as natural balanced diet and so are increasingly consumed by the urban populace in the continent. Therefore, they can no longer be ignored as they are among the main entry routes of AFM_1 into the human dietary system in Africa. The natural occurrence of AFM_1 in raw cow milk has been reported in quite a number of African countries as reflected in Table 1. A regional variation has been demonstrated in the data with lower concentrations occurring in the drier North Africa (Egypt and Morocco) than in the more humid SSA (Nigeria, Sudan, Kenya, Cameroon and South Africa). Many of the milk samples from the region (7.4, 18.7, 40, 52.6, 61.9 and 83.3 % of samples from Morocco, Kenya, Cameroon, Egypt, South Africa and Sudan, respectively, had AFM_1 contents above the legislated levels (0.05µg/L) of several countries including those in the EU. Similarly, their recorded mean levels according to JECFA (2001) are higher than those reported for European (0.023 µg/L), Latin America (0.022 µg/L), Far Eastern (0.36 µg/L), Middle Eastern (0.005 µg/L) and African diets (0.0018 µg/L). Since AFM_1 content in milk is a good indicator of AFB_1 contamination in feeds, hence at a transfer rate of 200:1 for cows it can be estimated that diary animals in Africa are exposed to between 3.6 – 414 µg/kg of AFB1 in their rations since the observed range of AFM_1 averages (Table 1) for Africa is 0.018 to 2.07 µg/L. Such high levels of AF in feeds in the continent have been shown in sections 3.1 and 3.2. The presence of AFM_1 in milk from breastfeeding mothers (Atanda et al. 2007) and other body fluids of diseased patients (Tchana et al. 2010) sometimes at above regulated levels, is confirmatory that humans are exposed to high AF levels in Africa and because milk is primary to infant nutrition it gives cause for considerable concern.

Raw milk is usually processed into dehydrated dairy products such as cream, butter, cheese and milk powder in order to extend its shelf-life. Yoghurt, another product of milk which has become part of human dietary system has a similar processing method as cheese making, only that the process is arrested before the curd forms. The separation of the components of milk during processing leads to distribution of AFM1 into dairy products with consequent lower levels of the toxins in the individual products than in milk. In spite of this anticipated loss, AFM_1 has been found in African dairy products at unacceptable levels. High prevalence (75%) of AFM1 with all positive samples having levels exceeding the legislated limit of 0.05 µg/L was seen in Libyan cheese (Elgerbi et al. (2004). Amer and Ibrahim (2010) found AFM_1 in 50/150 Egyptian cheese samples at levels between 0.051 to 0.182 µg/L and all the contaminated samples had levels beyond the EU regulated limit. Similarly, the toxin contaminates cheese, ice cream and yoghurt at intolerable amounts in Nigeria (Atanda et al. 2007). In order to determine how safe imported powdered milk samples are for human consumption, Makun et al. (2010) analyzed AFM_1 in 100 samples sold in the Lagos metropolis. As seen, 6 of the 19 positive samples had levels above the EU limit, however, all the tested samples were considered safe in Nigeria as levels of the toxin were below the country's maximum tolerable limit of 1 µg/kg. Clearly, the main strategy to reduce incidence of milk toxins is by feeding animals with AF free feeds.

Animal Tissues

Relative to such mycotoxins as OTA, a much smaller proportion of AF is absorbed into animal tissues. The transfer ratios into tissues for AFB1 can range between 1,000 and 14,000 (Smith and Moss, 1985). Any animal exposed to such high amount of mycotoxins would have shown some toxicity signs or even death and would likely not enter the human food chain in the developed world. Conversely, diseased animals are still eaten in Africa and as such, human exposure to AF via consumption of animal tissues can be a reality in the continent. There is a disparity between the AF content of beef and those of other edible organs with lower values obtained for the muscle tissue. This was shown by Oyero and Oyefolu (2010) when analysing fresh and sun-dried beef and edible organs (liver, kidney and heart). While, the dried animal tissues had lower levels of AFB (beef: 2.9, liver: 3.1, heart: 27.9, kidney: 75.8 ng/kg) and AFG (beef: 4.4, liver: 3.1, heart: 55, kidney: 141.3 ng/kg) than for fresh samples whereby AFB levels were for beef liver, heart, kidney were 21.7, 33.9, 55.9 and 85.2 ng/kg, respectively, and AFG levels in these organs recorded were respectively, 27, 41.4, 74.1 and 70.7 ng/kg. Accordingly, the levels of the toxins in the edible organs were consistently higher than those in beef with kidney being the most vulnerable. It is therefore, suggested that withdrawal of animals from contaminated feeds onto mycotoxin free diets for 3-4 weeks could have allowed for sufficient withholding period to clear the muscles and organs from the toxins. Despite the low carryover rates of the toxins into animal tissues, AF contaminate fresh and processed meat (especially liver and kidney) at toxicologically significant levels of up to 325µg/kg in Egypt (Aziz and Youssef, 1991; Abdelhamid, 2008). It is only proper to indicate further herein that consumption of animal visceral organs (kayan chiki) may constitute a major source of AF exposure than muscle. The garnishing of beef with peanut paste to produce dried beef product (kilishi) and roasted beef (Suya) is a common practice in West Africa that elevates the AF content of these processed meat products to over 194µg/kg (Jones et al. 2001; Chukwu and Imodiboh, 2009) which is beyond any known accepted maximum. Smoke-drying is the commonest method of preservation of fish, another major source of protein in the African continent. The inadequacy of the method with regards to preservation from contamination of smoke-dried fish by aflatoxigenic fungi and AF (Table 1) has been demonstrated (Adebayo-Tayo et al. 2009).

The carry-over of AF from feed to poultry by-products including meat and eggs has been investigated and found to be quite low varying with the product. In a feeding trial, Hussain et al. (2010) demonstrated AF transfer ratios of 1: 914 and 1: 1939 for the liver and muscle of broilers respectively. In hen, the transfer ratios are 1: 1103 for edible organs (gizzard, kidneys and liver) and 1: 33,100 for breast muscle (Wolzak et al. 1986). The carryover into eggs occurs at obviously high ratio of 1:6633 (Wolzak et al. 1985). The corresponding ratio of AFB_1 in feeds to residual levels in egg yolk and albumen were shown to be 1:4615 and 1:3846, respectively, in chicken hen (Bintvihok et al. 2002). In light of these high ratios, the presence of AF in eggs from Egypt (Abdallah et al., undated) and Cameroon (Tchana et al. 2010) at levels of up to 7.68 µg/kg (Table 1) indicates that poultry animals consume AF in feeds at alarming concentrations. Although the reported AF levels in both meat and meat products, and eggs from Africa seem insignificant, with chronic intake of such amount simultaneously occurring with other food borne toxicants can have deleterious health impact (Speijers and Speijers, 2004).

Fermented Products

Although fermentation reduces mycotoxins in contaminated food products (Hell and Mutegi, 2011), there is ample evidence to suggest that fermented products in Africa contain significant levels of AF. Kpodo et al. (1996) detected AF at levels as high as 289µg/kg in fermented maize dough in Ghana. The presence of AF was observed in all samples of fermented yams and plantains analyzed from the Southern region of Nigeria (Jonathan et al. 2011) at levels ranging between 37.67 – 96.34 µg/kg. Detectable (5.2 – 14.5 µg/kg) amounts of the toxins were also found in fermented cassava products from Cameroon (Essono et al. 2009). Sorghum based traditional opaque beer from Malawi contained AF at levels above the CODEX permissible limit of 10 µg/kg (Table 1) (Matumba et al. 2011). In their review on mycotoxin problem in Africa, Wagacha and Muthomi (2008) reported incriminating levels of AF (200, 000 – 400, 000 µg/l) in 33% of traditionally brewed beer in South Africa. Levels of up to 50µg/kg were found in sorghum based local beer from Lesotho (Sibanda et al. 1997).

With these unsafe levels in our fermented products, it will only be proper to adhere to the advice of Pietri et al. (2010) that if raw materials comply with the legislated limits, contribution of a moderate daily consumption of beer to AFB1 intake will not contribute significantly to exposure of the consumer.

Other Foods

Plants and plant products used as medicinal herbs, tea and spices may be commonly contaminated by AF at significant levels of up 2,230µg/kg especially in the case of liver curative herbal medicine sold in India (Trucksess and Scott, 2007). According to these authors, contamination of the toxins has been observed in ginger, garlic and capsicum. A survey conducted for aflatoxigenic fungi and AF in spices from Egypt (Aziz and Youssef, 1991) found black and white pepper contaminated with AF at unhealthy levels (range: 22-35 µg/kg). Zinedine et al. (2006) found natural presence of AFB1 in black pepper, ginger, red paprika and cumin from Morocco at average levels of 0.09, 0.63, 2.88 and 0.03µg/kg, respectively, with the highest level of contamination found in red paprika (9.68µg/kg). Although moulds are frequently isolated from African herbal plants (Bankole and Adebanjo, 2003), it seems the herbs are not prone to AF contamination (Katerere et al. 2008). Equally, a search on the literature failed to provide data on the incidence of AF in African tea. An attempt is made in this chapter to provide an extensive review on AF contamination of commodities in Africa and further information on the subject is referred to the review of Sibanda et al. (1997), Shephard (2003), Bankole and Adebanjo (2003) Bankole et al. (2006) and Wagacha and Muthomi (2008).

HUMAN AF EXPOSURE

An important part in elucidating the health effects of AF in humans is the estimation of exposure to the toxin that they receive. A tedious and rather inaccurate estimate of exposure level can be established by determining mean toxin level in the food people eat/amount eaten, either in the basic commodity, e.g. maize kernels and rice as in the case for Nigeria, or in its ready-to-eat form, e.g. porridge for those leaving in the rural area of Limpopo in South Africa. In order to approximate the level of exposure of Africans to AF, let's assume a Nigerian that eats 138kg of cereals annually (Bandyopadhyay et al. 2007) as a representative of Africa and whose main staples are maize and rice consumed in a ratio of about 3:1. From Table 1, the AFB_1 mean values for indigenous maize and rice destined for human consumption in the continent are approximately 258 and 83µg/kg, respectively. It can therefore be estimated that 29,500 µg/kg of AF is consumed yearly or let's say 81µg/kg daily by the subject from the two cereals. Even though food processing such as sorting, cooking and others factors will reduce the content (Hell and Mutegi, 2011), what might be left will still be incriminating causing certain chronic intoxications which are discussed subsequently in this chapter. Again, if 350 µg/kg is the mean level of AF in foods that can elicit acute symptoms (Azziz-Baumgartner et al., 2005), it can rationally be deduced that Africans are exposed to sub acute doses of AF in maize and rice amongst other food crops. While, the inferences made herein are generalized for Africa with regards to AF intake, it is not without flaws and exceptions, as it can also be reasonably inferred that Africans are exposed to the toxins from virtually all the foods they consume as seen in Table 1. This corroborates with high incidence of the toxin and its biomarkers in their body tissues and fluids (Gong et al. 2003, Tchana et al. 2010). This is also in agreement with the fact that in many regions of the continent, the estimate on the frequency of human exposure to AF is about 98% (Wild, 1996). More precise estimates of AF exposure among humans are best achieved by estimating the toxins and associated metabolites in bile, urine, faeces and hair as well as their distribution in blood, milk, liver, kidney and semen (Njobeh et al. 2010). Probably, the easiest one to use is the urine, which can conveniently be sampled and if a 24 hour sample is taken, it can give a fairly accurate content of AFB1 ingested over a period from the measurement of levels of AFM_1 (Nyathi et al. 1987) or adducts (Groopman et al. 1992).

Another useful method is to assay the amount in the blood and that bound to blood albumin. The latter can be regarded as a good biomarker for AFB1 exposure (Wild et al. 1990) and in effect, removes AFB1 from circulation. If one assumes that the half-life of albumin is about 20 days (Wild et al. 1990), then correlating the amount of conjugate with intake, an estimate of exposure that took place the previous week can be established. A similar case may be made for liver AFB_1-DNA adducts and this has been demonstrated in animals and by extrapolation, a relationship between the level of AFB1 albumin and AFB1-DNA conjugates is established (Wild et al. 1996).

HUMAN HEALTH IMPLICATIONS TO AF EXPOSURE WITH REFERENCE TO AFRICA

It is seen how there is extremely high degree of human exposure to AF in many parts of the African continent as previously stated in Section 4.0 of this chapter. Such level of exposure has enormous health and socioeconomic implications. Aspects of the effects of AF contamination of food and feed commodities in terms of the economy will also be highlighted in this review. In general, a disease caused by a mycotoxin is termed "mycotoxicosis" (aflatoxicosis in the case of AF). Aflatoxicosis seems a commoner in developing countries (Williams et al. 2004). Furthermore, for mycotoxins to exhibit disease symptoms of a chronic nature, they occur in feed or food at one part per billion (ppb)(μg/kg), the exception being FB1 which is present at three parts per million (ppm) (mg/kg) mark before symptoms in farm animals begin to be manifest (Bucci and Howard 1996). The difficulty when considering levels of mycotoxins for intoxication in human is that, except for "natural" cases of mycotoxicoses, disease causing levels of mycotoxins cannot be derived and hence, extrapolations from animal experiments have to be made. From known cases of human aflatoxicosis (Azziz-Baumgartner et al. 2005), it would seem that acute symptoms are found when levels in food ingested were at a mean value of 350ppb. Again the difficulty with such estimates, however, is gauging the exact amount of the toxin ingested per kg body weight, because this will depend not only on how much is ingested at a specific time, but for how long. Acute doses of mycotoxins may be ingested at one time or over a short period, whilst chronic levels may be consumed over long periods. An example in point of the latter is AFB_1 in groundnut, a staple where rural people ingest this mycotoxin most of their life time and as a consequence, this may increase the incidence of liver cancer (hepatocarcinoma) proportionately (Peers and Linsell, 1973) in the region. A similar situation is found where FB1 from maize as a staple is ingested but the consequence of a lifetime of exposure to this toxin is not as clearly defined as that of AFB_1, because FB_1 has a much lower toxicity but there are indications that it has a role in various diseases conditions (Dutton 2009). This is discussed subsequently in detail.

Toxicology of AFB_1

AFB_1 created great interest amongst the medical profession once it was established as a powerful carcinogen (Wogan, 1973). In fact it is claimed to be the most powerful naturally occurring carcinogen known. The toxin is also mutagenic and teratogenic (Raisuddin et al. 1993). Because of these toxic properties, several investigations have been carried out over the years since its discovery both in vitro and in vivo to elucidate its mode of action as a carcinogen, perhaps to the detriment of studying other toxic properties it exhibits as well as those of its congeners AFB_2, AFG_1 and AFG_2. To justify this obsession with AFB1, it is fair to point out that it is the most commonly occurring and at the highest levels as well as being the most potent when compared with the other AF. However, it might be worthwhile investigating the toxic properties of AFB2 more closely, as this has all the molecular attributes of AFB1 apart from the bishydro-furano double bond that confers the carcinogenic property. This double bond makes all the difference, as it allows detoxifying systems (cytochrome P_{450}) in the body to convert it to an epoxide by cytochrome P450s, specifically CYP3A4, CYP3A5, and/or CYP1A2 (Gallagher et al., 1994; Wang et al., 1999) to the exo-8, 9- epoxide, which ironically is the "activated" form of the molecule that can form adducts with deoxyribonucleic acid (DNA) leading to guanine nucleotide substitutions (Lilleberg et al., 1992). In passing it is also important to note that cytochrome P450 types also convert AFB1 to other derivatives e.g., AFM_1, AFP_1 and Q_1 (Campbell and Hayes 1976). The epoxide is a good alkylating agent and can react with bases such as those in DNA and RNA to form the AF– alkylated form. Obviously other factors are involved, as not all alkylating agents are carcinogens and further, AFB1 has a tendency to specifically attack guanine, one of the four DNA bases (Taylor 1992). Because AFB_1 is somewhat non-polar, it passes though membranes and other lipid barriers easily and also has a slight water solubility so it passes from the aqueous phase at low concentrations and accumulates in fat soluble phases such as adipose tissues. Furthermore, the molecule itself, because it is primarily aromatic in nature, the main core is a coumarin structure, which is rather flat and therefore, can intercalate into DNA (Jones et al. 1998). The epoxide of course can react with other nucleophiles, including those acting as part of the detoxification system, e.g., glutathione (Gopalan et al. 1992) and proteins such as blood albumin.

Because of the interaction of AFB_1 with DNA, it is reasonable to ascribe it to be its main toxicological action, certainly at low chronic level of exposure.

AFB_1 can form AFB_1-DNA adducts, DNA strand breaks, DNA base damage and oxidative damage that can lead to cancer (Wang and Groopman, 1999). This damage can be repaired by various mechanisms, e.g. base excision repair (Wood, 1999). However, certain mutations that occur due to AFB_1's action may interfere in these repair mechanisms, in particular, the xeroderma pigmentosum complementation gene group D (XPD) which is one of the groups encoding for groups in the nucleotide excision repair pathways. Recent findings suggest that two loci in this group are of particular importance in modulating the $_{AFB1}$ related development of liver cancer (Long et al. 2009). Probably of more direct importance is the action of AFB1 on the p53 gene where it causes an AGG to AGT transverse mutation at codon 249 (Bressac et al. 1991). Gene p53 is responsible for producing the p53 protein which has an important role in the regulation of cell cycle and in suppressing genome mutation (May E. and May P., 1999).

Disease Conditions in Africa Linked To AF Exposure

When considering mycotoxins and their effects in Africa there is a tendency to concentrate on those countries in SSA. The assumption being that those countries constituting the Sahara desert and border areas of the Mediterranean Sea are either more like European countries in terms of commercial food production and consumption; or the countries are so dry and unsuited to mass agrarian exploitation that they do not have the same problems of human mycotoxicoses as the rest within the continent. This is an erroneous supposition, as all foodstuffs at some stage may be contaminated with fungi and mycotoxins in these countries (Mokhles et al,. 2007, Ghali et al., 2010) but it is fair to say that in SSA, apart from the major cities, the African population is rural, mainly relying on subsistence agriculture. Because this type of activity is unregulated and often is insufficient to maintain a proper nutritional supply to these populations, it is not unreasonable to suppose that they are exposed routinely to mycotoxins and often have little natural resilience to their effects. Consequently, several diseases, which can be correlated to exposure to mycotoxins or can be in part attributed to them, are known. As AF is ubiquitous in African commodities (Nyathi et al. 1989), it is not surprising that several of such conditions are caused or exacerbated by these toxins. However, as indicated above, the main problem is AFB1 and these diseases will be discussed in terms of this toxin.

Acute Toxicity

Although most naturally occurring cases of acute toxicity caused by AFB1 are observed mainly in animals such as one that lead to the discovery of

AF in turkey poults (Blount, 1961), there are several cases of human acute aflatoxicoses (most of which often go unnoticed). Since 1982, deaths caused by AF-contaminated maize have repeatedly occurred in the Eastern Province of Kenya (Probst et al. 2007). Similar cases have also previously been recorded at various times in India (Krishnamachari et al. 1975) and Malaysia (Lye et al. 1995). A recent case reported in Kenya is as a result of consuming contaminated maize (Nyikal et al. 2004) with 125 deaths out of 317 cases (Azziz-Baumgartner et al., 2005) being recorded. The symptoms are anorexia, malaise and low grade fever leading to acute jaundice and lethal hepatitis. This outbreak was associated with high levels of AF in maize for human consumption (mean concentration of 355ppb) that led to increased levels of AF B1 albumin adduct and higher hepatitis B titres in the patients than controls (Azziz-Baumgartner et al., 2005). There seems to be little in the way of treatments for these cases, apart from straight forward strategies of antimicrobials and support for the damaged organs (Mwanza et al., 2005) as the toxin acts rapidly and can be lethal.

Conditions of the Liver

As the liver is the main organ of detoxification and the first major organ to be exposed to dietary intake of xenobiotics, it is not surprising to find that several liver conditions have been associated with AF, particularly in under developed countries, especially in Africa. The fact that many African crops, in particular, the staples such as ground nut and maize, can be routinely contaminated with AF, leads to an intuitive feeling that many diseases, including those of the liver, can be linked directly or indirectly, to AF, especially AFB_1. As already intimated, this is a dangerous supposition, as many chronic diseases are multi-factorial in nature, and as in any scientific hypothesis, needs strong evidence of support. This is not easily gathered in the rural areas of Africa where infrastructure is poor, health services varying from none to basic clinics and health centres with little time or inclination to gather usefully directed statistics. Even the analysis and quality control of staples is dependent upon external scientific studies, which often are one of investigations that merely provide a snapshot of a true situation. Consequently, the information available to investigators attempting to correlate disease conditions to mycotoxin exposure is sparse and patchy with respect to various African countries. In addition, the practise of comparing these conditions in Africa to those in developed countries may highlight interesting differences that can give clues to the aetiology of African disease but may also lead to western scientists dismissing explanations of such chronic diseases in Africa, because of well researched conditions in their own countries, which have explanations other than dietary ones. Clearly, people living in African rural areas have an environment that is completely different to those living in western cities and hence, elevated incidence of certain medical conditions may be explained by factors other than those appertaining let's say for example, in Europe or North America. In any case, there are readily available systems in developing countries particularly in Africa that could provide natural environments for human experimentations as it is in theory, possible to compare the chronic diseases in rural populations against those of urbanized populations of the same race group (Njobeh et al., 2010) elsewhere.

Hepatitis

Hendrickse (1991) lists five possible roles for AFB_1 in human disease including fatal AFB_1 poisonings that can "masquerade as hepatitis". This condition is an inflammation of the liver cells caused by various agents including viruses e.g., hepatitis B virus, (HBV) and may be self healing or in extended chronic cases, lead to cirrhosis. Several cases of hepatitis have been reported in the literature including one in India that was attended by high mortality (Krishnamachari et al. 1975). The outbreak was associated with maize contaminated by AF and it was concluded that it was as a result of aflatoxicosis.

Cirrhosis

Cirrhosis of the liver is a well known disease condition usually related to alcohol consumption. However, it may be caused by exposure of the organ to toxic principles other than ethanol and can often be found in children because of their higher susceptibility to toxin exposure. In certain countries in Africa (e.g. Ethiopia) (Tsega, 1977) and India (Yadgiri et al., 1970), the condition has been described and efforts made to correlate it with AF intake (Tandon et al., 1978). The whole issue is somewhat clouded because of the commoner occurrence of hepato-cellular carcinoma (HCC) together with cirrhosis, in areas of high AF exposure and hence, a relationship has been suggested by the two conditions; simply this would be that the onset of childhood and other cirrhoses leading to liver cancer (Lata, 2010). There is no real evidence, however, that there is a link (Kew and Popper 1984) and further, that the roles of HBV, hepatitis C (HVC) or AFB_1 is by no means clear. In some cases of cirrhosis, there is an over expression of the p53 gene and this has been related to simple cellular stress or mutation of the gene, which occurs in many of the subjects exposed to HBV and/or AFB_1 (Livni et al., 1995). More recent work on cirrhosis in The Gambia (Kuniholm et al. 2008) tentatively concludes that health effects due to exposure to AFB_1 could include this disease. It would seem that, because of various factors that African populations face, including alcohol abuse, it is difficult to tease out any particular one and ascribe a role to it, other than it does make a contribution to the overall condition.

Hepatocellular Carcinoma (HCC)

This chronic disease is a major global health problem, causing over 600,000 cases per annum (Ferenci et al. 2010) and accounts for over 70% of all liver carcinomas (Lata, 2010). In Africa, the most likely cause or promoter of such a cancer is AFB_1 (it has been estimated that AFB_1 may play a causative role in 4.6-28.2% of all global HCC cases) (Liu and Wu, 2010), which is a primary carcinoma in the case of areas where factors such as hepatitis and AF are found. In other situations, the cancer may be of secondary type, i.e., an infiltration of a metastasised type from other parts of the body. Thus the situation is, not as clear cut as might have been thought. Areas in Africa where groundnut is consumed regularly, which is one of the principal sources of the mycotoxin, such as Sudan (Omer et al. 1998); Mozambique and Kenya, have high levels of liver cancer which have been shown to be correlated to AFB1 levels (van Rensburg et al., 1985). These areas also have a high incidence of viral diseases that can also affect the liver, e.g., the highly contagious HBV plus HIV infection (Kew, 2010a); HCV (Ashfaq et al. 2011); and iron over load (Kew and Asare, 2007) which tends to cause the issue. Hence it has been argued that these viruses are the cause of the high liver cancer incidence with other factors such as alcoholism and mycotoxins being of secondary importance (Stoloff, 1969; Perz et al., 2006). Given the multi-factorial nature of cancer, these claims are dangerous ones to make and also apply to the school of thought that says HCC is mainly due to AFB_1. Nevertheless, hepatitis viruses do seem to play a major role in the development of HCC in African countries (Kirk et al., 2004; Ocama et al., 2009). It is, however, known that the risk posed by AFB_1 is independent of that of HBV (Blondski et al.2010) although where both factors occur together, the risk is increased. It has been suggested, for example that cirrhosis of the liver, as previously discussed, may be a precursor to liver carcinoma (Kew, 2010a; Lata, 2010). It seems, therefore that HCC is multi-factorial, although some factors may contribute more than others, depending upon circumstances, e.g., in Africa where AFB_1 is commonly found in food, it may play a bigger role here than elsewhere, where it is not found to any extent. Further, in simple terms one thinks of a cancer initiator and a cancer promoter. It may well be that the highest liver cancer areas have a combination of both cancer forming factors plus others such as genetic ones. This certainly is the case for AFB_1 as it has been seen that AFB_1 needs to be activated by the cytochrome P450 system in particular the CYP 3A4 version plus CYP3A5 and 3A7 (Kamdem et al., 2006). Hence those persons with genetics dictating less expression of these forms of cytochrome P450 would, in theory, have less chance of developing the cancer and vice versa. As earlier mentioned, AFB_1 may cause a missense mutation, where a guanine residue is converted to a thymine in the 249 codon of the p53 gene and this is considered to be an important marker in the promotion of HCC by AFB_1 (Bressac et al., 1991). In a study that was conducted on Chinese subjects, this mutation was detected in the blood and liver of HCC patients (Jackson et al., 2001). However, not all the patients had this mutation in their livers, indicating yet again a multi-factorial situation, although some of the liver negative subjects did have the mutation in their plasma samples. It was concluded that the detection of the codon 247 mutation may present a method of an early diagnosis of HCC.

Kwashiorkor

Kwashiorkor is a malnutrition condition which is in essence a protein deficiency of the young like marasmus but unlike it, the patient has sufficient calories. It is characterized by oedema, anorexia, dermatitis and an enlarged liver with fatty infiltrates (Bhattacharyya, 1986). Because of the liver's involvement, it was suggested that AFB_1 may be involved (Hendrickse et al., 1983). This appeared to be supported by other observation, e.g., relating peak prevalence for kwashiorkor with climatic conditions such as high humidity (de Vries and Hendrickse, 1988); general occurrence of AFB1 in children's excreta (de Vries et al. 1987, 1990); liver (Lamplugh and Hendrickes, 1982); serum (Coulter et al., 1986; Hatem et al., 2005); and its reduced product, aflatoxicol, in livers of Ghanaian children (Apeagyei et al., 1986).

Other work, however, does not support the hypothesis that AF have a role in Kwashiorkor (Househam and Hundt, 1991). What AFB1 exposure seems to influence in kwashiorkor patients is that their recovery time in hospital is lengthened and this is linked to a difference in the way AFB1 is metabolised as compared to control non-kwashiorkor infants (Ramjee et al., 1992). It could, therefore, be argued that kwashiorkor is protein deficiency in children who are receiving sufficient carbohydrate such as diet on maize gruel given to weaned children. Maize is not a balanced source of protein but has plenty of starch and hence its use as a staple would fit this hypothesis. Oedema would arise due to lack of sufficient blood proteins, i.e., hypoalbuminemia (Waterlow, 1984). This view has been questioned as it has been observed that the oedema symptoms can disappear before the blood albumin levels return to normal (Golden et al., 1980). An alternative mechanism suggested was that the condition was caused by oxidative species not being regulated by anti-oxidants, e.g., vitamin E, which were depleted because of the poor diet (Golden and Ramdath, 1987). In order to test this, children in Malawi were treated with antioxidants such as riboflavin, vitamin E and selenium (Ciliberto et al., 2005) but both the treated and control group given placebos showed similar levels as kwashiorkor as evidenced by oedema. More recent work seems to suggest that the incidence of kwashiorkor in Africa may be dropping, as evidenced by a Nigerian study (Oyelami and Ogunlesi, 2007). This was put down to better management of diarrhoeal diseases. An interesting point that was made in an earlier publication (Enwonwu, 1984) is that protein deficiency affects the cytochrome P450 system, which would allow for the accumulation of AFB1 in kwashiorkor children due to lack of conversion products. Because of the lack of the AFB_1 epoxide production, its carcinogenic property would be lost, which, ironically, would be re-instituted if protein supplementation was given. It would seem that the underlying causes of kwashiorkor still remains a mystery, is multi-factorial (Oyelami et al., 1995) and unrelated to the lack of any particular food or nutrient (Lin et al., 2007) although potassium supplement appeared to help (Manary and Brewster, 1997). However, it is still tempting to evoke some toxic factor in the diet of children that interacts with other factors to produce the condition. Kwashiorkor has been referred to at a certain time as a "maize disease" (Adhikari et al., 1994). This label has been used in other circumstances with reference to Fusarium mycotoxins, in particular the FB, i.e., FB_1 (Dutton, 2009). It is not impossible that these mycotoxins rather than AFB1 have some role in the development of kwashiorkor.

MISCELLANEOUS

Reye's Syndrome

This can be a fatal disease that affects mainly children and is typified by fatty liver, swelling of the brain (encephalopathy) and hypoglycemia. The liver becomes enlarged and firm but with no jaundice. There is also evidence of kidney damage. It is not particularly associated with Africa and indeed many cases were reported in first world countries (Harwig et al., 1975). When first described (Reye et al., 1963), the cause was unknown but it often occurred after a mild respiratory tract infection and, because of involvement of the liver, as in kwashiorkor and similar cases in Thailand, it was suggested that AFB_1 may be involved (Becroft, 1966; Olson et al., 1971). This was supported by further studies where materials resembling AF were isolated from the liver of a girl of 8 months of age (Becroft and Webster, 1972) and AFB_1 in the livers of 5 children (Stora et al. 1983) but other investigations showed that the presence of AF in Reye's syndrome livers was variable and probably, the syndrome was the result of multiple interrelated factors (Ryan et al., 1979; Rogan et al., 1985). Nevertheless, concern was raised with respect to the presence of AF in the serum and urine of children with Reye's syndrome and it was concluded that this was of general public health importance (Nelson et al., 1980). Because there was no clear cut evidence that AF played a central role in Reye's syndrome, other factors were considered. One of these inter alia, was the use of salicylates (Aspirin) in treating children with various infections, e.g., respiratory viral infections (Trauner, 1984). It has been claimed that since aspirin was abolished as a medication for influenza and similar diseases for patients under 18 years old, Reye's syndrome has become very rare (Kimura, 2011) and is considered a secondary mitochondrial disease.

Lung Cancer

Oyelami and co-workers (1997) found that children with various diseases, including pneumonia in Nigeria had high levels of AF in their lungs. The route of acquisition of this toxin was not clear but there is evidence to suggest that the lung can be exposed to AFB_1 by inhalation (Hayes et al., 1984) and that this may give rise to lung cancer (Dvorackova et al. 1981). Human lung microsomes cytochrome P_{450} does not seem to be well expressed, the CYP1A2 form is not expressed at all (Wheeler et al. 1990) although some CYP1A2 activity is (Kelly et al. 1997). The exposure of lung cells, however, to polycyclic aromatic hydrocarbons induces cytochrome P_{450}s that may activate AFB_1 (van Vleet et al., 2001). A study (Donnelly et al., 1996) was conducted on human lung tissue fractions to investigate the possible role of other oxidative systems in the activation of AFB_1 and its deactivation by glutathione Stransferase. The study concluded that AFB1 activation in human lung was primarily due to lipoxygenase and prostaglandin H synthase activity and that low conjugation activity contributed to human pulmonary susceptibility to AFB_1. Similar studies (van Vleet et al., 2002a) showed that cytochrome P_{450} in lung was capable of activating AFB_1 (van Vleet et al., 2002b) and CYP1A2 has greater importance in lung tissue in activating AFB_1. Later work has indicated that cytochrome P450 CYP2A13 is highly expressed in human bronchial cells, indicating that this cytochrome may be responsible for AFB1 activation in the lung (Zhu et al., 2006). Irrespective of what activates AFB_1, it seems that exposure of humans to AFB_1 inhalation may result in an increased risk of lung cancer (van Vleet et al., 2002b). In the case of rural Africans, this is then a possibility as they process their maize manually (partly by winnowing), often in enclosed conditions where they may be exposed to dust from contaminated cobs.

Immuno-suppression and Reproductive Problems

An often neglected aspect of chronic mycotoxin exposure in humans is the effect of these on the immune system. With the arrival of infections such as

HIV, which also attacks this system and is a major problem in SSA, the extra burden of immuno-suppressors may affect the course of the immune deficiency syndrome (Bondy and Pestka, 2000; Dutton, 2004; Jiang et al., 2008). There is a strong evidence to show that mycotoxins can promote "secondary infections" of organisms that are generally commensal in an animal due to immuno suppression, e.g. OTA in pigs (Stoev et al., 2000). Several mycotoxins have been shown to be immuno-suppressive or have a potential to be so (Sharma, 1993) and these include AFB1 (Cusumano et al., 1996). These effects may be compounded by the interaction of more than one mycotoxin, which may occur in substrates (Hussein and Brasel, 2001) that may cause synergism, although doubts have been expressed on assessing these effects (Speijers and Speijers, 2004). There is, however, evidence from in vitro experiments that this may occur (Creppy et al., 2004, Luongo et al., 2006; Del Rio Garcia et al., 2007; Orsi et al., 2007; Smith et al., 1997) and it is reasonable to suppose that these effects may be extrapolated to humans.

The action of AFB_1 on the human reproductive system and gestation and birth defects is experienced by animals (Ibeh and Saxena, 1997) so it not unreasonable to suppose that these may occur in humans and, from a recent review on the subject, it would seem that AFB_1 exposure in humans has several effects on reproduction (Shuaib et al., 2010a). Several studies have measured the blood albumin-AFB_1 conjugate levels in pregnant African women and have found substantial levels as might be expected (Turner et al., 2007). In one study, there was a correlation between these levels and anaemia in pregnant women in Ghana, if iron deficiency anaemia was excluded (Shuaib et al., 2010b). In the case of men, sperm abnormalities have been associated with AFB_1 in their semen (Ibeh et al., 1994). Other effects are low birth weight (de Vries et al., 1989) and jaundice in neonates (Abulu et al., 1998) which was correlated with AFB1 in cord blood and effects on their immune system (Turner et al., 2003). AFM_1 is a cytochrome P450 mediated hydroxylated product of AFB_1 found in most mammal secreted milks, previously exposed to AFB1 (Motawee et al., 2009). It does have toxic and carcinogenic properties, although not as marked as AFB1 (Hsieh et al., 1984). It has been found in breast milk of women from several African countries (Coulter et al., 1984; Zarba et al., 1992). Its presence in cow's milk, particularly in that produced in African rural areas (Mwanza, 2007; Tchana et al., 2010) is of great concern, considering its use in the nutrition of children.

To summarize on the health implications associated with human exposure to AF from the African viewpoint, it is but normal to state that of the AF, AFB_1 is considered to be the most important and much likely to be involved in human diseases. Although most work has concentrated on its role, sight must not be lost of the fact that its congeners, AFG_1 and M_1, which are also present in the environment, are capable of being converted to the active epoxide derivatives by the cytochrome P450 system. AFM_1 is of particular importance, because of its occurrence in milk from dairy cattle fed feeds contaminated with AFB_1. With AFB_1 itself, as discussed herein, it is very difficult to tease out its exact role in various human disease conditions. In animal trials, it has been shown to be a powerful carcinogen and there is no reason to suspect that it does not have a carcinogenic effect in humans. A strong claim that the role of mycotoxins in human disease has been ignored was recently made (Wild and Gong, 2010) and interestingly, AFB_1 and FB1 were cited as major culprits, because of their common occurrence in staple foods, often in combination. However, because of the complicating issues in the human environment and lack of direct experimentation on humans, it is difficult to assign a definitive role in any disease condition. In the case of HCC for example, at best we can claim that AFB1 is responsible for 5-28% of all cases in Africa, although the detection of specific point base mutations in liver cancer cells may allow for an estimation of the contribution of AFB_1 to human liver cancer. Whatever the precise medical role of mycotoxins in human diseases is , one cannot help but have a strong empathy with the pleas of Wild and Gong (2010) that they should be taken very seriously in Africa and other developing parts of the world.

POSSIBLE INTERVENTION CONTROL STRATEGIES FOR AF IN AFRICA

Considering the action and toxic nature of AF, it may have been thought that attempting to completely eliminate their actions in humans or perhaps, animals would be rather futile. However, several approaches have been or can be taken, either during pre or postexposure that can assist in alleviating or moderating the actions of AF, particularly those of AFB_1. Whatever actions are taken or have been adopted, do require financial considerations and because the current global financial situation is not in a good stead, make this even much more difficult, but of great importance. In order to assess "overall disease burden" (ODB), the concept of "disability-adjusted life year" (DALY) is used (Murray, 1994). It is defined as the number of years lost due to early death, ill health or disability and this is deduced by the addition of two variables, i.e. "years of life lost" (YLL) to "years lived with disability" (YLD). It follows that one DALY is equal to one year of healthy life lost. In order, therefore, to calculate the cost effectiveness of a programme to prevent HCC for example, the DALYs for this disease must be determined followed by computing the likely lowering of this figure. Against this, is the cost of reducing mycotoxins like AFB_1 and FB1 in a food staple such as maize. This in turn depends upon regulations governing the health risk factors, which in order to reduce risk to health, must become more stringent resulting in economic losses in the crop due to it not meeting the recommended standard. For example, it is claimed that export losses from AF in groundnut may exceed US$ 450 million, if the level were 4ppb AF (European Union regulation) as opposed to the level imposed in USA of 20ppb which would result to about $100 million loss (Wu, 2004). An important aspect of this approach is the fact that WHO has in the past not recognised AF as a high priority problem within the top 6 health problems but it has been argued that AF not only has an impact on HCC, but probably modulates several of the other top 6 problems (Williams et al., 2004). In order to determine the cost effectiveness of various interventions, it is necessary to establish the cost effectiveness ratio (CER), which is the gross domestic product (GDP) multiplied by DALY saved per unit cost. In a study involving AF in maize in Nigeria and groundnut in Guinea, two strategies were compared viz: pre-harvest control; and post harvest interventions (Wu and Khlangwiset, 2010a). Accordingly, it was shown that the cost of both interventions exceeded the monetised values of lives saved and quality of life gained by reducing HCC, if applied nationwide. Furthermore on this study of Wu and Khlangwiset (2010a), CER for biocontrol in Nigeria ranged from 5.1 to 24.8 and for post harvest intervention for groundnut in Guinea from 0.21- to 2.08. Any intervention with a CER 0.33 as just simply cost effective (Wu and Khlangwiset, 2010a). The implications of such calculations cannot be under estimated, as they do not only indicate to governments and world bodies their value, but also have an ethical dimension in terms of human and animal sufferings.

When considering interventions, several routes may be taken (Wu and Khlangwiset, 2010b). The best approach is that of prevention which is always better than cure. One such intervention is that of releasing non-aflatoxigenic strains of Aspergillus flavus into the agricultural environment and such a commercial product called Afla-Guard® is available commercially. This results in suppression of naturally occurring aflatoxigenic strains (Abbas et al., 2011). Another is the introduction of genetically modified variety of crops, e.g., genetically modified (GM) Bt maize which inhibits insect damage and hence fungal infection (Wu, 2006). Another preventive measure is feeding of animals with amino acids and vitamins particularly lysine and vitamin C that have protective actions against mycotoxins (Obidoa and Gugnani, 1992; Smith et al., 2000). A more traditional way is the use of fungicides and pesticides, although current preference is not in favour of this. The use of natural predators (cats and dogs) at fields and storage sites to deter rodents, birds and monkeys is a very practicable preventive control strategy for Africa.

Post harvest treatments are a little more difficult due to the persistence of AF in commodities even after processing. Early harvesting, effective drying (to moisture level of less than 14%), cleaning, removal of damaged produce (sorting), e.g., small and discoloured groundnuts (Chiou et al., 1994); good storage facilities with controlled humidity (Kew, 2010b) and packaging can all contribute to lowering the level of the mycotoxin in the final product. While these remain the most effective post harvest control measures for Africa, other alternative but less effective measures include reduction of storage time, use of chemical and botanical preservatives, and detoxification of contaminated produce. The most commonly used chemical preservatives are the organic acids; formic, acetic, propionic, sorbic and benzoic acids. Nonetheless, they are ineffective in foods that contain basic components that neutralize these acids (Smith and Moss, 1985). Alkalis, strong acids and oxidizing agents are quite effective in detoxifying AF but because they could drastically change the properties of the products, ammoniation is still the most preferred and developed detoxification procedure. But the changes in chemical compositions and organoleptic properties of ammoniated meals makes them unfit for human consumption nevertheless good enough for animals. Commercialization of the ammoniation procedure in Africa by governments and private companies as has been successfully done in the USA, could help provide livestock farmers in developing countries with relatively safer feeds in the face of highly contaminated feedstuffs and shortage of feeds. The toxic and 'off flavour' products of chemical preservation and detoxification processes has led scientists to search for natural, safer and environment-friendly fungicidal products. Among such African based studies, Lippia multiflora leaf extract has been shown to have fungistatic effect on A. flavus (Anjorin et al., 2008). More intense field trials of such promising plant products and their subsequent formulation into botanical fungicide would be impressive for the continent. Gamma irradiation of AF contaminated foods lowers both the toxicity (Ogbadu and Bassir, 1979) and production (Ogbadu, 1979; Ogbadu, 1980a; Ogbadu, 1981; Ogbadu, 1988) of the toxin in irradiated foods and so this could be a good post harvest, processing and packaging treatment option for African countries if suitable infrastructures are put in place. Hazard Analysis Critical Control Point (HACCP), a proactive management system in which food safety is maintained through the analysis and control of biological, chemical, and physical hazards from raw material production, procurement and handling, to manufacturing, distribution and consumption of the finished product has become a priceless tool for mycotoxin control (FAO, 2003).

Clinical applications in the control of conditions such as HCC have been applied with varying degree of success. This range from preventative measures such as the use of Novasil clay being added to the diet to bind AF (Afriyie-

Gyawu et al., 2008); and HVB immunisation (Kew 2005). The use of drugs can be considered in two parts, one that blocks cytochrome P_{450}s responsible for the activation of AFB1 to an active (its epoxide) form, e.g. oltipraz (Langouet et al., 1995; Wang et al., 1999) and natural foods, e.g. Brassicas (Manson et al., 1997); and those which may have some other non-clear cut effect such as the use of plant extracts as protective agents (Kotan et al., 2011); boric acid (Turkez and Geyikoglu, 2010); sorafenib, a blocker for signalling pathways involved in HCC (Dank, 2010).

LEGISLATION

In order to protect consumers against the hazards of mycotoxins, many countries including 15 of those from Africa (Sibanda et al., 1997; Fellinger, 2006; Njobeh et al., 2010) have instituted legislation against some mycotoxins notably AF. According to these authors, the maximum tolerable limits for AF in human foods in Africa is between 5-20 ppb, while for animal feeds is from 5 to 300 ppb with infant foods having the least regulated levels (0.05-10 ppb) (0.05 ppb for AFM1 in the case of South Africa). While these maximum allowable limits would protect citizens from the dangers of AF, the biggest challenge in regulating mycotoxins in the continent is the lack of enforcement of legislation partly attributed to the presence of informal food market systems operating in most countries. Under this market structure, raw agricultural produce from farms and storage barns are sold directly to consumers without being screened for mycotoxins neither are they subjected to inspection for spoilage. Furthermore, government agencies charged with the responsibility to regulate mycotoxins are non-existent in many of these countries and even when available, they are dysfunctional as they are composed of deplorable infrastructures and logistics. An effective mycotoxin surveillance and food quality control unit which ensures that all foods and feeds destined for human and animal consumption are devoid of mycotoxins at harmful levels must be in place to implement mycotoxin legislation in the continent.

PREVENTION BY SURVEILLANCE AND AWARENESS CAMPAIGN

Assessing the levels of mycotoxins and indeed other food toxicants is paramount to evaluating food safety. In line with the recommendation for effective mycotoxin survey and food and feed inspection for implementation of legislation for food safety, African governments must build or strengthen already existing regional laboratories to monitor mycotoxins in foods and feeds on regular basis. And to ensure that they are in compliance with set standards. Invariably, Africa must reinforce its food quality control agencies and this can only be achievable if professionals working in such establishments possess the academic as well as technical capacity for mycotoxin management which calls for the inclusion of courses on mycotoxins in the curricula for training of agriculturists, medical personnel and laboratory based scientists. Awareness on the adverse impact of mycotoxins should not be limited to professionals in the food and feed related industries, but to the entire consumers. Public awareness campaign on impact and prevention of mycotoxins especially the notorious AF via electronic and print media and other information dissemination modes is therefore an imperative. Such scientific and public enlightenment interventions require concerted national and international multidisciplinary strategies (WHO, 2006). It is but imperative to engage both national and international bodies to partner with one another to effectively manage mycotoxins. For example in Nigeria, experts in devised but related fields from academia and research institutions formed the Nigerian Mycotoxin Awareness and Study Network in 2005 (NMASNwww.ngmycotoxin.org) with a common goal of offering scientific and technical support towards managing mycotoxins in the country. In doing so, the network organizes annual workshops for stakeholders. Similarly this year (2011), the International Society of Mycotoxicology organised a world conference in Cape Town, South Africa on mycotoxin reduction (www.mycoredafrica2011.co.za) which brought scientists from all over the globe to not only share knowledge and expertise but to establish research collaborations towards strengthening the capacity of the African mycotoxicologists and laboratories. European Union (Leslie et al., 2008) and World Health Organisation (WHO, 2006) had earlier organized such international conferences in 2005 and 2006 in Ghana and Congo Brazzaville, respectively. It is only pertinent now to encourage scientists and institutions involved in mycotoxin research in Africa to collaboratively seek accessible research grants from the AU, EU and other foreign funding agencies for more effective investigations and control of mycotoxins.

SUMMARY

Aflatoxins are toxic secondary metabolites produced notably by Aspergillus flavus and A. parasiticus that frequently invade foods and feedstuffs before and after harvest. The four major aflatoxins include aflatoxin B_1, B_2, G_1 and G_2 with aflatoxin B_1 recognized as the most prevalent and toxic of all aflatoxins. Their presence in foods and feeds is inevitable and as such, humans and animals are exposed to them on a continuous basis leading to a wide array of health complications. Particularly aflatoxins B_1, they have been directly linked to hepatocarcinoma and deaths among humans and animals. Although this may be the case worldwide, the situation in sub-Saharan Africa is very severe as increased levels of exposure to this group of mycotoxins is a common phenomenon since it presents suitable environmental conditions for aflatoxin concentration in various food and feed materials. Again, the problem is further exacerbated by increased prevalence in the continent, of such endemic diseases as malaria, hepatitis and HIV/AIDS. In Africa recently, we have experienced the most fatal aflatoxinpoisoning outbreaks including two episodes in Kenya and one in Nigeria. In view of the significance therefore, of aflatoxin exposure, this chapter reviews the disparity in aflatoxin contamination of food and feeds worldwide with particular emphasis on Africa. It has also expounded briefly on those factors that influence the distribution of aflatoxins in various food and feeds. Additionally, an in-depth review is provided on the negative public health problems and the impact in the economy associated with this notorious group of secondary metabolites with particular reference made from the African context, while also discussing those control strategies possible within the continent's technological capacity.

REFERENCES

1.   Abalaka, J.A (1984). Aflatoxin distribution in edible oil-extracting plants and in poultry feed mills. Food and Chemical Toxicology 22 (6): 461-463

2.   Abbas, H. K., Zablotowicz, R. M., Horn, B. W., Phillips, N. A., Johnson, B. J., Jin, X. and Abel, C. A. (2011) Comparison of majo control strains of non-toxigenic Aspergillus flavus for the reduction of aflatoxins and cyclopiazonic acid in maize, Food Additives and Contaminants Part A 28, 198-208.

3.   Abdallah, M.I.M.; Manal, M.A.*; Dawoud, A.S. And Marouf, H.A. Occurrence of aflatoxins in table eggs sold in Damietta City regarding its health significance. Retrieved on 15th July, 2011 from www.nu.edu.sa/ uploads/sss/2/7/1/3/7.pdf

4.   Abdelhamid, A.M (2008). Thirty Years (1978 - 2008) of Mycotoxins Research at Faculty of Agriculture, Almansoura University, Egypt. Retrieved on 1st June, 2011 from www. engormix.com

5.   Abulu, E.O., Uriah, N., Aigbefo, H.S., Oboh, P.A. and Agbonlahor, D.E. (1998) Preliminary investigation on aflatoxin in cord blood of jaundiced neonates West African Journalof Medicine 17, 184-187.

6.   Adebayo-Tayo, B. C., Adegoke, A. A. and Akinjogunla, O. J. (2009). Microbial and physicochemical quality of powdered soymilk samples in Akwa Ibom, South Southern Nigeria. African Journal of Biotechnology 8 (13): 3066-3071

7.   Adebayo-Tayo, B. C. and Ettah, A E. (2010). Microbiological quality

and aflatoxin B1 level in poultry and livestock feeds. Nigerian Journal of Microbiology, 24(1): 2145 – 2152

8.   Adebayo-Tayo, B.C., Onilude, A.A., Ogunjobi, A.A., Gbolagade, J.S. and M.O. Oladapo(2006). Detection of fungi and aflatoxin in shelved bush mango seeds (Irvingia spp.) stored for sale in Uyo, Nigeria African Journal of Biotechnology 5 (19):1729-1732

9.   Adebayo-Tayo, B.C., Onilude, A.A. and Ukpe, G.P. (2008) Mycofloral of Smoke-Dried Fishes Sold in Uyo, Eastern Nigeria World Journal of Agricultural Sciences 4 (3): 346-350

10.   Adhikari, M., Ramjee, G. and Berjak, P. (1994) Aflatoxin, kwashiorkor and morbidity Natural Toxins 2, 1-3.

11.   Afla-guard, (2005). Aflatoxin in Africa. Retrieved on 11th May, 2011 fromwww.circleoneglobal.com/aflatoxin_africa.htm. Afriyie-Gyawu, E. Ankrah, N.A., Huebner, H.J., Ofosuhene, M., Kumi, J., Johnson, N.M., Tang,L., Xu, L., Jolly, P.E., Ellis, W.O., Ofori-Adjei, D., Williams, J.H., Wang, J.S., Phillips,

12.   T.D. (2008) Novasil clay intervention in Ghanaians at high risk for aflatoxicosis. I. Study design and clinical outcomes Food Additives and Contaminants Part A 25, 76-87.

13.   Agboola S.D (1992). Post harvest technologies to reduce mycotoxin contamination of food crops. In Z.S.C Okoye (ed) Book of proceedings of the first National Workshop on Mycotoxins held at University Jos, on the 29th November, 1990, 73-88.

14.   Akinmusire, O.O. (2011). Fungal species associated with the spoilage of some edible fruits inMaiduguri, Northern Eastern Nigeria Advances in Environmental Biology 5(1): 157-161

15.   Akrobortu, D.E. (2008). Aflatoxin contamination of maize from different storage locations in Ghana. An M.Sc. Thesis submitted to the Department of Agricultural Engineering, Kwame Nkrumah University of Science and Technology, Ghana. 27-32

16.   Amer, A.A. and Ibrahim, M.A.E. (2010) Determination of aflatoxin M1 in raw milk and traditional cheeses retailed in Egyptian markets Journal of Toxicology and Environmental Health Sciences 2 (4): 50-53

17.   Anjorin, S.T, Makun, H.A, and Iheneacho, H.E (2008). Effect of Lippia Multiflora leaf extract and Aspergillus flavus on germination and vigour indices of Sorghum Bicolor [L] (Moench). International Journal of Tropical Agriculture and Food System, 2 (1): 130 – 134.

18.   Apeagyei, F., Lamplugh, S. M., Hendrickse, R. G., Affram, K. and Lucas,

S. (1986) Aflatoxin in the livers of children with kwashiorkor in Ghana Tropical and GeographicalMedicine 38, 273-276.

19. Aroyeun, S.O and Jayeola, C.O (2005). Mycotoxins in cocoa. A paper presented at the Regional Workshop on Mycotoxins organized by National Agency for Food and Drug Administration and Control (NAFDAC) in collaboration with International Atomic Energy Agency (IAEA), Held at Meidan Hotels, Victoria Garden City, Lagos, Nigeria between 7th and 11th February, 2005.

20. Ashfaq, U. A., Javed, T., Rehman, S., Nawaz, Z. and Riazudding, S. (2011) An overview of HCV molecular biology, replication and immune response Virology Journal 8, In Press.

21. Atehnkeng, J., Ojiambo, P.S., Donner, M., Ikotun, K.,Sikora, R.A.,Cotty, P.J. and Bandypadhyay, R. (2008). Distribution and toxicity of Aspergillus species isolated from maize kernels from three agro-ecological zones of Nigeria. International Journal Food Microbiology. 122, 74–84.

22. Atanda, O., Oguntubo, A., Adejumo, O., Ikeorah, J. and Akpan, I. (2007). Aflatoxin M1 contamination of milk and ice cream in Abeokuta and Odeda local governments of Ogun State, Nigeria Chemosphere 68, 1455–1458

23. Aziz, N.H. and Youssef, Y.A (1991). Occurrence of aflatoxins and aflatoxin-producing moulds in fresh and processed meat in Egypt Food Additives and Contaminants. 8 (3):321-31

24. Azziz-Baumgartner, E., Lindblade, K., Gieseker, K., Rogers, H. S., Kieszak, S., Njapau, H., Schleicher, R., McCoy, L. F., Misore, A., DeCock, K., Rubin, C. and Slutsker, L.(2006) Case control study of an acute aflatoxicosis outbreak, Kenya 2004 Environmental Health Perspectives 113, 1779-1783.

25. Baiyewu, R.A., Amusa, N. A., Ayoola, O.A. and Babalola, O.O. (2007). Survey of the postharvest diseases and aflatoxin contamination of marketed pawpaw fruit (Caricapapaya L) in South Western Nigeria African Journal of Agricultural Research 2 (4):178-181.

26. Bandyopadhyay, R., Kumar, M. and Leslie, J.F. (2007). Relative severity of aflatoxin contamination of cereal crops in West Africa Food Additives and Contaminants 24(10):1109-14.

27. Bankole, S.A and Adebanjo, A. (2003). Mycotoxins in food in West Africa: current situation and possibilities of controlling it. African Journal of Biotechnology 2 (9): 254-263.

28. Bankole, S.A. and Mabekoje, O.O (2004). Mycoflora and occurrence of aflatoxin B1 in dried yam chips from markets in Ogun and Oyo States,

Nigeria Mycopathologia. 157(1):111-5

29. Bankole, S.A., Ogunsanwo, B.M. and Eseigbe, D. A (2005). Aflatoxins in Nigerian dryroasted groundnuts. Food Chemistry 89: 503–506.

30. Bankole, S.A., Ogunsanwo, B.M., Osho, A. and Adewuyi, G.A (2006). Fungal contamination and aflatoxin B1 of _egusi_melon seeds in Nigeria. Food Control 17: 814–818.

31. Bankole,S Schollenberger, M. and Drochner, W. (2006). Mycotoxins in food systems in Sub Saharan Africa: A review Mycotoxin Research 22 (3): 163-169

32. Barkai-Golan, R. and Paster, N. (2008). Mycotoxins in Fruits and Vegetables. AcademicPress, San Diego, USA. 3-11

33. Bassa, S., Mestres, C., Champiat, D., Hell, K., Vernier, P. and Cardwell, K. (2009) First report of aflatoxin in dried yam chips in Benin. Plant Disease 85 (9): 1032

34. Becroft, D.M.O. (1966) Syndrome of encephalopathy and fatty degeneration of viscera in New Zealand children British Medical Journal 2, 1351.

35. Becroft, D. M. O. and Webster, D. R. (1972) Aflatoxins and Reye's disease British Medical Journal 14th October, p117.

36. Beucker, S., Raters, M. and Matissek, R. (2005). Mycotoxin Study III: MycoDONA Deoxynivalenol, Ochratoxin A and Aflatoxins in Cocoa and Cocoa-containing Products Analysis, Situation Assessment, and Monitoring. Retrieved on 15th July, 2011 from http://www.kakao-stiftung.de/pdf/Projekt_38e.pdf

37. Bhattacharyya, A. K. (1986) Protein-energy malnutrition (Kwashiorkor-Marasmus syndrome): terminology, classification and evolution World Review of Nutrition and Dietetics 47, 80-133.

38. Bilgrami, K.S., Prasad, T., Misra, R.S. and Sinha, K.K.1981. Aflatoxin contamination in maize under field conditions.Indian Phytopathology. 34, 67–68.

39. Bintvihok, A., Thiengnin, S., Doi, K. and Kumagai, S. (2002). Residues of aflatoxins in the liver, muscles and eggs of domestic fowls Journal of Veterinary Medical Science 64(11): 1035-1037

40. Blondski, W., Wojciech, Kotlyar, D. S. and Forde, K. A. (2010) Non-viral causes ofhepatocellular carcinoma World Journal of Gastroenterology 16, 3603-3615.

41. Blount W.P. (1961) Turkey X disease Turkeys (Journal of the British Turkey Association) 9, 55-58. Bondy, G. S. and Peska, J. J. (2000)

Immunomodulation by fungal toxins Journal of Toxicologyand Environmental Health B 3, 109-143.

42.   Bressac, B., Kew, M., Wanda, J. and Ozturk, M. (1991) Selective G to G mutations of p53gene in hepatocellular carcinoma from southern Africa Nature 350, 429-431.

43.   Bucci, T. J. and Howard, P. C. (1996) Effect of fumonisin mycotoxins in animals Journal of Toxicology Toxin Reviews 15, 293-302.

44.   Brimer, L (2011). Chemical food safety of traditional grains. Retrieved on 1st June, 2011 from http://www.sik.se/traditionalgrains/workshop/proceedings/Leon_Brimer_proc.pdf

45.   Campbell, T.C. and Hayes, J.R. (1976) The role of aflatoxin metabolism in its toxic lesions Toxicology and Allied Pharmacology 35, 199-222.

46.   Center for Disease Control and Prevention (CDC), 2004. Outbreak of aflatoxin poisoning— Eastern and Central provinces, Kenya, January–July, 2004. Retrieved on 20th June, 2011 from http://www.cdc.gov/mmwr/preview/mmwrhtml/mm5334a4.htm.

47.   Chandrashekar, A., Bandyopadhyay, R., and Hall, A.J. (eds.). 2000. Technical and institutional options for sorghum grain mold management: proceedings of an international consultation, 18-19 May 2000, ICRISAT, Patancheru, India. (In En. Summaries in En, Fr.) Patancheru, 502324, Andhra Pradesh, India: InternationalCrops Research Institute for the Semi- Arid Tropics. 299 pp. ISBN 92-9066-428-2. Order code CPE 129.

48.   Chiou, R. Y. Y., Wu, P. Y. and Yen, Y. H. (1994) Color sorting of lightly roasted and deskinned peanut kernels to diminish aflatoxin contamination in commercial lots Journal of Agricultural Food Chemistry 42, 2156-2160.

49.   Chukwu, O. and Imodiboh, L.I. (2009) Influence of Storage Conditions on Shelf-Life of Dried Beef Product (Kilishi) World Journal of Agricultural Sciences 5 (1): 34-39

50.   Ciliberto, H., Ciliberto, M., Briend, A., Ashorn, P., Bier, D. and Manary, M. (2005) Antioxidant supplementation for the prevention of kwashiorkor in Malawian children: randomised, double blind, placebo controlled trial British Medical Journal 330, 1109-1114.

51.   Codex Alimentarius Commission. 2004. Code of Practice for the Prevention and Reduction of Aflatoxin Contamination in Peanuts. Retrieved 13th May, 2011 from http://www.codexalimentarius.net/download/standards/10084/CXC_055_2004e. pdf. Codex Alimentarius Commission. (2011). Discussion paper on mycotoxins in Sorghum. Joint FAO/WHO Food Standards Programme CODEX Committee on

contaminants in foods' 5th Session held in The Hague, The Netherlands on 21 – 25 March 2011.

52. Coulter JB, Lamplugh SM, Suliman GI, Omer MI and Hendrickse RG. (1984) Afltoxins in human breast milk Annals of Tropical Paediatrics 4, 61-66.

53. Coulter, J. B., Hendrickse, R. G., Lamplugh, S. M., MacFarlane, S. B., Moody, J. B., Omer, M. I., Suliman, G. I., and Williams, T. E. (1986) Aflatoxins and kwashiorkor: clinical studies in Sudanese children, Transactions of the Royal Society of Tropical Medicine and Hygiene 80, 945-951.

54. Creppy, E. E., Chiarappa, P., Baudrimont, I., Borracci, P., Moukha, S. and Carratu, M. R. (2004) Synergistic effects of fumonisin B1 and ochratoxin A: are in vitro cytotoxicity data predicitive of in vivo acute toxicity? Toxicology 201, 115-123.

55. Cusumano, V., Rossano, F., Merendino, R. A., Arena, A., Costa, G. B., Mancuso, G., Baroni, A. and Losi, E. (1996) Immunobiological activities of mould products: functional impairment of human monocytes exposed to aflatoxin B1 Research in Microbiology 147, 385-391.

56. Dank, M. (2010) Treatment of primary hepatocellular carcinoma Orv Hetil 151, 1445-1449. (In Hungarian; abstract PubMed 20739261).

57. Del Rio Garcia, J. C., Moreno, R., C., Pinton, P., Mendoza, E. S. and Oswald, I. P. (2007) Evaluation of the cytotoxicity of aflatoxin and fumonisins on swine intestinal cells Rev Iberoam Micol 24, 136-141. (In Spanish; abstract PubMed 17604433).

58. de Vries, H. R. and Hendrickse, R. G. (1988) Climatic conditions and kwashiorkor in Mumias: a retrospective analysis over a 5-year period Annals of Tropical Paediatrics 8, 268-270.

59. de Vries, H. R., Lamplugh, S. M. and Hendrickse, R. G. (1987) Aflatoxins and kwashiorkor in Kenya: a hospital based study Annals of Tropical Paediatrics 7, 249-251.

60. de Vries, H.R., Maxwell S.M. and Hendrickse, R.G. (1989) Foetal and neonatal exposure to aflatoxins Acta Paediatric Scandanavia 78, 373-378.

61. de Vries, H. R., Maxwell, S. M. and Hendrickse, R. G. (1990) Aflatoxin excretion in children with kwashiorkor or marasmic kwashiorkor - a clinical investigation Mycopathologia 110, 1-9.

62. Donnelly, P.J., Stewart, R.K., Ali, S.L., Conlan, A.A., Reid, K.R., Petsikas, D. and Massey, T.E.(1996) Biotransformation of aflatoxin B1 in human lung Carcinogenesis 17, 2487-2494.

63. Dvorakova, I., Stora, C. and Ayraud, N. (1981) Evidence for aflatoxin B1 in two cases of lung cancer in man Journal of Cancer Research and Clinical Oncology, 100 221-224.

64. Dutton, M. F (2004) Chapter 13 Fumonisin B1 in animal and human health. In Recent Researches on Fungi Ed. R.K.S. Kushwaha, Scientific Publishers (India) Jodhpur. Dutton, M. F. (2009) The African Fusarium/ maize disease, Mycotoxin Research 25, 29-39.

65. Dutton, M., Mwanza, M., de Kock, S., Khilosia, L. (2012). Mycotoxins in South African foods: a case study on aflatoxin in milk. Mycotoxin Research 28: 19-25

66. Elgerbi, A. M., Aidoo, K. E. Candlish, A. A. G. and Tester, R. F. (2004). Occurrence ofaflatoxin M1 in randomly selected North African milk and cheese samples Food Additives & Contaminants: Part A, 21 (6) : 592 - 597

67. Elzupir AO, Younis M.H, Himmat Fadul M, Elhussein AM (2009) Determination of Aflatoxins in Animal Feed in Khartoum State, Sudan. Journal of Animal and Veterinary Advances 8 (5): 1000-1003

68. Enwonwu, C. O. (1984) The role of dietary aflatoxin in the genesis of hepatocellular carcinoma in developing countries, Lancet 2 (8409), 956-958.

69. Essono, G., Ayodele, M., Akoa, A., Foko, J., Filtenborg, O. and Olembo, S. (2009). Aflatoxinproducing Aspergillus spp. and aflatoxin levels in stored cassava chips as affectedby processing practice Food Control 20 648–654

70. Fapohunda, S.O (2011).Impact of Mycotoxins on Sub-Saharan Africa : Nigeria as a Case Study. European Mycotoxin Awareness Network Retrieved on 7th July, 2011 fromhttp://www. services.leatherheadfood. com/mycotoxins/index.asp

71. F.A.O. (2003) Manual on the application of HACCP system to mycotoxin control. Retrievedon 13th May, 2011 from http://www.fao. org/docrep/005/y1390e/y1390e00.htm).

72. F.A.O (1983). Post harvest losses in quality of food grains. Food and AgricultureOrganisation (Food and Nutrition Paper No 29, pg. 103

73. FAOSTAT (2006). Production statistics. Retrieved on July, 2010 from www.faostat.fao.orgFAOSTAT, (2010) Crops primary equivalent. Retrieved on 16th May, 2011 fromwww.faostat.fao.org

74. FDA (2011), Hazard Analysis Critical Control Point. Retrieved on 13th May, 2011 fromhttp://www.fda.gov/food/foodsafety/

hazardanalysiscriticalcontrolpointshaccp/ default.htm)

75. Fellinger, A. (2006). Worldwide mycotoxin regulations and analytical challenges. WorldGrain Summit: Foods and Beverages, September 17–20, 2006, San Francisco,California, USA.

76. Ferenci, P., Fried ,M., Labrecque, D., Bruix, J., Sherman, M., Omata, M., Heathcote, J., Piratsivuth, T., Kew, M., Otegbayo, J.A., Zheng, S.S., Sarin, S., Hamid, S., Modawi, S.B., Fleig, W., Fedail, S., Thomson, A., Khan, A., Malfertheiner, P., Lau, G., Carillo, F.J., Krabshuis, J., Le Mair, A., World Gastroenterology Organisation Guidelinesand Publications Committee (2010) World gasteroenterology organisation guideline. Hepatocellular carcinoma (HCC): a global perspective, Journal of Gastrointestinal Liver Disease 19, 311-317.

77. Fernández-Cruz, M.L., Mansilla, M.L. and Tadeo, J.L. (2010). Mycotoxins in fruits and theirprocessed products: Analysis, occurrence and health implications Journal of Advanced Research 1, 113–122

78. Frisvad, J.C. and Samson, R.A (2004). Emericella venezuelensis, a New Species with Stellate Ascospores Producing Sterigmatocystin and Aflatoxin B1 Systematic and Applied Microbiology. 27, 672-680.

79. Frivad, J.C., Samson, R.A. and Smedsgaard, J. (2004) Emericella astellata, a new producer of aflatoxin B1, B2 and sterigmatocystin. Letters in Applied Microbiology 38, 440–445

80. Frisvad, J. C., Skouboe, P., and Samson, R. A. (2005). Taxonomic comparison of three different groups of aflatoxin producers and a new efficient producer of aflatoxin B1, sterigmatocystin and 3-O-methylsterigmatocystin, Aspergillus rambellii sp. nov. Systematic AppliedMicrobiology, 28, 442–453.

81. Gallagher, E.P., Wiekers, L.C., Stapleton, P.L., Kunze, K.L and Eaton, D.L. (1994) Role of human microsomal and human complementary DNA expressed cytochromes P4501A2 and 3A4 in the bioactivation of aflatoxin B1 Cancer Research 54, 101-108.

82. Gbodi, T.A. (1986). Studies of mycoflora and mycotoxins in Acha, maize and cotton seed in plateau state, Nigeria. A Ph. D thesis, submitted to Department of Physiology and Pharmacology, Faculty of Veterinary Medicine, A.B.U, Zaria, pg. 1-213.

83. Ghali, R., Belouaer, I., Hdiri, S., Ghorbel, H., Maaroufi, K. and Hedilli, A. (2009) Simultaneous HPLC determination of aflatoxins B1, B2, G1 and G2 in Tunisian sorghum and pistachios. Journal of Food Composition and Analysis 22, 751–755

84. Ghali, R., Khlifa, K. H., Ghorbel, H., Maaroufi, K. and Hedilli, A. (2008).

Incidence of aflatoxins, ochratoxin A and zearalenone in Tunisian foods. Food Control 19, 921–924

85. Ghali, R., Khlifa, K. H., Ghorbel, H., Maaroufi, K. and Hedilli, A. (2010) Aflatoxin determination in commonly consumed foods in Tunisia, Journal of the Science of Foodand Agriculture 90, 2347-2351.

86. Golden, M.H.N., Golden, B.E. and Jackson, A.A., (1980) Kwashiorkor and nutritionaloedema Lancet 1 (8180), 114-116.

87. Golden, M.H.N. and Ramdath, D. (1987) Free radicals in the pathogenesis of kwashiorkor Proceedings of Nutrition Society 46, 53-68.

88. Gong, Y.Y., Egal, S., Hounsa, S., Hall, A.J., Cardwell, K.F., Wild, C.P., (2003). Determinants of aflatoxin exposure in young children from Benin and Togo, West Africa: the critical role of weaning. International Journal of Epidemiology 32, 556–662.

89. Gong, Y., Hounsa, A., Egal, S., Sutcliffe, A.E., Hall, A.J., Cardwell, K.F., Wild, C.P., (2004).Postweaning exposure to aflatoxin results in impaired child growth: a longitudinal study in Benin, West Africa. Environmental Health Perspectives 112, 1334–1338.

90. Gopalan, P., Jensen, D. E. and Lotikar, P. D. (1992) Glutathione conjugation of microsome mediated and synthetic aflatoxin B1-8,9-oxide by purified glutathione S-transferase from rats Cancer Letters 64 225-233.

91. Gnonlonfin, G.J.B., Hell, K., Fandohan, P. and Siame, A.B. (2008). Mycoflora and natural occurrence of aflatoxins and fumonisin B1 in cassava and yam chips from Benin West Africa International Journal of Food Microbiology

92. Groopman, J. D., Hall, A. J., Whittle, H., Hudson, G. J., Wogan, G. N., Montesano, R. and Wild, C. P. (1992) Molecular dosimetry of aflatoxin-N7-guanine in human urine obtained in the Gambia, West Africa Cancer Epidemiological Biomarkers and Prevention 1, 221-227.

93. Harwig, J., Przybylski, W. and Moodie, C. A. (1975) A link between Reye's syndrome and aflatoxins? Canadian Medical Association Journal 113, 281.

94. Hatem, N. L., Hassab, H. M., Abd Al-Rahman, E. M., El-Deeb, S. A. and El-Sayed Ahmed, R. L. (2005) Prevalence of aflatoxins in blood and urine of Egyptian infants Food and Nutrition Bulletin 26, 49-56.

95. Hayes, R.B., Van Nieuwenhuize, J.P., Raatgever, J.W. and Ten Kate, F.J.W. (1984) Aflatoxin exposures in industrial setting: an epidemiological study of mortality Food Chemistry and Toxicology 22, 39-43.

96. Hell, k. and Mutegi, C. (2011). Aflatoxin control and prevention strategies in key crops of Sub Saharan Africa. African Journal of Microbiology Research 5 (5):459-466

97. Hell, K., Cardwell, K.F., Setamou, M., Poehling, H.M., (2000). The influence of storagepractices on aflatoxin contamination in maize in four agroecological zones of Benin, West Africa. Journal of Stored Products Research 36,365–382.

98. Hell, K., Gnonlonfin, .G.J., Kodjogbe, G., Lamboni, Y. and Abdourhamane, I.K. (2009). Mycoflora and occurrence of aflatoxin in dried vegetables in Benin, Mali and Togo, West Africa International Journal of Food Microbiology 135, 99–104

99. Hendrickse, R. G. (1991) Clinical implications of food contaminated by aflatoxins Annals of Academic Medicine Singapore 20, 84-90.

100. Hendrickse, R. G., Coulter, J. B., Lamplugh, S. M., MacFarlane, S. B., Williams, T. E., Omer, M. I., Suliman, M. I., and El-Zorganui, G. A. (1983) Aflatoxins and kwashiorkor. Epidemiology and clinical studies in Sudanese children and findings in the autopsy liver samples from Nigeria and South Africa Bulletin of the Exotic Pathology Society 76, 559-566

101. Holbrook, C.C., Guo, B.Z., Wilson, D.M., Kvien, C., (2004). Effect of drought tolerance on preharvest aflatoxin contamination in peanut. Proceedings of the 4th International Crop Science Congress Brisbane, Australia, 26 Sep–1 Oct 2004

102. Househam, K. C. and Hundt, H. K. (1991) Aflatoxin exposure and its relationship to kwashiorkor in African children Journal of Tropical Pediatrics 37, 300-302.

103. Hsieh, D. P. H., Cullen, J. M. and Ruebner, B. H. (1984) Comparative hepatocarcinogenicity of aflatoxins B1 and M1 in the rat Food and Chemical Toxicology 22, 1027-1028.

104. Hussain, Z., Khan, M.Z., Khan, A., Javed, I., Saleemi, M.K., Mahmood, S. and Asi, M.R.(2010). Residues of aflatoxin B1 in broiler meat: Effect of age and dietary aflatoxin B1 levels Food and Chemical Toxicology 48 (12): 3304-3307

105. Hussein, H. S., and Brasel, J. M. (2001) Toxicity, metabolism and impact of mycotoxins onhumans and animals, Toxicology 167, 101-134.

106. International Crops Research Institute for the Semi-Arid Tropics (ICRISAT) (2010). Assessing occurrence nd distribution of aflatoxins in Malawi. Project Final Report (Grant No. 08-598). Retrieve on 1st June, 2011 from http://mcknight.ccrp.cornell.edu/program_docs/

project_documents/SAF_0642_g   roundnut_breeding/Assessing%20
Occurrence%20and%20Distribution%20of%20Aflatoxin%20%20
final2.pdf.

107. Ibeh, I. N. and Saxena, B. N. (1997) Aflatoxin B1 and reproduction. I. Reproductive performance in female rats, African Journal of Reproductive Health 1, 79-84.

108. Ibeh IN, Uraih N, Ogonar JI. (1994) Dietary exposure to aflatoxin in human male infertility in Benin City, Nigeria. International Journal of Fertility and Menopausal Studies 39, 208-214

109. IPCC, (2007) International Panel on Climate Change. Fourth Assessment Report. Retrieved on 16th May, 2011 from www.ipcc.ch/publications_ and_data/publications_and_data_reports.shtml

110. Ito, Y., Peterson, S. W., Wicklow, D. T., & Goto, T. (2001). Aspergillus pseudotamarii, a new Aflatoxin producing species in Aspergillus section Flavi. Mycological Research, 105, 233–239.

111. Jackson, P. E., Qian, G.-S., Friesen, M. D., Zhu, Y.-R., Lu, P., Wang, J.-B., Wu, Y., Kensler, T. W., Vogelstein, B. and Groopman, J. D. (2001) Specific p53 mutations detected in plasma and tumors of hepatocellular carcinoma patients by electrospray ionization mass spectrometry, Cancer Research 61, 33-35.

112. Jarvis, B (1971). Factors affecting the production of mycotoxins Journal of Applied Bacteriolog. 34(1):199-213.

113. Jarvis, B. (1976). Mycotoxins in food. In: Skinner, F.A and Carr, J.G (Eds), Microbiology inAgriculture, Fisheries and Food, Academic Press, London, pp. 251-267.

114. JECFA, (2001). WHO Food additives series:47 safety evaluation of certain food additives and contaminants: Aflatoxin M1. Joint FAO/WHO Expert Committee on Food Additives

115. Jiang, Y., Jolly, P. E., Preko, P., Wang, J. S., Ellis, W. O., Phillips, T. D. and Williams, J. H. (2008) Aflatoxin related immune dysfunction in health and in human immunodeficiency virus disease, Clinical and Developmental and Immunology 2008, 12 pages. (On Line Free access).

116. Jimoh, K.O. and Kolapo, A.L. (2008). Mycoflora and aflatoxin production in market samples of some selected Nigerian foodstuffs. Research Journal of Microbiology 3 (3): 169-174

117. Jonathan, G., Ajayi, I. and Omitade, Y.(2011). Nutritional compositions, fungi and aflatoxins detection in stored 'gbodo' (fermented Dioscorea rotundata) and 'elubo ogede' (fermented Musa parasidiaca) from South

western Nigeria. African Journal of Food Science 5(2): 105 – 110.

118. Jonathan, S. G. and Esho E.O. (2010) Fungi and aflatoxin detection in two stored oyster mushroom (Pleurotus ostreatus and Pleurotus pulmonarius) from Nigeria Electronic Journal of Environmental, Agricultural and Food Chemistry 9 (11): 1722-1736

119. Jones, W. R., Johnston, D. S. and Stone, M. P. (1998) Refined structure of the doubly intercalated d(TATafbGCATA)2 aflatoxin B1 adduct Chemical Research in Toxicology 11, 873-881.

120. Jones, M. J., Tanya, V. N., Mbofiing, C. M.F., Fonkem, D. N. and Silverside, D. E. (2001) A Microbiological and nutritional evaluation of the West African dried meat product, Kilishi The Journal of Food Technology in Africa, 6 (4), 126-129

121. Juan, C., Zinedine, A., Molto, J.C., Idrissi, L. and Man~es, J. (2008). Aflatoxins levels in dried fruits and nuts from Rabat-Sale´ area, Morocco Food Control 19, 849–853

122. Kaaya, N.A and Kyamuhangire, W. (2006). The effect of storage time and agroecological zone on mould incidence and aflatoxin contamination of maize from traders in Uganda. International Journal of Food Microbiology 110, 217–223

123. Kaaya, N.A. and Warren, H.L. (2005). A review of past and present research on aflatoxin in Uganda African Journal of Food Agriculture Nutrition and Development 2005, 18 pages (On Line Free Access)

124. Kamdem, L. K., Meineke, I., Godtel-Armbrust, U., Brockmoller, J. and Wojnowski, L. (2006) Dominant contribution of P450 3A4 to the hepatic activation of aflatoxin B1 Chemical Research in Toxicology 19, 577-586.

125. Kamika, I. and Takoy, L.L (2011). Natural occurrence of Aflatoxin B1 in peanut collected from Kinshasa, Democratic Republic of Congo. Food Control. In press.

126. Kang.ethe, E.K. and Lang.a, K.A. (2009). Aflatoxin B1 and M1 contamination of animal feeds and milk from urban centers in Kenya. African Health Sciences 9 (4): 218-226

127. Kastner, S., Kandler, H., Hotzb, K., Bleisch, M., Lacroix, C, Meile, L.(2010). Screening for mycotoxins in the inoculum used for production of attie´ke´ a traditional Ivorian cassava product. Food Science and Technology 43: 1160–1163

128. Katerere, D.R., Stockenstrom, S., Thembo, K.M., Rheeder, J.P., Shephard, G.S. and Vismer,H.S. (2008). A preliminary survey of mycological and fumonisin and aflatoxin contamination of African traditional herbal

medicines sold in South Africa Human and Experimental Toxicology 27 (11): 793-798

129. Kelly, J.D., Eaton, D.L., Guengerich, F.P. and Colombe (1997) Aflatoxin B1 activation in human lung Toxicology and Applied Pharmacology 144, 88-95.

130. Kew, M. C. (2005) Prevention of hepatocellular carcinoma, HPB 7, 16-25.

131. Kew, M. C. (2010a) Hepatocellular carcinoma in African blacks: recent progress in etiology and pathogenesis World Journal of Hepatology 2, 65-73.

132. Kew, M. C. (2010b) Prevention of hepatocellular carcinoma Annals of Hepatology 9, 120 - 132.Kew, M. C. and Asare, G.A. (2007) Dietary iron overload in the African and hepatocellular carcinoma Liver International 27, 735-741. Kew, M. C. and Popper, H. (1984) Relationship between hepatocellular carcinoma and cirrhosis Seminars in Liver Disease 4, 136-146.

133. Kimura, A. (2011) Reye's syndrome and Reye-like syndrome, Nippon Rinsho 69, 455-459. (In Japanese, PubMed 21400838).

134. Kirk, G. D., Lesi, O. A., Mendy, M., Akano, A. O., Sam, O., Goedert, J. J., Hainaut, P., Hall, A. J., Whittle, H. and Montesano, R. (2004) The Gambia liver cancer study: infection

135. with hepatitis B and the risk of hepatocellular carcinoma in west Africa Hepatology39, 211-219.

136. Kotan, E., Alpsoy, L., Anar, M., Aslan, A. and Agar, G. (2011) Protective role of methanolextracts of Cetrartia islandica against oxidative stress and genotoxic effects of aflatoxinB1 in human lymphocytes in vitro Toxicology and Industrial Health 2011, In press.

137. Kpodo, K., Sorensen, A.K. and Jakobsen, M.(1996). The occurrence of mycotoxins infermented maize products Food Chemistry 56 (2):147-153.

138. Krishnamachari, K. A., Bhat, R. V., Nagarajan, V. and Tilak, T. B. (1975) Hepatitis due toaflatoxicosis. An outbreak in Western India Lancet 1(7915), 1061-1063.

139. Kuniholm, M. H., Lesi, O. A., Mendy, M., Akano, A. O., Sam, O., Hall, A. J., Whittle, H., Bah,E., Goedert, J. J., Hainaut, P. and Kirk, G. D. (2008) Aflatoxin exposure and viralhepatitis in the etiology of liver cirrhosis in the Gambia West Africa Environmental Health Perspectives 116, 1553-1557.

140. Lamplugh, S. M. and Hendrickse, R. G. (1982) Aflatoxins in the livers

of children with kwashiorkor Annals of Tropical Paediatrics 2, 101-104.

141. Lane, K.S, (2005). New support for FDA regulation of tobacco. Retrieved on July, 2006 from www.Tobacco.org.

142. Langouet, S., Coles, B., Morel, F., Becquemonet, L., Beaune, P., Guengrich, P., F., Ketterer, B. and Guillouzo, A. (1995) Inhibition of CYP1A2 and CYP3A4 by olitpraz results in reduction of aflatoxin B1 metabolism in human hepatocytes in primary culture Cancer Research 55, 5574-5579.

143. Lata, J. (2010) Chronic liver disease as tumor precursors, Digestive Diseases 28, 596-599. Leslie, J.F., Bandyopadhyay, R. and Visconti, A. (2008), Mycotoxins: Detection methods, management, public health and agricultural trade. Cromwell Press, Trowbridge, UK. Pp 476

144. Lilleberg, S. L., Cabonce, M. A., Raju, N. R., Wagner, L. M. and Kier, L. D. (1992) Alterations in the structural gene and expression of p53 in rat liver tumors induced by aflatoxin B1 Molecular Carcinogenesis 6, 159-172.

145. Lin, C. A., Boslaugh, S., Ciliberto, H. M., Maleta, K., Ashorn, P., Briend, A. and Manary, M. J. (2007) A prospective assessment of food and nutrient intake in a population of Malawian children at risk for kwashiorkor Journal of Paediatric and

146. Gastroenterological Nutrition 44, 487-493. Liu, Y. and Wu, F. (2010) Global burden of aflatoxin-induced hepatocellular carcinoma: a risk assessment Environmental Health Perspectives 118, 818-824.

147. Livni, N., Eid, A., Ilan, Y., Rivkind, A., Rosenmann, E., Blendis, L. M., Shouval, D. and Galun, E. (1995) p53 expression in patients with cirrhosis with and without hepatocellular carcinoma Cancer 75, 2420-2426.

148. Long, X. D., Ma, Y., Zhou, Y. F., Yao, J. G., Ban, F. Z., Huang, Y. Z. and Huang, B. G. (2009) XPD codon 312 and 751 polymorphisms and aflatoxin B1 exposure and hepatocellular carcinoma risk BMC Cancer 9, 400-409.

149. Lubick, N (2010) Examining DDT's urogenital effects. Environmental Health Perspectives 118,A18

150. Luongo, D., Severino, L., Bergamo, P., De Luna, R., Lucisano, A. and Rossi, M. (2006) Interactive effects of fumonisin B1 and alpha-zearalenol on proliferation and cytokine expression in Jurkat T cells Toxicology in Vitro 20, 1403-1410.

151. Lye, M. S., Ghazali, A. A., Mohan, J., Alwin, N. and Nair, R. C. (1995) An outbreak of acute hepatic encephalopathy due to severe aflatoxicosis

in Malaysia American Journal of Tropical Medical Hygiene 53, 68-72.

152. Maenetje, P.W. and Dutton, M.F (2007). The incidence of fungi and mycotoxins in South African barley and barley products Journal of Environmental Science and Health, Part B: 42, (2) 229 - 236

153. Makun, H.A, Anjorin, S.T., Moronfoye, B., Adejo, F.O., Afolabi, O.A., Fagbayibo, G., Balogun, B.O. and Surajudeen, A.A. (2010). Fungal and aflatoxin contaminations of some human food commodities in Nigeria. African Journal of Food Sciences. 4 (4): 127 – 135

154. Makun, H. A., Dutton, M.F., Njobeh, P.B., Mwanza, M. and Kabiru A.Y. (2011) Natural multi- mycotoxin occurrence in rice from Niger State, Nigeria Mycotoxin Research. 27 (2): 97-104.

155. Makun HA, Gbodi TA, Akanya HO, Sakalo AE, Ogbadu HG (2007) Fungi and somemycotoxins contaminating rice (Oryza sativa) in Niger state, Nigeria. African Journal of Biotechnology 6(2):99–108

156. Makun HA, Gbodi TA, Akanya HO, Salako EA, Ogbadu GH (2009) Fungi and some mycotoxins found in mouldy Sorghum in Niger State, Nigeria. World Journal of Agricultural Sciences 5(1):5–17

157. Manary, M. J. and Brewster, D. R. (1997) Potassium supplement in kwashiorkor Journal of Paediatric and Gasteroenterological Nutrition 24, 194-201.

158. Manson, M. M., Ball, H. W. L., Barrett, M. C., Clark, H. L., Jujah, D. J., Williamson, G. and Neal, G. E. (1997) Mechanism of action of dietary chemoprotective agents in rat liver: induction of phase I and II drug metabolizing enzymes and aflatoxin B1 metabolism Carcinogenesis 18, 1729-1738.

159. Matumba, L., Monjerezi, M., Khonga, E. B., Lakudzala, D.D. (2011). Aflatoxins in sorghum, sorghum malt and traditional opaque beer in southern Malawi. Food Control 22,266-268

160. Mashinini, K. and Dutton, M.F. (2006) The incidence of fungi and mycotoxins in South Africa wheat and wheat-based products Journal of Environmental Science and Health Part B, 41:285-296

161. Maxwell, D., Levin, C., Armar-Klemesu, M., Ruel, M., Morris, S., Ahiadeke, C. (2000). Urban Livelihoods and Food and Nutrition Security in Greater Accra, Ghana. IFPRI, Washington, p. 172.

162. May, P. and May, E. (1999) Twenty years of p53 research: structural and functional aspects of the p53 protein Oncogene, 18 7621–36.

163. Martin, P.M.D, and Gilman, G.A. (1976). A consideration of the mycotoxin hypothesis with special reference to the mycoflora of maize,

sorghum, wheat and groundnut. Rep. Trop. Prod. Inst. G105, Vil=112pg. 1-63.

164. Mazen, M.B, El-Kady, I.A and Saber, S.M (1990). Survey of the mycoflora and mycotoxins of cotton seeds and cotton seed products in Egypt. Mycopathologia 110 (3):133-138.

165. MRC (Medical Research Council) (2006). Report on aflatoxins in groundnuts and peanuts products retrieved on 14th July, 2011 from www. doh.gov.za/department/foodcontrol/docs/nmp.html

166. Mestres, C., Bassa, S., Fagbohoun, E., Nago, M., Hell, K., Vernier, P., Champiat, D., Hounhouigan, J. and K. F. Cardwell. (2004). Yam chip food sub-sector: hazardous practices and presence of aflatoxins in Benin. Journal of Stored Products Research 40 (5): 575-585

167. Miller, J.D. (1995). Fungi and mycotoxins in grains: Implications for stored product research. J. Stored. Prod. Res. 31 (1): 1-16

168. Miller, J.D. (1996). Mycotoxins. In: Cardwell, K.F. (Ed.), Proceedings of the Workshop onMycotoxins in Food in Africa. November 6–10, 1995, Cotonou, Benin. International Institute of Tropical Agriculture, Cotonou,Benin, pp. 18–22.

169. Mircea, C., Poiata, A., Tuchilus, C., Agoroae, L., Butnaru, E. and Stanescu, U. (2008) Aflatoxigenic fungi isolated from medicinal herbs Toxicology Letters. 180: 32-246

170. Motawee, M.M., Bauer, J. and McMahon, D.J. (2009) Survey of aflatoxin M1 in cow, goat, buffalo and camel milks in Ismailia-Egypt. Bulletin of the Environmental and Contamination Toxicology, 83, 766-769.

171. Mokhles, M., Abd El Wahhab, M. A., Tawfik, M., Ezzat, W., Gamil, K. and Ibrahim, M. (2007) Detection of aflatoxin among hepatocellular carcinoma in patients in Egypt Pakistan Journal of Biological Sciences 10, 1422-1429.

172. Mngadi, P.T., Govinden, R. and Odhav, B. (2008). Co-occurring mycotoxins in animal feeds. African Journal of Biotechnology 7 (13): 2239-2243.

173. Muhammad, S., Shehu, K., Amusa, N.A., (2004). Survey of the market diseases and aflatoxin contamination of tomato (Lycopersicon esculentum Mill) fruits in Sokoto, northwestern Nigeria. Nutrition and Food Science 34, 72–76.

174. Murray, C. J. L. (1994) Quantifying the burden of disease: the technical basis for disabilityadjustedlife ears Bulletin of the World Health Organisation 72, 429-445.

175. Mutegi, C.K., Ngugi, H.K., Hendriks, S.L. and Jones, R.B.(2009). Prevalence and factors associated with aflatoxin contamination of peanuts from Western Kenya International Journal of Food Microbiology 130, 27–34.

176. Muthomi, J.W., Ndung'u, J.K., Gathumbi, J.K., Mutitu, E.W. and Wagacha, J.M. (2008). The occurrence of Fusarium species and mycotoxins in Kenyan wheat. Crop Protection 27, 1215– 1219.

177. Muzanila, Y.C., Brennan, J.G. and King, R.D. (2000) Residual cyanogens, chemical composition and aflatoxins in cassava ⁻our from Tanzanian villages Food Chemistry 70, 45-49

178. Mwanza, M. (2007) A survey of fungi and mycotoxins with respect to South African domestic animals in the Limpopo Province. Master of Technology Dissertation, University of Johannesburg. http://152.106.6.200:8080/dspace/bitstream/10210/884/1/ Mwanza%20M%20tech%20dissertation. pdf

179. Mwanza, O. W., Otieno, C. F. and Omonge, E. (2005) Acute aflatoxicosis: case report East African Medical Journal 82, 320-324.

180. Nelson, D. B., Kimbrough, R., Landrigan, P. S., Hayes, A. W., Yang, G. C. and Benanides, J. (1980) Aflatoxin and Reye's syndrome: a case study Pediatrics 66, 865-869.

181. Njobeh, B.P., Dutton, M.F., Koch, S.H. and Chuturgoon, A. (2009) Contamination with storage fungi of human foods from Cameroon International Journal of Food Microbiology 135, 193-198.

182. Njobeh B. P., Dutton F. M., Makun, H.A (2010). Mycotoxins and human health: Significance,prevention and control In: Ajay K. Mishra, Ashutosh Tiwari, and Shivani B. Mishra (Eds) 'Smart Biomolecules in Medicine' VBRI Press, India 132-177

183. Nyathi, C. B., Mutiro, C. P., Hasler, J. A. and Chetsanga, C. J. (1987) A survey of urinary aflatoxin in Zimbabwe International Journal of Epidemiology 16, 516-519.

184. Nyathi, C. B., Mutiro, C. F., Hasler, J. A. and Chetsang, C. J. (1989) Human exposure to aflatoxins in Zimbabwa Central African Journal of Medicine 35, 542-545.

185. Nyikal, J., et al. (2004) Outbreak of aflatoxin poisoning - Eastern and Central Provinces, Kenya January-July 2004, Morbidity and Mortality Weekly Control 53, 790-793.

186. Obidoa, O and Gugnani, H.C. (1992). Mycotoxins in Nigerian foods: causes, Consequences and remedial measures. In Z.S.C Okoye (ed) Book

of proceedings of the first National Workshop on Mycotoxins held at University Jos, on the 29th November, 1990, 95-114.

187. Obilana AB. 2002. 'Overview: importance of millets in Africa', Retrieved on 1st June, 2011 from Website: www.afripro.org.uk/papers/ Paper02Obilana.pdf

188. Ocama, P., Nambooze, S., Opio, C. K., Shields, M. S., Wabinga, H. R. and Kirk, G. D. (2009) Trends in the incidence of primary liver cancer in central Uganda. 1960-1980 and 1991-2005 British Journal of Cancer 100, 799-802.

189. Odoemelam, S. A and Osu, C.I (2009). Aflatoxin B1 contamination of some edible grains marketed in Nigeria. E-Journal of Chemistry 6 (2):308-314.

190. Ogbadu, G. (1979) Effect of low gamma irradiation on the production of aflatoxin B1 by Aspergillus flavus growing on Capsicum annuum Microbios letters 10, 139-142.

191. Ogbadu, G. and Bassir, O. (1979) Toxicological study of γ-irradiated aflatoxins using the chicken embryo Toxicology and Applied Pharmacology 51, 379-382.

192. Ogbadu G (1980a) Influence of gamma irradiation of aflatoxin B1 production by Aspergillus flavus growing on some Nigerian foodstuffs. Microbios.; 27(107):19-26.

193. Ogbadu, G. (1981) Ultra structural changes in gamma-irradiated Aspergillus flavus spores Cytobios 30, 167-171.

194. Ogbadu, G. (1988) Use of gamma irradiation to prevent aflatoxin B1 production in smoked dried fish International Journal of Applied Instrumentation 31, 207-207.

195. Ogunsanwo, B.M., Faboya, O.O., Idowu, O.R., Lawal, O.S. and Bankole, S.A (2004). Effect of roasting on the aflatoxin contents of Nigerian peanut seeds. African Journal of Biotechnology 3 (9): 451-455

196. Okoye, Z. S. C. (1987). Carryover of aflatoxin B1 in contaminated substrate corn into Nigerian native beer. Bulletin of Environmental Contamination and Toxicology 37 (4) 482-489

197. Olson, L.C., Bourgeois, C.H., Cotton, R.B., Harikul, S., Grossman, R.A. and Smith, T.J. (1971) Encephalopathy and fatty degeneration of the viscera in northeastern Thailand. Clinical syndrome and epidemiology Paediatrics 47, 707- 716.

198. Omer, R. E., Bakker, M. I., van't Veer, P., Hoogenboom, R. L., Polman, T. H., Alink, G. M., Idris, M. O., Kadaru, A. M. and Kok, F. J. (1998)

Aflatoxin and liver cancer in Sudan Nutrition and Cancer 32, 174-180.

199. Ominski, K.H., Marquardi, R.R., Sinha, R.N and Abramson, D (1994). Ecological aspects of growth and mycotoxin production by storage fungi. In: Miller, J.D and Trenholm, H.L (1994). Mycotoxins in grains: Compounds other than aflatoxins. Eagan Press, St. Paul Minnesota, USA. 287-314.

200. Opadokun, J.S (1992). Occurrence of aflatoxin in Nigerian food crops In Z.S.C Okoye (ed)Book of proceedings of the first National Workshop on Mycotoxins held at University Jos, on the 29th November, 1990, 95-114.

201. Orsi, R. B., Oliveira, A. A., Dilkin, P., Xavier, J. G., Direito, G. M. and Correa, B. (2007) Effects of oral administration of aflatoxin B1 and fumonisin B1 in rabbits(Oryctiolagus cuniculus) Chemical and Biological Interactions 170, 201-208.

202. Otsuki, T., Wilson, J.S. and Sewadeh, M. (2001) What price precaution? European harmonization of aflatoxin regulations and African groundnut exports. European Review of Agricultural Economics 28: 263-283.

203. Oyelamin, O. A., and Ogunlesi, T. A. (2007) Kwashiorkor - is it a dying disease?, South African Medical Journal 97, 65-68.

204. Oyelami, O. A., Maxwell, S. M., Aladekomo, T. A. and Adelusola, K. A. (1995) Two unusual cases of kwashiorkor: can protein deficiency explain the mystery? Annals of Tropical Paediatrics 15, 217-219.

205. Oyelami, O. A., Maxwell, S. M., Adelusola, K. A. and Oyelese, A. O. (1997) Aflatoxins in the lungs of children with kwashiorkor and children with miscellaneous diseases in Nigeria Journal of Toxicology and Environmental Health 51, 623-628.

206. Oyero, G.O and Oyefolu, A.B (2010) Natural occurrence of aflatoxin residues in fresh and sun-dried meat in Nigeria Pan African Medical Journal. 7:14

207. Paterson, R.R.M and Lima, N (2010) How will climate change affect mycotoxins in foods? Food Research International 43, 1902 - 1914

208. Peers, F. G. and Linsell, C. A. (1973) Dietary aflatoxins and human liver cancer. A population based study in Kenya British Journal of Cancer 27, 473-484.

209. Perz, J. F., Armstrong, G. L., Farrington, L. A., Hutin, Y. J. and Bell, B. P. (2006) The contribution of hepatitis B virus and hepatitis C virus infection to cirrhosis and primary liver cancer world wide Journal of Hepatology 45, 529-538.

210. Peterson, S.W., Ito, Y., Horn, B.W., Goto, T. (2001) Aspergillus bombycis,

a new aflatoxigenic species and genetic variation in its sibling species, A. nomius. Mycologia 93, 689–703

211. Pietri, A., Bertuzzi, T., Agosti, B. and Donadini, G. (2010). Transfer of aflatoxin B1 and fumonisin B1 from naturally contaminated raw materials to beer during an industrial brewing process. Food Additives & Contaminants: Part A: Chemistry, Analysis, Control, Exposure & Risk Assessment. 27 (10): 1431 - 1439

212. Pildain, M.B., Frisvad, J.C., Vaamonde, G., Cabral, D., Varga, J. and Samson, R.A. (2008).

213. Two novel aflatoxin-producing Aspergillus species from Argentinean peanuts International Journal of Systematic and Evolutionary Microbiology 58, 725-735.

214. Probst, C., Njapau, H., and Cotty, P. J. (2007) Outbreak of an acute aflatoxicosis in Kenya in 2004: identification of the causal agent, Applied and Environmental Microbiology 73, 2762-2764.

215. Rehana, F. and Basappa, S.C. (1990). Detoxification of aflatoxin B1 in maize by different cooking methods. Journal Food Science Technology. 27: 379-399.

216. Raisuddin, S., Singh, K. P., Zaidi, S. I. A., Paul, B. N. and Ray, P. K. (1993) Immunosuppressive effects of aflatoxin in growing rats Mycopathologia 124, 189-194

217. Ramjee, G., Berjak, P., Adhikari, M. and Dutton, M. F. (1992) Aflatoxins and kwashiorkor in Durban, South Africa Annals of Tropical Paediatrics 12, 241-247.

218. Reiter EV, Dutton MF, Mwanza M, Agus A, Prawano D, Häggblom P, Razzazi-Fazeli E, Zentek J, Andersson G, Njobeh PB (2011) Quality control of sampling for aflatoxins in animal feedingstuffs: Application of the Eurachem/CITAC guidelines. Analyst.136 (19): 4059-4069

219. Reye, R.D.K., Morgan, G. and Baral, J. (1963) Encephalopathy and fatty degeneration of the viscera: a disease entity in childhood Lancet 2: 749-752.

220. Riba, A., Bouras, N., Mokrane, S., Mathieu, F., Lebrihi, A. and Sabaou, N (2010). Aspergillus section Flavi and aflatoxins in Algerian wheat and derived products. Food and Chemical Toxicology 48, 2772–2777

221. Robertson, A. (2005). Risk of aflatoxin contamination increases with hot and dry growing conditions. Integrated Crop Management 185–186. Retrieved on 11th May, 2011 from http://www.ipm.iastate.edu/ipm/icm/2005/9-19/aflatoxin.html.

222. Rogan, W. J., Yang, G. C. and Kimborough, R. D. (1985) Aflatoxin and Reye's syndrome: a study of livers from deceased cases Archives of Environmental Health 40, 91-95.

223. Ryan, N. J., Hogan, G. R., Hayes, A. W., Unger, P. D. and Sirai, M. Y. (1979) Aflatoxin B1; its role in the etiology of Reye's syndrome Paediatrics 64, 71-75.

224. Sahar, N., Ahmed, M., Parveen, Z., Ilyas, A. and Bhutto, A. (2009). Screening of mycotoxinsin wheat, fruits and vegetables grown in Sindh, Pakistan Pakistan Journal Botany.,41(1): 337-341

225. Sánchez-Hervás, M., Gil, J.V., Bisbal, F., Ramón, D and Martínez-Culebras, P.V (2008).Mycobiota and mycotoxin producing fungi from cocoa beans. International Journalof Food Microbiology 125:336–340.

226. Schindler, A.F. (1977) Temperature limits for production of aflatoxin by twenty-five isolatesof Aspergillus flavus and Aspergillus parasiticus. Journal of Food Protection. 40:39–40.

227. Sharma, R. P. (1993) Immunotoxicity of mycotoxins Journal of Dairy Science 76, 892-897.

228. Shuaib, F. M., Ehiri, J., Abdullahi, A., Williams, J. H., and Jolly, P. E. (2010a) Reproductivehealth effects of aflatoxins: a review of the literature, Reproductive Toxicology 29, 262-270.

229. Shuaib, F. M., Jolly, P. E., Ehiri, J., Jiang, Y., Ellis, W. O., Stiles, J. K., Yatich, N. J.,Funkhouser, E., Person, S. D., Wilson, C. and Williams, J. H. (2010b) Associationbetween anemia and aflatoxin B1 biomarker levels among pregnant women,American Journal of Tropical Medical Hygiene 83, 1077-1083.

230. Sibanda L, Marovatsanga LT, Pestka JJ (1997) Review of mycotoxin work in sub-SaharanAfrica. Food Control 8:21–29 647.

231. Sinha, R.N (1984). Journal of Economic Entomology 77: 1463-1488.

232. Smith, T.K., Mehrdad, M. and Ewen, J.M. (2000) Biotechnology in the Feed Industry.Proceedings of Alltech's 16th Annual Symposium, Pp 383–390.

233. Smith, T. K., McMillan, E. G. and Castillo, J. B. (1997) Effect of feeding blends of Fusariummycotoxin-contaminated grains containing deoxynivalenol and fusaric acid ongrowth and feed consumption of immature swine Journal of Animal Science 75, 2184-2191.

234. Smith, J.E and Moss M.O. (1985). Mycotoxins: formation, analysis and significance. JohnWiley & sons. Chichester, Britain, 83-103.

235. South African Department of Health (2004a) Government Gazette 6th

March 2009.Regulations governing tolerance for fungus-produced toxins in foodstuffs. Foodstuff, Cosmetics and Disinfectants Act 1972 (Act 54 of 1972).

236. South African Department of Health (2004b). Government Gazette 6th March 2009. Regulations governing tolerance for fungus-produced toxins in foodstuffs.Foodstuff, Cosmetics and Disinfectants Act 1972 (Act 54 of 1972) 1.

237. Speijers, G. J. and Speijers, M. H. (2004) Combined toxic effects of mycotoxins ToxicologyLetters 153, 91-98.

238. Stoev, S. D., Goundasheva, D., Mirtcheva, T. and Mantle, P. G. (2000) Susceptibility to secondary bacterial infections in growing pigs as an early response in ochratoxicosis Experimental Toxicology and Pathology 52, 287-296.

239. Stoloff, L. (1989) Aflatoxin is not a probable human carcinogen: the published evidence issufficient Regulatory Toxicology and Pharmacology 10, 272-283.

240. Stora, C., Dvorackova, I. and Ayraud, N. (1983) Aflatoxin and Rey's syndrome Journal of Medicine 14, 47-54.

241. Stössel, P. (1986). Aflatoxin contamination in soybeans: role of proteinase inhibitors, zinc availability, and seed coat integrity. Applied and Environmental Microbiology 52, 68–72.

242. Sylla, A., Diallo, M.S., Castegnaro, J., Wild, C.P (1999). Interaction between hepatitis V virus and exposure to aflatoxin in the development of hepatocellular carcinoma: a molecular epidemiological approach. Mutation Research 428:187-196

243. Tandon, H. D., Tandon, B. N. and Ramalingaswami, V. (1978) Epidemic toxic hepatitis in India of possible mycotoxin origin Archives of Pathology and Laboratory Medicine 102, 372-376.

244. Taylor, E. R. (1992) Aflatoxin B1 and DNA adducts. Proposed model for surface noncovalentand covalent complexes with N7 of guanine Journal of Biomolecular Structure and Dynamics 10, 533-550.

245. Tchana, A. N., Moundipa, P. F. and Tchouanguep, F. M. (2010) Aflatoxin contamination in food and body fluids in relation to malnutrition and cancer status in Cameroon International Journal of Environmental Research and Public Health 7, 178-188.

246. Trauner, D. A. (1984) Reye's syndrome The Western Journal of Medicine 141, 206-209.

247. Trenk, H.L., and Hartman, P.A. (1970) Effects of moisture content and

temperature on aflatoxin production in corn. Applied Microbiology. 19:781–784.

248. Trucksess, M. W. and Scott, P. M.(2008) 'Mycotoxins in botanicals and dried fruits: A review', Food Additives & Contaminants: Part A, 25: 2, 181 — 192

249. Tsega, E. (1977) Current views on liver diseases in Ethiopia Ethiopian Medical Journal 15, 75-82.

250. Turkez, H., and Geyikoglu, F. (2010) Boric acid: a potential chemoprotective agent against aflatoxin B1 toxicity in human blood, Cytotechnology 62, 157-165.

251. Turner, P.C., Moore S.E., Hall, A.J., Prentice A.M. and Wild C.P. (2003) Modification of immune function through exposure to dietary aflatoxin in Gambian children Environmental Health Perspectives 111, 217-220.

252. Turner, P.C., Collinson, A.C., Cheung, Y.B., Gong, Y., Hall, A.J., Prentice, A.M. and Wild, C.P. (2007) Aflatoxin exposure in utero causes growth faltering in Gambian infants International Journal of Epidemiology 36, 1119-1125.

253. Udoh, J.M., Cardwel, K.F., Ikotun, T. (2000) Storage structures and aflatoxin content of maize in five agro-ecological zones of Nigeria. Journal of Stored Products Research 36, 187–201. US Grain Council (2000) Manual pp 1-42

254. Van Rensburg, S. J., Cook-Mozaffari, P., Van Schalkwyk, J. J., Van der Watt, J. J., Vincent, T. J. and Purchase, I. F. (1985) Hepatocellular carcinoma and dietary aflatoxins in Mozambique and Transkei, British Journal of Cancer 51, 713-726.

255. Van Vleet, T. R., Klein, P. J. and Coulombe, R. A. (2001) Metabolism of aflatoxin B1 by normal bronchial epithelial cells Journal of Toxicology and Environmental Health 63, 525-540.

256. Van Vleet, T. R., Klein, P. J. and Coulombe, R. A. (2002a) Metabolism and cytotoxicity ofaflatoxin B1 in cytochrome P450 expressing human lung cells Journal of Toxicology and Environmental Health A 65, 853-867.

257. Van Vleet, T. R., Mace, K. and Coulombe, R. A. (2002b) Comparative aflatoxin B1 activation and cytotoxicity in human bronchial cells expressing cytochromes P450 1A2 and 3A4Cancer Research 62, 105-112.

258. Wagacha, J.M., Muthomi, J.W. (2008) Mycotoxin problem in Africa: current status, implications to food safety and health and possible

management strategies. International Journal Food Microbiology. 124, 1–12.

259. Wang, J. S. and Groopman, J. D. (1999) DNA damage by mycotoxin Mutation Research 424, 167-181.

260. Wang, J.-S., Shen, X., He, X. H., Zhu, Y.-R., Zhang, B.-C., Wang, J.-B., Qian, G.-S., Kuang, S.- Y., Zarba, A., Egner, P. A., Jacobson, L. P., Munoz, A., Helzlsouer, Groopman, J. D. and Kensler, T. W. (1999) Protective alterations in Phase 1 and 2 metabolism of aflatoxin B1 by oltipraz in residents of Qidong, People's republic of China Journal of the National Cancer Institute 91, 347-354.

261. Waterlow, J.C. (1984) Kwashiorkor revisited: the pathogenesis of oedema in kwashiorkor and its significance. Transactions of the Royal Society of Tropical Medical Hygiene 78, 436-441.

262. Wheeler, C.W., Park, S.S. and Guenthner, T.M. (1990) Immunological analysis of cytochrome P450 1A1 homologue in human lung microsomes Molecular Pharmacology 38, 634-643.

263. World Health Organization (WHO) (2006) Mycotoxins in African foods: implications to food safety and health. AFRO Food Safety Newsletter. World Health Organization Food safety (FOS), Issue No. July 2006. www.afro.who.int/des.

264. Wild, C. P. (1996). Summary of data on aflatoxin exposure in West Africa. In K. F. Cardwell (Ed.), Proceedings of the workshop on mycotoxins in food in Africa, November 6– 10, 1995 (p. 26). Cotonou, Benin: International Institute of Tropical Agriculture

265. Wild, C. P. and Gong, H. Z. (2010) Mycotoxins and human disease: a largely ignored global health, Carcinogenesis 31, 71-82.

266. Wild, C. P., Jiang, Y. Z., Sabbioni, G., Chapot, B. and Montesano, R. (1990) Evaluation of methods for quantitation of aflatoxin-albumin adducts and their application to human exposure assessment Cancer Research 50, 245-251.

267. Wild, C. P., Hasegawa, R., Barraud, L., Chutimataewin, S., Chapot, B., Ito, N. and Montesano, R. (1996) Aflatoxin-albumin adducts: a basis for comparative carcinogenesis between animals and humans Cancer Epidemiology, Biomarkers and Prevention 5, 179-189.

268. Williams, J. H., Phillips, T. D., Jolly, P. E., Stiles, J. K., Jolly, C. M. and Aggarwal, D. (2004) Human aflatoxicosis in developing countries a review of toxicology, exposure potential health consequences and intervention American Journal of Clinical Nutrition 80, 1106-1122.

269. Wogan, G. (1973) Aflatoxin carcinogenesis Methods in Cancer Research 7, 309-344. Wolzak,A., Pearson, . A. M., Coleman, T. H., Pestka, J. J. and Gray, J. I. (1985). Aflatoxin deposition and clearance in the eggs of laying hens Food and Chemical Toxicology, 23 (12): 1057-1061

270. Wolzak, A., Pearson, A. M., Coleman, T. H., Pestka, J. J., Gray, J. I. and Chen, C. (1986) Aflatoxin carryover and clearance from tissues of laying hens Food and Chemical Toxicology 24, 37-41

271. Wood, R.D. (1999) DNA damage recognition during nucleotide excision repair in mammalian cells, Biochemie 81, 39-44.

272. Wu, F. (2004) Mycotoxin risk assessment for the purpose of setting international regulatory standards Environmental Science and Technology 38, 4049-4055.

273. Wu, F. (2006) Mycotoxin reduction in Bt corn: potential economic, health and regulatory impacts Transgenic Research 15, 277-289.

274. Wu, F. and Khlangwiset, P. (2010a) Health economic impacts and cost-effectiveness of aflatoxin reduction strategies in Africa: case studies in biocontrol and post harvest interventions Food Additives and Contaminants 27, 496-509.

275. Wu, F. and Khlangwiset, P. (2010b) Evaluating the technical feasibility of aflatoxin risk reduction strategies in Africa Food Additives and Contaminants 27, 658-676.

276. Yadgiri, B., Reddy, V., Tulpule, P. G., Srikantia, S. G. and Goplan, C. (1970) Aflatoxin and Indian childhood cirrhosis The American Journal of Clinical Nutrition 23, 94-98.

277. Yousef, A.E. and Marth, E.H. (1985) Degradation of aflatoxin M1 in milk by ultraviolet energy. Journal of Food Protection, 48, 697–698.

278. Zain, M.E (2010). Impact of mycotoxins on humans and animals. Journal of Saudi Chemical Society. Retrived on 16th

279. Zarba A., Wild C.P., Hall A.J., Montesano R, Hudson G.J. and Groopman JD. (1992) Aflatoxin M1 in human breast milk from The Gambia, west Africa, quantified by combined monoclonal antibody immunoaffinity chromatography and HPLC Carcinogenesis 13, 891-894.

280. Zhu, L. R., Thomas, P. E., Lu, G., Reuhl, K. R., Yang, G. Y., Wang, L. D., Wang, S. L., Yang, C. S., He, X. Y. and Yan, H. J. (2006) CYP2A13 in human respiratory tissues and lung cancers: an immunohistochemical study with a new peptide specific antibody Drug Metabolism and Deposition 34, 1672-1676.

281. Zinedine, A., Juan, C., Soriano, J.M., Moltó, J.C., Idrissi, L. and Mañes, J. (2007a). Limited survey for the occurrence of aflatoxins in cereals and poultry feeds from Rabat, Morocco International Journal of Food Microbiology 115, 124–127

282. Zinedine, A., González-Osnaya, L., Soriano, J.M., Moltó, J.C., Idrissi, L., Mañes, J. (2007b). Presence of aflatoxin M1 in pasteurized milk from Morocco International Journal of Food Microbiology 114, 25–29

283. Zinedine, A., Brera, C., Elakhdari, S., Catano, C., Debegnach, F., Angelini, S., De Santis, B.,Faid, M., Benlemlih, M., Minardi, V., Miraglia, M. (2006) Natural occurrence of mycotoxins in cereals and spices commercialized in Morocco. Food Control 17, 868–874.

Chapter 4

NEW CHEESE-LIKE FOOD PRODUCTION FROM SOY MILK — UTILITY OF SOY MILK CURDLING YEAST

Makoto Kanauchi[1], Sakiko Hatanaka[2] and Makoto Shimoyamada[3]

[1]Department of Food Management, Miyagi University, Japan

[2]Industrial Technology Institute, Miyagi Prefectural Government, Japan

[3]School of Food and Nutritional Sciences, University of Shizuoka, Japan

ABSTRACT

Soybeans are a traditional food in eastern Asia, particularly in Japan and China. They were eaten in 100 BC in China. The beans can be processed into Tofu, soy milk, fermented seasonings, soy sauce or Miso paste, and Natto and green beans. Soybeans have rich nutrition, protein lipid, and other functional substances such as isoflavones. However, soybeans are difficult to process for use as food because of tissue and cell wall hardness. Therefore, soybeans are conducted to do some treatments, e.g., boiling, steaming, roasting, crushing/grinding, and some enzyme treating, to eat soy protein easily. Soy storage proteins mainly comprise two proteins as 7S globulin composed with β-conglycinin and 11S globulin containing glycinin composed of 5 subunits. β-Conglycinin, included in 7S globulin, is composed of three subunits. To modify the physical properties of soy protein, a new type of enzyme for curdling soybean milk enzyme was purified as an extract from yeast. Yeast producing curdling soybean milk enzyme, the SCY003 strain, was isolated from 1345 yeast strains. According to the morphology, physiology, and molecular and characteristics, SCY003 was identified as Saccharomyces bayanus. The soy milk curdling enzyme having proteolytic activity was approximately 45 kDa and monomer protein. The optimum pH for the protease activity was pH 7.5;

the optimum temperature was 50°C.The enzyme cleaved the β-conglycinin as α–, α'-, and part of glycinin as A3 A4, A1b, and A2 in soy protein by endoproteolysis. Soybean protein became loosely curdled with the addition of other proteases from microorganisms or plants. Soybean milk curdled after cleaving endoproteolysis enzyme in SCY003 strain.The breaking point of curd curdled by enzyme was 58.4% strain. Their breaking stress was 10,900 (N·m–2). The brittleness point is 81.2% and 10,200 (N·m–2), and the brittleness of this curd produced using the enzyme was 727 (N·m–2). Brittleness of the curd produced by the enzyme was less. Their breaking point was greater than that of the curd produced using the glucono-δ-lactone (GDL). Furthermore, the curd had sticky and chewy texture. The curd made by enzyme has resilience more than normal Tofu. It is considered that the curd produced by enzyme was not like Tofu rheologically.

INTRODUCTION

Utility of Soybeans

Soybeans have been used as traditional foods from ancient times. They are rich in nutrients such as rich proteins, lipids, and others. Furthermore, soybeans can be eaten after processing in various ways.

Eastern Asian people and particularly Japanese people have eaten soybeans after various stages of processing. Soybean seedlings are eaten in many dishes as bean sprouts. Furthermore, soybeans are eaten as green beans in the pod after boiling, as *Edamame*. Soybean flour made from roasted soybeans is eaten as *Kinako* powder. After boiling soybeans, the beans can be fermented can using molds to produce *Chi* or *Tempe*. Furthermore, soy sauce is made from a molded mixture with boiled beans and roasted wheat, and salted water. Soy paste is made from fermented boiled soybeans and *Koji* with salt. The resultant umami taste is an extract from the bean, facilitated by an enzyme reaction because it is thought that umami components are stored as proteins in hard tissue.

After hard tissues in boiled soybeans are crushed and ground physically, the soluble fraction is extracted as soy milk. Soy milk is processed as *Yuba* from a soybean sheet, and *Tofu* is produced. Regarding the insoluble fraction, spent soy is also eaten as *Okara*. Finally, compressed soybeans produce oil that is widely used as cooking oil. The residue of oil pressing can then be used for soy sauce production or soy protein for food manufacture. Comparing soy products to milk casein, which is eaten as cheese, soy products are not used as widely as food. One reason is that soybean curd such as *Tofu* lacks taste and has less elastic properties and texture compared to cheese. Therefore, to make rich nutrition and produce a food that has good taste and texture, soybean protein is modified by enzymes and is extracted by microorganisms [1].

History of Soybeans

Theories about the origin of soybean use and cultivation remain controversial in their details. By some accounts, soybeans used as food originated in the area of Manchuria in China and Siberia in Russia. Alternatively, their use as food originated in southern China [1]. Yet another possible history holds that soybeans were bred from wild soybeans as *Glycine soja* in China. In fact, the bean was present in ancient Japan: beans were found in the bottom of an earthen vessel produced in the middle Jomon period (3000–2000 B.C.) in Japan. However, soybeans have not been recorded as eaten in that period.

In China, the first literal record can be found in a Chinese dictionary published in 100 B.C. The dictionary inserted "*Chi*," representing fermenting soybeans with salt. Furthermore, archaic *Miso* paste and soy sauce fermented by soybeans were described in the Chinese text "Qi-min-yao-shu," published in the sixth century.

In Japan also, "*Chi*" fermented soybeans with salt was recorded in the Taiho Code in A.D. 701. It is considered that soybean fermentation practices diffused from China. Tofu was recorded in the tenth century in China. It was recorded in the same period in Japan. In Japan, it was initially eaten as a vegetarian dish for Buddhists because, during the Kamakura period, people gradually became more and more vegetarian in their eating practices. In the Edo Era, a *Tofu* recipe book was published, with 283 recipes explained in it.

Consumption Worldwide and in Japan

Soybeans were produced only in eastern Asia for a long time. In contrast, other cereals such as rice, barley, wheat, and corn have diffused throughout the world. Moreover, it is considered that some other endemic bean or pea or pulse had become cultivated in each area already [1].

For instance, in central Asia, broad beans were cultivated, as were chick peas in India, shell peas in western Asia, and kidney beans and ground peas in North America. Nevertheless, soybeans have been cultivated in the United States as oil seed crops since the 1920s. Furthermore, the crop has begun to be cultivated in Canada and South America. In 2012, approximately 82 million tons of soybeans were harvested, with the United States accounting for 34% of the world production. Brazil harvested approximately 66 million tons that year, accounting for 27% of world production. Argentina harvested approximately 40 million tons, or 17% of world production. Therefore, most soybeans (over

80%) consumed worldwide are now produced and harvested in North and South America [2] (Table 1).

In Japan, soybean production was sufficient to provide for domestic consumption until the Taisho Era [1]. Soybean consumption in Japan has been high, but it decreased after the Taisho Era. In 2013, 30 million tons were consumed, but only 240 thousand tons were harvested domestically. That figure is less than 0.1% of the world production amount. Soybeans used domestically account for 104 thousand tons for feed, 6 thousand tons for seed, and 1.9 million tons for oilseeds, all together accounting for 70% of the 30 million tons consumed. Furthermore, those figures indicate that only 30% of soybeans are used as food. The self-sufficiency ratio of soybeans was 97% in 1947. It decreased gradually to 28% in 1959, 11% in 1965, and 7% in 2013 [3]. The ratios of soybeans used for food are 49% used for *Tofu*, 13% used for *Miso* paste and *Natto*, 4% used for soy milk, and 3.5% used for soy sauce production. As mentioned earlier, some soy sauce production companies have used soy meal after oil pressing to produce soy sauce (Table 2).

Table 1: Soybeans production

	Production 2012 (10 thousand tons)	Share of world production (%)
United States	8205	33.9
Brazil	6585	27.2
Argentina	4010	16.6
India	1467	6.1
China	1305	5.4
Canada	509	2.1
Paraguay	434	1.8
Uruguay	300	1.2
Ukraine	241	1.0
Bolivia	206	0.9
Russia	181	0.7
Indonesia	84	0.3
South Africa	65	0.3
Nigeria	58	0.2
North Korea	35	0.1

Japan	24	0.1
Myanmar	21	0.1
Others	454	1.9
World total	24,184	100

[i] - Source: http://www.maff.go.jp/j/seisan/ryutu/daizu/d_data/pdf/014.pdf

Table 2: Changes in amount of soybeans for applications

	2003	2013
Miso paste	138	123
Soy sauce	38	33
Tofu and fried tofu	494	454
Natto	137	125
Frozen *tofu*	30	20
Soy milk	19	40
Delicatessen of soybean	33	30
Kinako soy powder	17	18
Other	128	93
Total	1034	936

(Unit: thousand tons)

Ref. http://www.maff.go.jp/j/seisan/ryutu/daizu/d_data/

NUTRITION

Soybean Protein

Soybeans contain 35% protein as storage protein, which is used for nutrition during germination. That storage protein is stored in granules, called protein bodies, of about 5–8 μm diameter. Soluble soy protein is extracted from insoluble protein bodies that are burst during soy milk and *Tofu* production. All soy protein is stored in the protein body [1]. Other proteins exist as nonstorage proteins, containing important physiological proteins such as trypsin inhibitors. When an animal ingests a trypsin inhibitor, a digestive enzyme, trypsin activity is inhibited by combination of the trypsin inhibitor specifically with trypsin. Consequently, the pancreas works excessively to secrete and supplement trypsin activity [1]. However, after heating, the inhibitor loses its inhibitory activity and does not bind with trypsin.

Protein digestibility-corrected amino acid score (PDCAAS) values are used to evaluate the protein quality based on the amino acid requirements of humans and their ability to digest it. It was long thought that the amount of amino acid requirements of humans dictate a low score for soybeans because methionine and cysteine residues, sulfur amino acids, in soybean storage proteins have a low composition. The score was only 86 points based on the amino acid requirements of a developing rat. However, in 1985, the score was modified to 100 points, the same as milk and eggs, based on the amino acid requirements of humans. Soy storage proteins are rich in nutrition for human needs [4].

Throughout the world, the recently improving healthy image of soy protein is interesting. In particular, the health benefits of soy foods attract attention in the United States. Health claims are authorized by the Food and Drug Administration (FDA) in the United States: foods containing 6.25 g of soy protein or more can be said by manufacturers to reduce the risk of heart disease if a consumer ingests 25 g/day of soy protein [5]. In Japan, some soybean containing foods are manufactured as *Tokuho*: government-approved foods for specified health benefits, as for hypocholesterolemic activity in this case.

The taxonomy of soybean storage protein has been conducted according to the sedimentation coefficient by an ultracentrifugal fraction as 2S globulin, 7S globulin, 11S globulin, and 15S globulin. Yamauchi [1] reported details of soybean proteins: 2S globulin contained α-conglycinin, 7S globulin composed with β-conglycinin and γ-conglycinin, and 11S globulin-containing glycinin. In addition, the 11S globulin composes hexamer. It is a 350,000 Da protein. Furthermore, their proteins are composed with five subunits as G_1–G_5; their subunit was 10 polypeptides as A_{1a}, A_2, A_{1b}, A_3, A_5A_4, A_4, B_2, B_{1b}, B_4, and B_3. Their polypeptides are combined specifically as $A_{1a}B_{1b}$, $A_{1b}B_2$, A_2B_{1a}, A_3B_4, and $A_5A_4B_3$. In fact, β-conglycinin, called 7S globulin, combines a dimer protein and a monomer protein, which are 150,000–200,000 Da, or an average of 180,000 Da. They are composed of three subunits: an α-subunit of 63,000 Da, an α'-subunit of 67,000 Da, and a β-subunit of 48,000 Da. The protein has a low concentration of sulfur amino acids. In particular, the β-subunit does not contain methionine, cysteine, and tryptophan [6].

Lipid

Soybeans have 20% lipids. The lipid concentration varies among harvested regions. Soybeans harvested in the United States have more lipids than those in China [1]. A main reason is that soybeans there have long been bred and modified to contain high oil concentrations as oilseed.

Table 3: Components of fatty acids in foods (%)

	Palmitic acid	Stearic acid	Oleic acid	Linoleic acid	Linolenic acid
Pork fat	26.2	13.5	42.9	9	0.3
Beef fat	38.7	3.8	42.1	2.3	2.4
Milk lipid	31.1	9.2	21.7	1.6	0.4
Soybean oil	10.5	3.2	22.3	54.5	8.3
Sunflower oil	trace	4	27.6	58.3	trace
Cotton oil	27.3	3.1	16.7	50.4	trace
Safflower oil	6	3.4	12.2	77	0.3

[i] - Source: Yamauchi and Ookubo (1992).

Components of fatty acids in foods shows Table 3. Soy oil comprises a small amount of saturated fatty acids, such as palmitic acid and stearic acid, and large amount of unsaturated fatty acids such as oleic acid, linoleic acid, and linolenic acid. Polyunsaturated fatty acids (PUFAs) containing more than unsaturated bonds are important nutrition as necessary lipids for humans. Soybeans have over 60% PUFA. In particular, one kind of PUFA as linoleic acid contained approximately 54.5%.

Actually, PUFAs in animal lipids have low concentration. Therefore, they are insufficient nutritionally. Saturated fats and unsaturated fats are ideally in the following ratio: saturated–unsaturated (1:2) [1]. Soybean lipids were well known to be much stable against oxidation because they are covered as oil body particle by oleosin and other proteins.

Isoflavone

Isoflavone is one kind of flavonoid (Fig. 1). Fabaceae sp. contain high concentrations (Fig. 1).

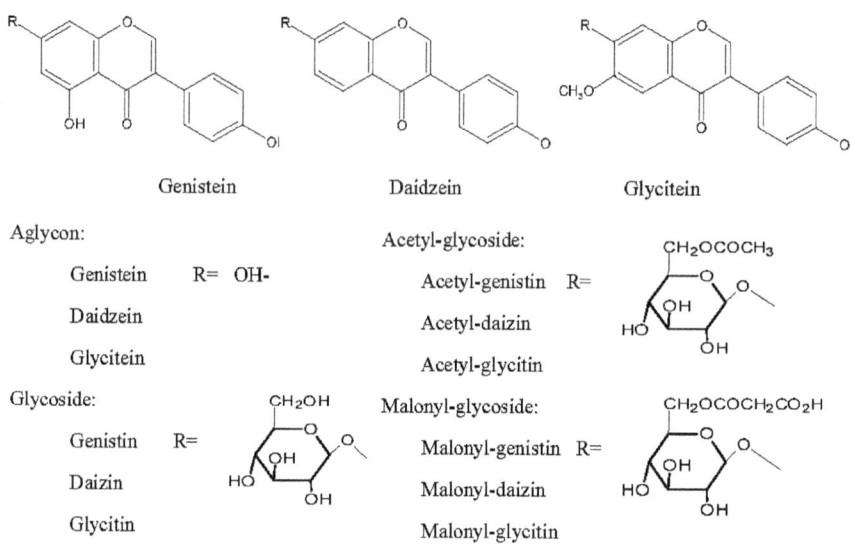

Figure 1: Isoflavone structure.

Generally, soybeans have totally 12 isoflavones in 3 aglycones, and they have three types of glycosides as glucoside, acetyl-glycoside and malonyl-glycosides: genistein, daidzein, glystein, genistin, daidzin, glycitin, acetyl-genistin, acetyl-daidzin, acetyl-glycitin, malonyl-genistin, malonyl-daidzin and malonyl-glycitin [7]. After soybean consumption, glycoside isoflavone, which is contained in food as soy milk or*Tofu*, hydrolyzes aglycon and glycoside by bacteria in intestines. Their aglycon are absorbed by the body. Genistein and daidzein have estrogenic effects and hormone-like activity. The isoflavone binding with estrogen receptor reacts as an estrogen agonist in the human body. Their substances from plants having estrogenic effects are called "plant estrogen." *Miso* paste has a high level of isoflavone because glucosyl isoflavone is hydrolyzed to their related aglycons.

Many Japanese and throughout eastern Asia intake isoflavone from soybeans. Some researchers have reported negative opinions about plant estrogen [8]. Isoflavones are produced via phenyl–propanoid pathway from phenylalanine in plants. Two intermediate substances, naringenins, are converted to genistein by two specific enzymes in soybeans: isoflavone synthetase and dehydrogenase. Chalcones are converted to daidzein by three specific enzymes in soybean: chalcone reductase, chalcone isomerase, and isoflavone synthetase. Isoflavone and a similar substance, phytoalexin, are used as antibacterial substances against phytopathogenic fungi and bacteria. In addition, they grow well as root nodule bacteria at the root for nitrogen fixation [9, 10].

Other beans and peas, legumes, have isoflavones: chickpeas have biochanin A [11]; alfalfa has formononetin and coumestrol [12]; and ground peas have genistein [13].

Isoflavones in many plants store glucosyl, malonyl-glucosyl, and acetyl-glucosyl conjugate as hydrophilic substances. After invasion of phytopathogenic fungus and bacteria, the glucosyl conjugate isoflavone is transferred to infested wounds, where it hydrolyzes for phytoalexin [14]. Isoflavone has health functions against climacteric disturbances and type 2 diabetes. In Japan, soybean isoflavones are a *Tokuho* (government-approved food for specified health purpose) for the prevention of osteoporosis. The Ministry of Health, Labour and Welfare in Japan alerts consumers to avoid overdosing on isoflavones. The amount of isoflavone intake was 30 mg a day, omitting isoflavone intake from meals and by supplements.

FOOD PROCESSING

Formation of TOFU Curdling

Soy proteins have properties that produce curd to add specific metal ions. The property is applied for*Tofu* production. Tofu, soybean curd from soybean milk, is consumed throughout Asia. It was eaten in the tenth century in China and Japan. *Tofu* is traditionally consumed after it is produced with a combination of magnesium dichloride ($MgCl_2$) as *Nigari* and calcium dichloride ($CaCl_2$) as *Sumashi-ko*. More recently, glucono-δ-lactone (GDL) has been added to it for commercial production. Tofu resembles cheese or yogurt made from milk curd of cows or other mammals. It is made from soy milk. It curdles by *Nigari* or a coagulant agent [15]. The forming system of *Tofu* curdling is shown in Fig. 2.

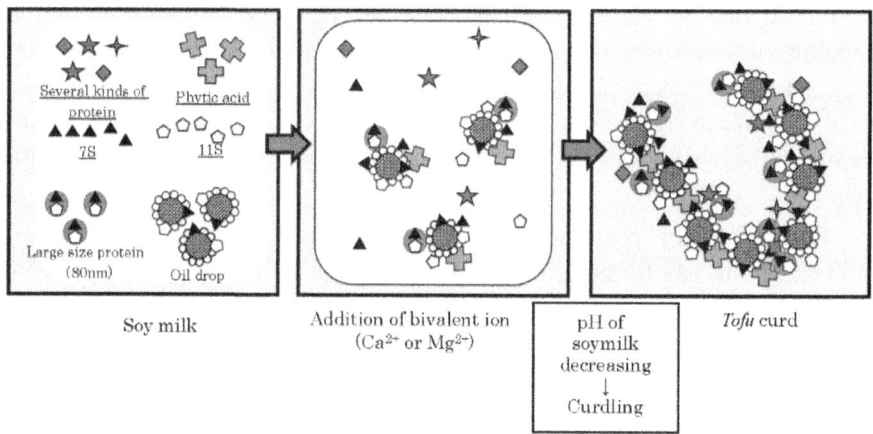

Figure 2: Formation of *Tofu* curdling [15].

Soy milk is an emulsion composing lipid and soy protein, mainly glycinin and β-conglycinin. Actually, 60% of the protein in soy milk protein is composed of these two proteins, and a large size of the protein body was constructed with their proteins and others. Lipid is a triacylglyceride composed of linoleic acid, oleic acid, and phospholipids. Before treating, triacylglyceride is stored in oil bodies in soybeans. During soy milk production, and of course during *Tofu* production, oil drops are suspended, forming an emulsion after crushing of soybeans. The oil drop is formed and held stably by lecithin, phospholipid, and oleosin, all proteins forming oil bodies in soybeans. Furthermore, an outside layer is covered by soy protein. Phytic acid (*myo*-inositol-1,2,3,4,5,6-hexaphosphate) and mineral and oligosaccharide are in the soluble component of soy milk [16]. When some bivalent ions such as calcium ion and magnesium ion are added to soy milk, the ions combine with phytic acid. As a result, decreasing the pH of the soybeans immobilizes the protein in it [17].

This phenomenon induces the charge of protein to dissipate by a combination of phytic acid and bivalent ion. Repulsion among their proteins is decreased. Moreover, the immobilized protein combines with outside layer protein easily around an oil drop [18]. Consequently, the steric network structure of soy protein is formed to gather oil drops with intermediary soy protein. Becoming low pH in soy milk, soluble soy protein becoming dissoluble is taken into the network. The curd produces a hard gel that is water retentive, such that moisture is trapped in the network.

SOY MILK CURDLING YEAST AND CHARACTERISTICS OF CURDLING ENZYMES

Soy Milk Curdling

As discussed above, soybean proteins provide rich nutrition [4, 19]. In fact, soymilk consumption is increasing quickly throughout the world because of its health benefits. Moreover, differently from bovine milk, it contains no cholesterol. Yogurt-like foods and cheese-like foods made from soybeans can be consumed by people who are concerned about health issues or allergies related to bovine milk. Tofu, soybean curd from soybean milk, resembles cheese or yogurt made from milk curd of cows or other mammals. However, the cheese mouthful sense and physical properties are not identical to those of casein protein. Unlike the protein casein in bovine milk, enzymatic curdling of soybean milk produces poor flavor and texture. It is not yet a viable alternative to dairy foods. For that reason, the commercial use of enzymes such as bromelain, ficin, and papain for the curdling of soybean milk has been unsuccessful [20–22]. The authors have reported that physical properties of soy protein modified by enzyme reactions such as germinated proteolysis in soybeans. Therefore, in this section, along with a report of yeast containing soybean curdling enzyme [23], this investigation was undertaken to screen and identify specific food yeast strains (*Saccharomyces* sp.) that produce a soybean milk curdling enzyme and to purify the enzyme using chromatographic procedures.

Screening of Yeast Producing Curdling Soybean Milk Enzyme

The yeast strains (1345 strains) stored in the laboratory were screened using soybean milk agar plate medium. The strains were inoculated by streaking on a plate surface. Then they were incubated at 30°C for 7 days. After cultivation, the clear zone diameter was measured using calipers. Yeast strains that produced a clear zone were selected. Results show that 1242 yeast strains among all 1345 yeast strains produced no clear zone on the plate medium. The yeast strains (42 strains) produced less than 1 mm of a clear zone. Also, 57 yeast strains produced 1.0–5.0 mm; 4 yeast strains produced more than 6 mm.

In the second screening of curdling soybean milk enzyme-producing yeast, the screened strains (103 strains) were inoculated to the soybean milk liquid medium. Purchased soybean milk was added to them aseptically.

The soybean milk medium was incubated at 30°C for 24 h. When curdling occurred, the pH of whey was measured using a pH meter (Horiba Ltd.). Results show that three yeasts curdled at pH greater than 5.90. The media were pH 5.90 (SCY 001), pH 6.05 (SCY 002), and pH 6.38 (SCY003) (Table 4).

Table 4: Curdling soybean milk condition by screened yeast

pH	Curdling soy milk			
	(++)	(+)	(–)	
≥6.50	0	0	7	
6.49–5.90	1	2	43	
5.89–5.50	0	17	17	
5.49–5.00	2	11	0	
≤4.99	3	0	0	

[i] - Initial pH was 6.70. ++, coagulating very well; +, coagulating; –, noncoagulating.

The curd activity of strain SCY003 was the highest among the strains. Therefore, the SCY003 strain was finally screened. Isolated yeasts were classified taxonomically and were identified according to methods described in earlier studies [24].

The morphology was observed by microscope. Their 1.5- to 6.5-μm-long cells were short and ovaloid. The yeast, which buds by multibudding reproduction, does not form pseudomycelia or pellicles on the liquid medium. It forms ascospores. It was identified as *Saccharomyces* sp.

For researching physiological characteristics of the strain, the yeast was inoculated into a yeast nitrogen base medium (Difco Laboratories) adding 0.5% of each carbon source: sugar or organic acid as glucose, galactose, sucrose, maltose, raffinose, trehalose, lactose, melibiose, cellobiose, melezitose, starch, D-xylose, L-arabinose, D-ribose, L-rhamnose, erythritol, D-mannitol, salicin, inositol, dulcitol, ethanol, D-sorbitol, disodium succinate, and trisodium citrate. The yeast was inoculated into yeast carbon base medium (Difco Laboratories) adding sodium nitrate solution.

The glucose, galactose, sucrose, maltose, and raffinose in the medium were fermented as carbon sources using strain SCY003. The yeasts grew in a vitamin-free medium. Furthermore, the strain did not grow in 0.01% cycloheximide. According to the morphological, physiological, and molecular characteristics, it was identified as *Saccharomyces bayanus*.

For researching molecular biological characteristics of the strain, primers were used for amplification and sequencing of 18S-rRNA-encoding genes. The PCR products were sequenced using a kit (ABI Prism Big Dye Terminator Cycle Sequencing Ready Reaction; Applied Biosystems). Analyses of

DNA sequence reactions were performed using a sequencer (3130; Applied Biosystems). The 18S rRNA coding DNA was sequenced. Homology was assessed using the Basic Local Alignment Search Tool (BLAST; http://www.ncbi.nlm.nih.gov/BLAST/).

As a result, the yeast showed homology of 99% with *S. bayanus* (accession no. AY046227). It was identified as *S. bayanus* through homology research and phenotypic testing.

Purification of The Protease As A Soybean Milk Curdling Enzyme

Enzyme extraction, intercellular, in *S. bayanus* SCY003, soybean milk curdling test/activity was conducted using the method described [25-27] for modified soybean milk from bovine milk-curdling activity. The mixture was centrifuged at 400×g for 10 min. The supernatant was removed gently using a Pasteur pipette. The weight of the precipitate was measured using a chemical balance.

Generally, commercial soy milk has dispersion stability attributable to the presence of oleosomes or forming aggregate formation of soy proteins on it [28, 29]. Therefore, no precipitate is produced from commercial soybean milk by low centrifugal gravity as 400×g. However, precipitation ratios increase with the enzyme reaction period.

The precipitation ratio was related with the reaction period, and with the enzyme solution volume. They are mutually correlated: $R^2 = 0.9978$. Therefore, one curdling unit expressed the ratio of curdling (%) from 1 mL of soybean milk at 40°C for 1 h. The precipitation ratio from soybean milk was assayed efficiently to propose a new curdling method. The precipitation ratio was assayed (Fig. 3).

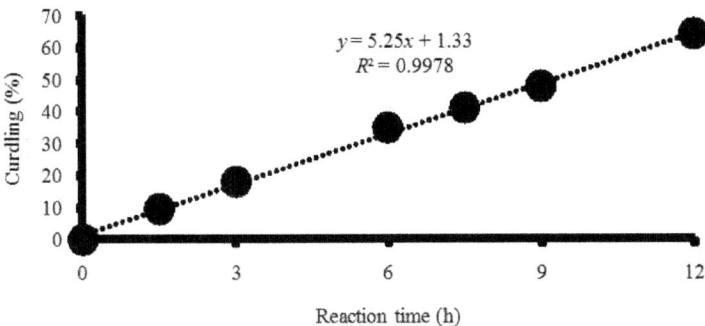

Figure 3: Curdling soybean milk enzyme activity.

After crude extraction of the enzyme, the enzyme protein was purified using chromatography. After crushing cells and extracting the enzyme, the enzyme was precipitated to between 30% and 40% saturation of ammonium sulfate. After redissolving the precipitate, the solution was dialyzed overnight at 4°C, and they carried out ion-exchange chromatography using a column (25 mm × 300 mm) of DEAE gel. The protein was eluted using a 0- to 1-M NaCl linear gradient. Proteolytic activity and curdling were assayed each fraction. Proteolytic activity was measured in duplicate using a commercial kit fluorescein isothiocyanate-labeled casein (FTC). Fluorescence was measured using a 485-nm excitation wave and a 535-nm emission wave. The proteolytic activity was decided for one unit expressed equal amount of trypsin (1 ng·mL^{-1}) doing proteolysis of FTC solution.

After ion chromatography, fractions containing the highest level of activity were pooled and reprecipitated using 80% saturation of ammonium sulfate. After redissolving the precipitate, gel filtration chromatography (10 mm × 350 mm, P-100 gel; Bio-Rad Laboratories Inc., CA, USA) was carried out. Then the molecular weight of the enzyme was analyzed using also the chromatography as various molecular weight standards (myosin, 200 kDa; serum albumin, 66.2 kDa; ovalbumin, 45.0 kDa; trypsin inhibitor, 21.5 kDa).

The results are portrayed in Figs. 4a and 4b. One main curdling activity peak was identified using ion-exchange chromatography. The peak (fraction no. 49) agreed with protein and activity. Furthermore, curdling activity agreed with the same fractions presenting protease activity (fraction no. 49).

As a result, the larger peak of proteolysis activity was found around fraction number 25, and a small peak was found at fraction number 49. The fraction of the proteolysis enzyme around number 25 was not representative of curdling activity. It is considered that the former fractions are attributable to intense proteolysis enzymes and that the latter fractions are attributable to soybean milk-curdling enzymes.

After reprecipitation, the sample was analyzed using gel filtration chromatography. A peak was found at fraction numbers 11–14. Their fractions agreed with soybean milk curdling activity, proteolytic activity, and protein. This result on their chromatograms demonstrates that protease and soybean milk curdling enzyme have some mutual relation of activity.

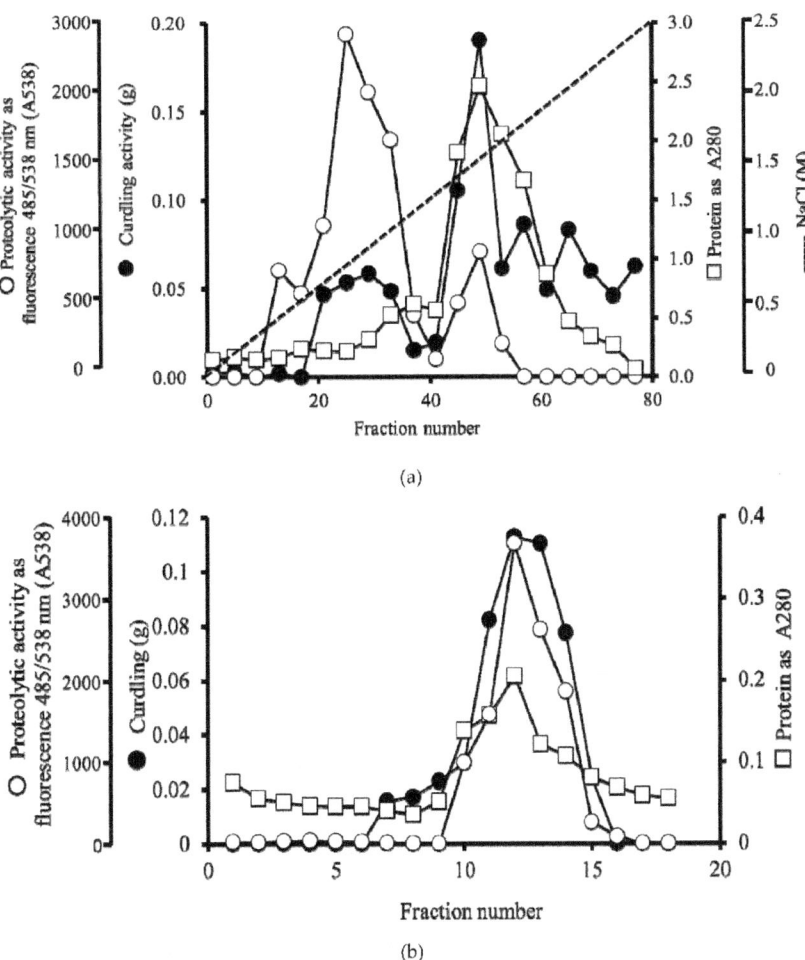

Figure 4: (a). Ion-exchange chromatography of soybean milk curdling. enzyme. (b). Gel filtration chromatography of soybean milk curdling. enzyme.

After purification, the enzyme protein band was approximately 45 kDa (Fig. 5a), which agrees with data of other proteases. The protease molecular weight was measured using gel filtration chromatography (Fig. 5b). The molecular mass is about 45 kDa. The protease was inferred.

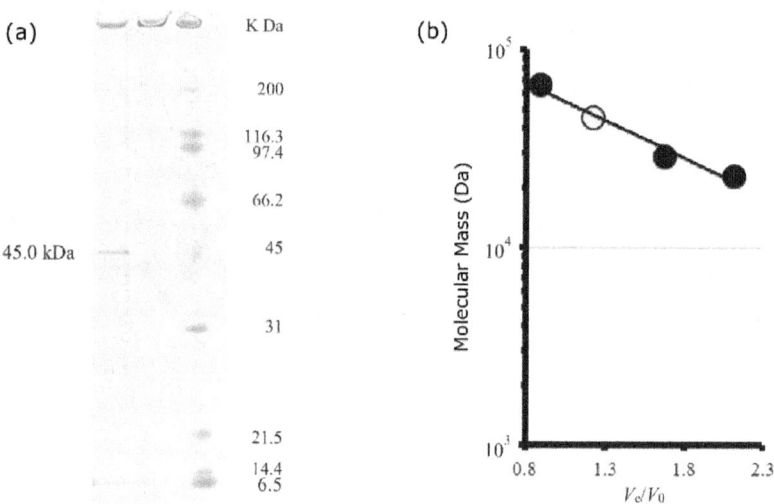

Figure 5: (a). Photograph of SDS-PAGE of soybean curding enzyme. (b). Measurement of molecular weight.

The soy milk curdling enzyme has proteolytic activity. Results suggest that the soy milk curdling enzyme was a proteolysis enzyme. Many researchers have reported protease produced by yeasts as*Candida albicans* [30, 31, 32], *Candida humicola* [33], and *Saccharomyces cerevisiae* [34]. Extracellular proteases produced by yeasts as *Candida* spp. are 42–45 kDa [35, 32]; those by bacteria are 21 kDa [36].

By contrast, few reports describe intracellular protease producing *Saccharomyces cerevisiae*, although many intracellular proteases in the vacuole or other organelles are known to be related to proteinase A, which is 42 k Da [34]. The molecular weight of curdling soy protein enzyme protease agreed with protease produced by other yeast as Ascomycota. However, the *Mucor* sp. enzyme, which curdles bovine milk, produced a 49-kDa protease [37], which is larger than those produced by these yeasts.

Characteristics of The Protease As A Soybean Milk Curdling Enzyme

Optimum pH, temperature, and stability of the enzyme are presented in Figs. 6a and 6b. Optimum pH was assayed at pH 4.0–8.0 using 50 mM phosphate-citric buffer or phosphate-NaOH. For optimum pH of the enzyme assaying, 0.1 mL of a fluorescein isothiocyanate-labeled casein (FTC) solutions, which dissolved in each phosphate buffer (50 mM, pH 4.0–8.0), and 20 μL of enzyme solution were reacted at 40°C for 60 min. For optimum temperature of the enzyme assaying, FTC solution at pH 7.5 (0.1 mL) and 20 µL of enzyme solution were reacted at 15°C–70°C for 60 min to find optimum temperature. After reaction, fluorescence was measured using a 485-nm excitation wave and a 535-nm emission wave.

The optimum pH for the protease activity as curdling was pH 7.5; the optimum temperature was 50°C. The optimum pH of a bovine milk curdling protease, *Mucor pusillus*, is pH 5.0. The optimum pH values of many commercially available proteases are pH 5.9–6.7. However, soymilk curdling activity decreases concomitantly with increasing alkalinity.

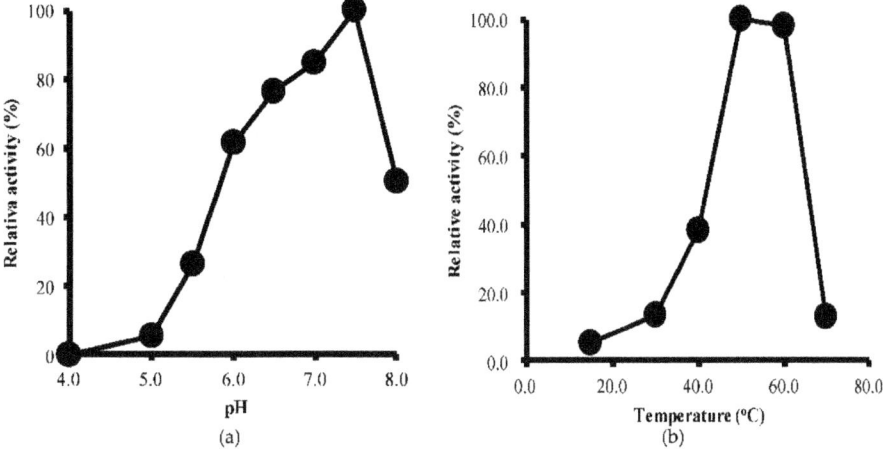

Figure 6: (a). Optimum pH of soybean curdling enzyme. (b). Optimum temperature of soybean curdling enzyme.

Park *et al.* [38] reported that the optimum pH of soybean milk curdling protease produced by *Bacillus* was pH 6.0. Regarding this enzyme, the optimum pH of the protease was pH 7.5. The optimum temperature of the protease was 50°C. The enzyme also curdled soybean milk at pH 7.5 and 50°C. Commercial soybean milks sold in Japan are pH 7.0–7.2. The pH range agrees with their optimum pH range.

Effects of metal ions and inhibitors on protease are presented in Table 5. In fact, zinc, copper, and mercury all inhibit protease activity. The amino acid of the active site contains cysteine residue because of inhibition by mercury [39]. Furthermore, EGTA (10 mM) inhibited protease activity (62.0% of relative activity).

Table 5: Effects of metal ion and inhibitor of the protease

Metal ion and inhibitor	Concentrations	Relative activity (%)
$N^{a}+$	1 mM	101.5
K_+	1 mM	105.9
$M^{n2}+$	1 mM	101.9
$M^{g2}+$	1 mM	46.2
$F^{e2}+$	1 mM	95.4
$Z^{n2}+$	1 mM	23.6
$C^{o2}+$	1 mM	47.9
$C^{u2}+$	1 mM	17.9
$C^{a2}+$	1 mM	100.9
Iodo acetate	1 mM	74.1
$H^{g2}+$	1 mM	34.1
EGTA	1 mM	101.6
EGTA	10 mM	62.0
EDTA	1 mM	96.1
EDTA	10 mM	89.8
NEM	1 mM	98.5
Azid	1 mM	100.4
NBS	1 mM	97.6
Cont.	–	100.0

The activity was not activated by metal ions, but it was inactivated by mercury. Soybean milk-curdling enzyme [38] was inhibited by zinc ions and mercury ions. These results agree with our data related to zinc and mercury. Its survival activity was 18% by mercury. The protease was not activated by metal ions, which indicates that the protease is not a metalloprotease: a metal-dependent enzyme. The amino acid of active site contains cysteine residue.

The mechanisms of curdling soybean milk protease were investigated. At first, The curdled soybean milk samples with added protease and without protease were treated with sample buffer solution.

Soybean milk was poured into a glass vessel (inner diameter 32 mm, height 45 mm). After 0.1 mL of enzyme solution adding to the soybean milk, or without enzyme 0.1 mL D.W., the mixtures were incubated at 40°C, and they were sampled sequentially; between 4- and 24-h. samples (0.01 mL) were added to 0.01 mL of sample buffer and then heated at 100°C for 3 min. Then samples (10 μl) were added in each well. The samples were electrophoresed on a 12.5% uniform gel at 20 mA.

They were subsequently examined using SDS–PAGE (Fig. 7) of curdled soybeans. The left side lane shows the standard of protein size. The next lane (0 h.) shows soybean milk protein without reaction of protease. The other lanes show soybean milk protein decomposed for 4, 8, 12, 16, and 24 h. Lane 0 h shows the α'- and α-subunits of β-conglycinin (approximately 84–73 kDa), the A_3 acidic subunit (approximately 40 kDa), other acidic subunits as A_4, A_{1a}, A_{1b}, and A_2 (approximately 30–42 kDa) of glycinin, the β subunit (approximately 50 kDa), and basic subunits as B_3, B_{1a}, B_{1b}, and B_4(approximately 20 kDa) [40, 41].

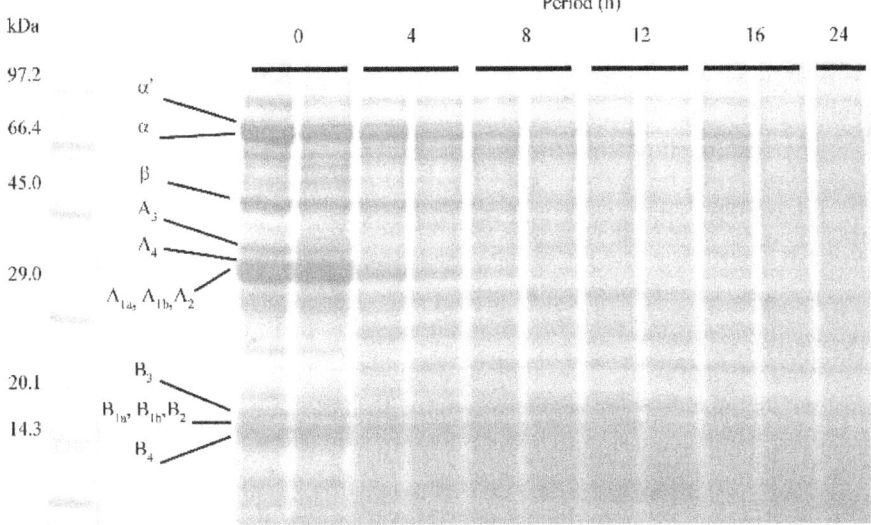

Figure 7: Digestion of soy protein during curding by soybean milk curdling enzyme.

The two bands shown as α' and α disappeared gradually after the reaction, showing the protein band from curd making the protease. In the glycinin subunits, the band of the A_3 acidic subunit disappeared completely after 4 h. Furthermore, A_4, A_{1a}, A_{1b}, and A_2 disappeared to a partial degree same as A_3acidic subunit. Peptides smaller than 20 kDa were detected on the gel. The β-conglycinin as α, α', and part of glycinin as A_3 A_4, A_{1b}, and A_2 were decomposed. Soybean protein became loosely curdled with the addition of other proteases from microorganisms or plants. The protein was decomposed. The low-molecular-weight peptides increased on the polyacrylamide gel. Generally, 11S glycinin was related to the formation of a stiffer gel. Furthermore, Ono *et al.* [42] reported hydrophobic bonding and hydrogen bonding related to curdling *Tofu*. Utsumi *et al.* [43] reported that the basic subunit and β-subunit formed macro complexes by heating. The complexes were regarded as forming cores for *Tofu* coagulation. The complexes were reportedly wrapped in α-, α'-, and acidic subunits [42].

According to our data, after the curdling soy milk by enzyme, α- and α'-subunits cleaved by the protease easily, whereas basic and β-subunit remained. It is considered that surface proteins as α- and α'-subunits were decomposed easily. Some decomposed subunits such as α, α', A_3, acidic, and basic subunits are regarded as related to the curdling soybean milk.

The enzyme of mechanisms for proteolysis was searched that the enzyme had the peptidase activity as exotype proteolysis activity and protease as endotype proteolysis activity. The synthesis peptide substrates, Z-glutamyl-tyrosine, and casein, FTC, were reacted by the enzymes. Peptidase (carboxypeptidase) activity was determined by the increase in ninhydrin after hydrolysis of benzyloxycarbonyl-glutamyl-tyrosine (pH 8.0) at 40°C.

The results show that the enzyme had 0.14 $U \cdot mg^{-1}$ protein as peptidase activity and 0.55 $U \cdot mg^{-1}$ protein protease as endotype proteolysis activity (data not shown). Results also show that the soybean milk-curdling enzyme as a proteolysis enzyme had endotype proteolysis activity.

Jones *et al.* [34] reported proteolysis enzymes of three types in yeast classes: cytosolic protease, vacuolar proteases, and proteases located within the secretory pathway. They belong to aspartic type, serine, or metallo-type proteolysis enzyme cleaved substrates with endotype or exotype. Generally, metallo-type enzyme requires metal ions such as zinc. The optimum condition of aspartic protease is an acidic condition. The enzyme did not require ion metal. Its optimum pH was 7.5, which is weakly alkaline. It is therefore considered that the enzyme is a serine protease of one kind. These results agreed with serine protease from *Bacillus* sp. curdling soy protein [44]. For future studies, we will ascertain the amino acid sequence in a substrate cleaved by enzyme using synthesis substrate.

Characteristics of Curd Curdling By Enzyme

As mechanisms that are closely involved in curdling soybeans, curdled soybean milk by enzyme and glucono-δ-lactone (GDL, 3.0% solution) as control samples were dissolved in chemical solutions. The enzyme solution (0.1 mL) was added to the soybean milk. The mixture was incubated at 40°C for 4 h., or 0.1 mL of glucono-δ-lactone (GDL, 3.0% solution) was added to the soybean milk (1.0 g) and incubated at 80°C for 1 h as control sample. To each of the two curdled soybean milk samples, 9 mL of chemical solution as 2% SDS solution,

4 M urea solution, and 10 mM 2-mercaptoethanol were added. They were held for 1 h at room temperature and were then centrifuged at 4000×g for 10 min. Protein in the supernatant was assayed using the Lowry method.

Results show the relative ratio of protein (%); that is, 100% relative ratio represents the amount of protein dissolving no-curdling soy milk in each solution (Fig. 8).

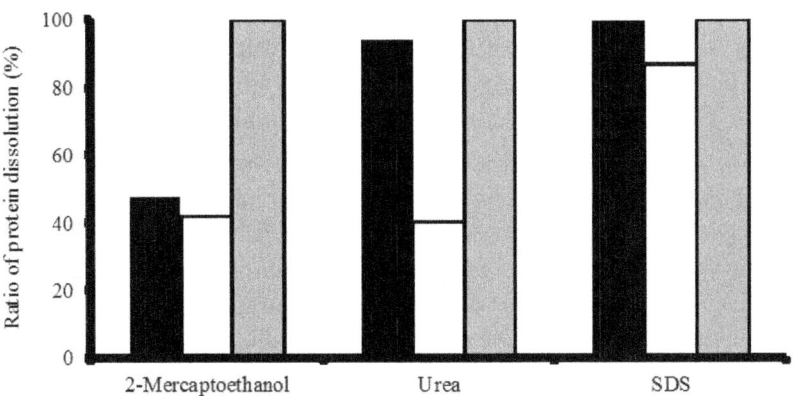

■, SCE; □, GDL; ■, no-curdled soy milk.

Figure 8: Curdled soy bean milk dissolved in chemical solutions. SCE shows soymilk curdled by enzymes; GDL shows soymilk curdled by glucono δ-lactone.

After dissolving the solutions, curd produced by the enzyme dissolved 47.3% of relative ratio from curd to the 2-mercaptoethanol solutions. That curdled by GDL dissolved 41.6% of relative ratio from the curd to the urea solutions. With urea solution, the curd making the enzyme dissolved 93.8% of relative ratio and GDL dissolved 40.3% of relative ratio. The curd produced by the enzyme can dissolve with urea. Both the curd produced by the enzyme and GDL dissolved with SDS solution. Actually, 2-mercaptoethanol solution cleaves the disulfide bond in protein. The urea solution cleaves the hydrogen bond, and the SDS solution cleaves the hydrophobic bond. That inference agrees with results reported by Yasuda *et al.* [44] that serine protease from *Bacillus* sp. curdled soybean milk and produced a protein bond through mutual hydrophobic bonding.

Next, the effects of the proteolysis enzyme against two protein in soybean, glycinin and β-conglycinin, were researched. Glycinin and β-conglycinin were extracted from commercial soy protein according a process described by Nagano *et al.* [44]. The soy protein (100 g) suspended 1500 mL of distilled water at pH 7.5 adjusted 0.1 M NaOH. From their extraction, the glycinin was carried out to do isoelectric precipitation at pH to 6.4. Moreover, β-conglycinin was also precipitated at pH 4.8. The two fractions were freeze-dried. Each fraction as glycinin fraction and β-conglycinin (50 mg) was resolved to 1 mL of 50 mM phosphate buffer (pH 7.5). Furthermore, enzyme solution (0.1 mL) was added. The mixture were incubated at 40°C. After reaction, soybean milk curdling activity was assayed according the preceding method. Curdling activity was assayed against two substrates: glycinin and β-conglycinin (Fig. 9). Glycinin was curdled strongly: soybean curdling activity was 86.9 (U·mL^{-1}·min^{-1}). However, β-conglycinin was curdled weakly: 38.0 (U·mL^{-1}·min^{-1}).

The data agree with reports in the relevant literature [45, 46, 47]. Bromelain decomposes 11S globulin to curdling. The entire band of acidic subunits and most basic subunits disappeared [46]. Glycinin-rich soybean milk was curdled strongly [47]. However, the enzyme made glycinin curdle without metal ion or GDL. Generally, glycinin is known to contain more sulfur amino acid than β-conglycinin does. According to Fig. 7, soymilk was curdled by a hydrogen bond or hydrophobic bond. Furthermore, some alkaline protease as subtilisin and chymotrypsin cleaves hydrophobic amino acid residue.

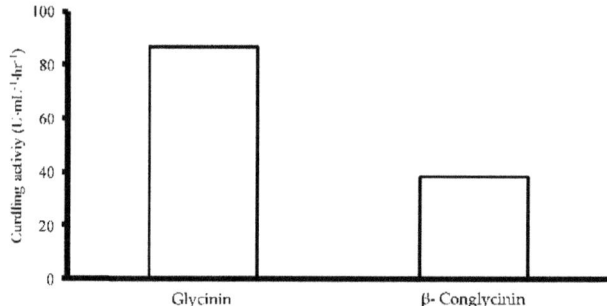

Figure 9: Curdling activity against two soy protein as glycinin and β-conglycinin.

The enzyme made curd from soymilk mainly by an enzyme reaction against glycinin. Choi *et al.* [48] reported that the α-subunit in β-conglycinin contains a hydrophobic sequence.

Soybean milk (10 mL) with 0.01 mL of anti-foam (KM-72F) was poured into a glass cup (32 mm inner diameter, 45 mm height). Then 1.5 mL of enzyme solution was added to the soybean milk. The mixture was incubated at 40°C for 4 h. As a control sample, glucono-δ-lactone (GDL, 0.3%) was added to soybean milk (10 mL) with 0.01 mL of anti-foam (KM-72F) added. Then the mixture was incubated at 80°C for 1 h. The curdled soybean milk samples were held at room temperature for 30 min. The rheological characteristics of enzyme curdling soybean milk were measured directly using a creep meter with a 16-mm-diameter plunger compressing 1 mm s^{-1} with 80.0%. As a control sample, soybean milk was curdled by GDL, 0.3% at 80°C for 1 h.

The rupture strength of the curdled soybean milk in the cups was measured directly using a creep meter (RE-3305; Yamaden Co. Ltd., Tokyo, Japan) with a 16-mm-diameter plunger compressing 1 mm s^{-1}with 80.0%.

The stress–strain curves of curdled soybean milk are presented in Fig. 10. The vertical axis shows stress (N·m^{-2}), which represents internal forces of the sample curd pushing back against the strain. The horizontal axis shows the strain of the curd. The sample curd strained by the plunger is broken by a force that exceeds a certain force: the breaking point. The breaking load represents the hardness or softness of the curd sample. The breaking strain represents the resilience of the sample curd: a large value signifies a high-resilience sample. Pressure by the plunger was loaded more. Then the curd was broken more heavily. After strain loading, the stress value decreased partly. The brittleness shows a different breaking point with the local minimal value. A large brittleness load shows brittle sample curd. The soybean milk was poured into a glass vessel and then curdled using the respective methods. After curdling, rheological analysis of the sample was conducted using a creep meter without taking out.

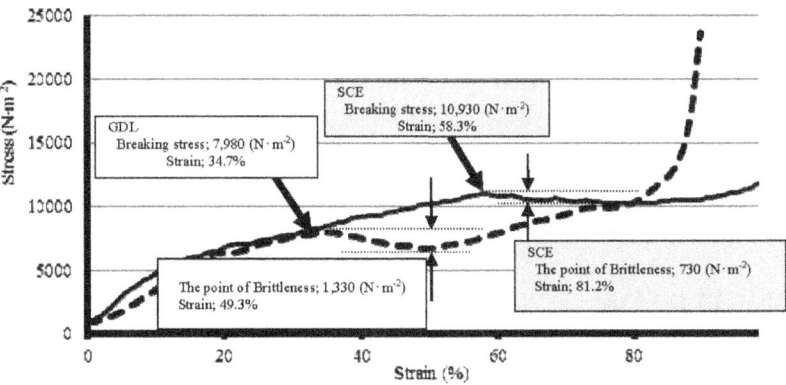

Figure 10: Stress–strain curves of curdled soybean milk.

The breaking point of curd curdled using GDL, which is used by many Japanese *Tofu* production companies to produce commercial *Tofu*, was 34.7% strain. The breaking stress was 7980 (N·m^{-2}). The point of brittleness was 49.3% and 6650 (N·m^{-2}). The brittleness of this curd produced using GDL was 1330 (N·m^{-2}). In fact, their *Tofu* curdled by GDL same as the Japanese *Tofu* is soft and brittle.

By contrast, the breaking point of curd curdled by soy curdling enzymes (SCE) was 58.4% strain. Their breaking stress was 10,930 ($N·m^{-2}$). The brittleness point is 81.2% and 10,200 ($N·m^{-2}$). The brittleness of this curd produced using the enzyme was 730 ($N·m^{-2}$). The curd that used SCE had 1.4 times greater breaking point, shown hardness, than that of the curd produced using the GDL same as the Japanese*Tofu*. Moreover, the curd that used SCE had 1/2 times smaller brittleness than that of GDL. Great breaking point shows hard curd and elasticity. It showed that the curd produced by SCE had hard, but soft, springy and sticky curd.

Heretofore, Yasuda *et al.* [45] reported soy milk curdled by bacteria protease. As a result, the curd produced using the bacteria protease is too soft to measure rheological characteristics, but they also reported that curd produced using the bacteria protease with calcium ion had $2–3×10^5$ ($N·m^{-2}$) of the breaking stress. Their curd was as hard as 20–30 times of the curd produced using this yeast enzyme. The curd produced using the bacteria protease with calcium was broken during fermentation and aging for long time. However, the curd produced using the bacteria protease had not springy and sticky texture.

It is considered that the curd produced using yeast enzymes was not like *Tofu* or the curd produced using the bacteria protease of its main characteristics or rheology.

Guo and Ono [47] and Toyokawa *et al.* [49] reported that the breaking stress of normal *Tofu* showed a relationship to soy milk conditions, such as their concentrations of glycinin, proteins, or temperature. Future investigations will examine if the condition of soy milk has a relationship with the breaking stress for the enzyme curdling.

According to this report, the new protease from *S. bayanus* SCY 003 produced a new texture of soy food that is applicable for new healthy foods, anti-milk allergy foods, and others.

CONCLUSION

Soybeans are a traditional food in eastern Asia, particularly in Japan and China. They were eaten in 100 BC in China. The beans can be processed into *Tofu*, soy milk, fermented seasonings, soy sauce or *Miso*paste, and *Natto* and green beans. Soybeans have rich nutrition, protein lipid, and other functional substances such as isoflavones. However, soybeans are difficult to process for use as food because of tissue and cell wall hardness. Therefore, soybeans are conducted to do some treatments, *e.g.*, boiling, steaming, roasting, crushing/grinding, and some enzyme treating, to eat soy protein easily.

Soy storage proteins mainly comprise two proteins as 7S globulin composed of β-conglycinin and γ-conglycinin and 11S globulin containing glycinin composed of 5 subunits. β-Conglycinin, included in 7S globulin, is composed of three subunits.

To modify the physical properties of soy protein, a new type of enzyme for curdling soybean milk enzyme was purified as an extract from yeast. Yeast producing curdling soybean milk enzyme, the SCY003 strain, was isolated from 1345 yeast strains. According to the morphology, physiology, and molecular and characteristics, SCY003 was identified as *S. bayanus*. The soy milk curdling enzyme having proteolytic activity was approximately 45 kDa and a monomer protein. The optimum pH for the protease activity was pH 7.5; the optimum temperature was 50°C.

The enzyme cleaved the β-conglycinin as α and α', and part of glycinin as A_3 A_4, A_{1b}, and A_2 in soy protein by endoproteolysis. Soybean protein became loosely curdled with the addition of other proteases from microorganisms or plants. Soybean milk curdled after cleaving endoproteolysis enzyme in SCY003 strain. The rheological characteristics of enzyme curdling soybean milk, the breaking point, was 58.4% strain; their breaking stress was 10,900 $(N \cdot m^{-2})$; the brittleness point is 81.2% and 10,200 $(N \cdot m^{-2})$. The brittleness of the curd produced using the enzyme was 727 $(N \cdot m^{-2})$. The curd had a sticky and chewy texture and did not resemble *Tofu* rheologically.

In this way, some properties of soy protein were modified by enzymes, such as decomposing specific subunits in soybeans and making soy milk curdle, which are expected to be applicable for new healthy foods, anti-milk allergy foods, and others.

ACKNOWLEDGEMENTS

We thank A-STEP, Adaptable and Seamless Technology, Transfer Program through Target Driven R&D by Japan Science and Technology Agency (JST) (AS231Z00291E) for financial support.

REFERENCES

1. Yamauchi F, Okubo K. editors. Daizu no Kagaku: Science of Soy Bean. Asakura Publishing, Tokyo: 1992 (in Japanese).

2. The yield amount of soy bean; http://www.maff.go.jp/j/seisan/ryutu/daizu/d_data/pdf/014.pdf

3. Trend in demand of soy bean; http://www.maff.go.jp/j/seisan/ryutu/daizu/d_data/pdf/011.pdf

4.   Endres JG. In: Soy Protein Products: Characteristics, Nutritional Aspects, and Utilization. Champaign, Illinois: AOCS Publishing; pp.10–18 (2001).

5.   Food and Drug Administration (FDA) in U.S. Soy Protein and Coronary Heart Disease. In: Federal Register. 1999:57700–57733.

6.   CMC Publishing. β-Conglycinin. Bio industry, 27 (5): 68–69 (2010) (in Japanese).

7.   Yamori Y, Ohta S, Watanabe S. Soybean Isoflavones. Tokyo: Saiwai Shobo; 2001 (in Japanese).

8.   Fritz H, Seely D, Flower G, Skidmore B, Fernandes R, Vadeboncoeur S, Kennedy D, Cooley K, Wong R, Sagar S, Sabri E, Fergusson D. Soy, red clover, and isoflavones and breast cancer: a systematic review. PLoS One 2013;8(11): e81968.

9.   Simons R, Vincken JP, Roidos N, Bovee TF, van Iersel M, Verbruggen MA, Gruppen H. Increasing soy isoflavonoid content and diversity by simultaneous malting and challenging by a fungus to modulate estrogenicity. J Agric Food Chem 2011;59(12): 6748–58.

10. Zimmermann MC, Tilghman SL, Boué SM, Salvo VA, Elliott S, Williams KY, Skripnikova EV, Ashe H, Payton-Stewart F, Vanhoy-Rhodes L, Fonseca JP, Corbitt C, Collins-Burow BM, Howell MH, Lacey M, Shih BY, Carter-Wientjes C, Cleveland TE, McLachlan JA, Wiese TE, Beckman BS, Burow ME. Glyceollin I, a novel antiestrogenic phytoalexin isolated from activated soy. J Pharmacol Exp Ther 2010;332(1): 35–45.

11. Kraft B, Barz W. Degradation of the isoflavone biochanin A and its glucoside conjugates by *Ascochyta rabiei*. Appl Environ Microbiol 1985;50(1): 45–8.

12. Tikhonovich IA. Nitrogen Fixation: Fundamentals and Applications: Fundamentals and Applications. Proceedings of the 10th International Congress on Nitrogen Fixation, St. Petersburg, Russia; 1995.

13. Takashima M, Nara K, Niki E, Yoshida Y, Hagihara Y, Stowe M, Horie M. Evaluation of biological activities of a groundnut (*Apios americana*Medik) extract containing a novel isoflavone. Food Chem 2013;138(1): 298–305.

14. Lin LZ, He XG, Lindenmaier M, Yang J, Cleary M, Qiu SX, Cordell GA. LC-ESI-MS study of the flavonoid glycoside malonates of red clover (*Trifolium pratense*). J Agric Food Chem 2000;2(48): 354–65. doi:10.1021/jf991002

15. Shimoyamada M. Processing characteristics of soy protein—curdling property of soy milk. J Cookery Sci Japan 2007;40(1): 37–40 (in

Japanese).

16. Ono T, Takeda M, Guo ST. Interaction of protein particles with lipids in soybean milk. Biosci Biotechnol Biochem 1996;60: 1165–9.

17. Ono T, Katho S, Mochizuki K. Influences of calcium and pH on protein solubility in soybean milk. Biosci Biotechnol Biochem 1993;57: 24–8.

18. Guo ST, Ono T, Mikami M. Incorporation of soy milk lipid into protein coagulum by addition of calcium chloride. J Agric Food Chem 1999;47: 901–5.

19. Larkin T, Price WE, Astheimer L. The key importance of soy isoflavone bioavailability to understanding health benefits. Crit Rev Food Sci Nutr 2008;48(69): 538–52.

20. Park YW, Kusakabe I, Kobayashi H, Murakami K. Production and properties of a soymilk-clotting enzyme system from a microorganism. Agric Biol Chem 1985;49(11): 3215–9.

21. Park YW, Kobayashi H, Kusakabe I, Yoshida S, Murakami K. Action of soymilk-clotting enzyme from *Bacillus* sp. K-295G-7 on the acidic subunit of soybean 11S globulin. Agric Biol Chem 1989;53(8): 2289–90.

22. Qua DV, Shimizu U, Taga N. Purification and some properties of halophilic protease produced using a moderately halophilic marine *Pseudomonas* sp. Can J Microbiol 1981;27: 505–10.

23. Hatanaka S, Maegawa M, Kanauchi M, Kasahara S, Shimoyamada M, Ishida M. Characteristics and purification of soybean milk curdling enzyme-producing yeast *Saccharomyces bayanus* SCY003. Food Sci Technol Res 2014;20(5): 927–38.

24. Kurtzman CP, Fell JW. In: The Yeast, A Taxonomic Study. 4th ed.. Amsterdam, The Netherlands: Elsevier Science Publishers B.V., pp.360–361 and 891–913, 1998.

25. Arima K, Yu J, Iwasaki S, Tamura G. Milk-clotting enzyme from microorganisms. V. Purification and crystallization of Mucor rennin from *Mucor pusillus* var. Lindt. J Appl Microbiol 1968;16: 1727–33.

26. Khan MR, Blain JA, Patterson, JDE. Extracellular protease of *Mucor pusillus*. Appl Environ Microbiol 1979;37(4): 719–24.

27. Nouani A, Belhamiche N, Slamani R, Belbraout S, Fazouane F, Bellal MM. Extracellular protease from *Mucor pusillus*: purification and characterization. Int J Dairy Technol 2009;6(1): 112–7.

28. Waschatko G, Junghans A, Vilgis TA. Soy milk oleosome behaviour at the air–water interface. Faraday Discuss 2012;158: 157–69.

29. Shimoyamada M, Tsushima N, Tsuzuki K, Asao H, Yamauchi R. Effect

of heat treatment on dispersion stability of soymilk and heat denaturation of soymilk protein. Food Sci Technol Res 2008;14: 32–8.

30. Remold H, Fasold H, Staib F. Purification and characterization of a proteolytic enzyme from *Candida albicans*. Biochim Biophys Acta 1968; 167: 399–406.

31. Ruchel R. Properties of a purified proteinase from the yeast *Candida albicans*. Biochim Biophys Acta 1968;659: 99–113.

32. Negi M, Tsuboi R, Matsui T, Ogawa H. Isolation and characterization of proteinase from *Candida albicans*: substrate specificity. J Invest Dermatol 1984;83: 32–6.

33. Ray MK, Uma DK, Seshu KG. Shivajo, S. Extracellular protease from the Antarctic yeast *Candida humicola*. Appl Environ Microbiol 1992;88(6): 1918–23.

34. Jones EW. Three proteolytic systems in the yeast *Saccharomyces cerevisiae*. J Biol Chem 1991;266(13): 7963–6.

35. Ruchel R. Properties of a purified proteinase from the yeast *Candida albicans*. Biochim Biophys Acta 1968;659: 99–113.

36. Cowan DA, Daniel RM. Purification and some properties of an extracellular protease (caldolysin) from an extreme thermophile. Biochim Biophys Acta 1982;705: 293–305.

37. Nouani A, Belhamiche N, Slamani R, Belbraout S, Fazouane F, Bellal MM. Extracellular protease from *Mucor pusillus*: purification and characterization. Int J Dairy Technol 2009;6(1): 112–7.

38. Park YW, Kobayashi H, Kusakabe I, Murakami K. Purification and characterization of soymilk-clotting enzymes from *Bacillus* sp. K-295G-7. Agric Biol Chem 1987;51(9): 2343–9.

39. Springham DG, Moses V, Cape RE. In: Biotechnology, the Science and the Business. New York: Harwood Academic Publishers; 1999.

40. Thanh VH, Shibasaki K. Heterogeneity of b-conglycinin from soybean seeds. Biochim Biophys Acta 1976a;439: 326–38.

41. Thanh VH, Shibasaki K. Major proteins of soybean seeds: Subunit structure of b-conglycinin. J Agric Food Chem 1978;26: 692–5.

42. Ono T, Wada T, Imai A. The structure of Tofu for preventing the change of lipid. Daizu Tanpakushitsu Kenkyu 2004;7: 42–7 (in Japanese).

43. Utsumi S, Damodaran S, Kinsella JE. Heat-induced interactions between soybean proteins: preferential association of 11S basic subunits and β subunits of 7S. J Agric Food Chem 1984; 32: 1406–12.

44. Nagano T, Hirotsuka M, Mori H, Kohyama K, Nishinari K. Dynamic viscoelastic study of the gelation of 7S globulin from soybeans. J Agric Food Chem 1992;40: 941–4.

45. Yasuda M, Kuba M, Tachibana S, Aoyama M. Analysis for mechanism of soybean-milk-coagulation by bacterial protease and utilization of the enzyme to the food processing. Daizu Tanpakushitsu Kenkyu 2002;5: 36–40 (in Japanese)

46. Mohri M, Matsushita S. Improvement of water absorption of soybean protein by treatment with bromelain. J Agric Food Chem 1984;32(3): 486–90.

47. Guo S, Ono T. The role of composition and content of protein particles in soymilk on Tofu curding by glucono-δ-lactone or calcium sulfate. Food Chem Toxicol 2005;70(4): 258–62.

48. Choi SK, Adachi M, Utsumi S. Improved bile acid-binding ability of soybean glycinin A1a polypeptide by the introduction of a bile acid binding peptide (VAWWMY). Biosci Biotechnol Biochem 2005;68: 1980–3.

49. Toyokawa T, Uehara M, Mochizuki T, Tamamura T, Higa K. Comparisons of physiological and chemical characteristics with Okinawa Tofu and Japanese Tofu. Report of Okinawa Industrial Technology Center. Okinawa, Japan: Okinawa Industrial Technology Center; 2008;11: 7–11 (in Japanese).

Chapter 5

ANTIBACTERIAL EFFECTS OF CINNAMON: FROM FARM TO FOOD, COSMETIC AND PHARMACEUTICAL INDUSTRIES

Seyed Fazel Nabavi [1], Arianna Di Lorenzo [2], Morteza Izadi [3], Eduardo Sobarzo-Sánchez [4], Maria Daglia [2,] and Seyed Mohammad Nabavi [1,]

[1]Applied Biotechnology Research Center, Baqiyatallah University of Medical Sciences, P.O. Box 19395-5487, Tehran 14359-16471, Iran

[2]Department of Drug Sciences, Medicinal Chemistry and Pharmaceutical Technology Section, University of Pavia, Pavia 27100, Italy

[3]Health Research Center, Baqiyatallah University of Medical Sciences, Tehran 14359-16471, Iran

[4]Laboratorio de Química Farmacéutica, Facultad de Farmacia, Universidad de Santiago de Compostela, Santiago de Compostela 15782, Spain

ABSTRACT

Herbs and spices have been used since ancient times, because of their antimicrobial properties increasing the safety and shelf life of food products by acting against foodborne pathogens and spoilage bacteria. Plants have historically been used in traditional medicine as sources of natural antimicrobial substances for the treatment of infectious disease. Therefore, much attention has been paid to medicinal plants as a source of alternative antimicrobial strategies. Moreover, due to the growing demand for preservative-free cosmetics, herbal extracts with antimicrobial activity have recently been used in the cosmetic industry to reduce the risk of allergies connected to the presence of methylparabens. Some species belonging to the genus *Cinnamomum*, commonly used as spices, contain many antibacterial compounds. This paper reviews the literature published over the last five years regarding the antibacterial effects of cinnamon. In addition, a brief summary of the history, traditional uses, phytochemical constituents, and clinical impact of cinnamon is provided.

INTRODUCTION

Herbs and spices have been used since ancient times, not only as antioxidants and flavoring agents, but also for their antimicrobial activity against degradation induced by foodborne pathogens and food spoilage bacteria. Many plants used in traditional medicine represent rich sources of natural bioactive substances with health-promoting effects and no side effects. Nowadays, over 65% of the world population relies on traditional medicine for health care [1,2,3,4]. Recently, a large demand has risen for preservative-free cosmetics and antimicrobial herbal extracts, aimed at reducing the risk of allergies connected to synthetic preservatives such as methylparabens [5].

During the last two decades, growing evidence shows that plants are rich sources of different classes of antimicrobial substances acting as defense systems to protect them against biotic (living) and abiotic (non-living) stresses [6]. Among these secondary metabolites, polyphenols, terpenoids, alkaloids, lectins, polypeptides, and polyacetylenes are known to be antimicrobial agents; most of these metabolites are also approved as a GRAS (Generally Recognised as Safe) material for food products, showing negligible side effects. These properties give them special economic importance [7]. There are many edible and medicinal plants with high antimicrobial effects, such as thyme (*Thymus vulgaris* L.), tea (*Camellia sinensis* L.), garlic (*Allium sativum* L.), turmeric (*Curcuma longa* L.), berries belonging to Rosaceae family, and cinnamon (species belonging to *Cinnamomun* genus) [6,8].

The genus *Cinnamomum* (family Lauraceae) contains more than 300 evergreen aromatic trees and shrubs [9]. Four species have great economic importance for their multiple culinary uses as common spices worldwide: *Cinnamon zeylanicum* Blume (a synonym of *Cinnamon verum* J. Presl, known as Sri Lanka cinnamon), *Cinnamon loureiroi* Nees (known as Vietnamese cinnamon), *Cinnamon burmanni* (Nees & T. Nees) Blume (known as Indonesian cinnamon) and *Cinnamon aromaticum* Nees (a synonym of *Cinnamon cassia* (L.). J. Presl, known as Chinese cinnamon) [10]. The term cinnamon commonly refers to the dried bark of *C. zeylanicum* and *C. aromaticum* [11] used for the preparation of different types of chocolate, beverages, spicy candies and liquors [12]. Moreover, cinnamon is used in various savory dishes, pickles, soups, and Persian sweets. Cinnamon bark, leaves, flowers and fruits are used to prepare essential oils, which are destined for use in cosmetics or food products. Moreover, according to traditional Chinese medicine (dating roughly 4000 years), cinnamon has been used as a neuroprotective agent [13] and for the treatment of diabetes [14]. Cinnamon has also been used as a health-promoting agent for the treatment of diseases such as inflammation, gastrointestinal disorders and urinary infections [15,16].

Another potential medical use of cinnamon would be with regards to its antimicrobial properties, especially antibacterial activity. It is well known that infection is one of the leading causes of morbidity and mortality worldwide. According to the World Health Organization reports, in 2011, there were more than 55 million deaths worldwide with infection being responsible for one-third of all deaths [17]. The high prevalence of infection and long-term exposure to antibiotics has lead to the antibiotic resistance of microorganisms. Therefore, much attention has been paid to the discovery and development of new antimicrobial agents that might act against these resistant microorganisms, and cinnamon could be an interesting candidate [6,18].

The aim of this review is to analyze the available scientific data, published over the last five years, regarding the antibacterial effects of cinnamon and its active constituents such as cinnamaldehyde and eugenol. In addition, a brief summary on the history, cultivation, chemical composition, traditional uses, and clinical impacts of cinnamon is provided.

HISTORY

For thousands of years, cinnamon has been known as one of the most common spices, with multiple culinary usages [19]. In Ayurvedic medicine it has been used as antiemetic, anti-diarrheal, anti-flatulent, and stimulant agent [20]. Furthermore, it was used for embalming by the ancient Egyptian people [21]. In the 16th century, Portuguese conquistadors discovered *C. zeylanicum* growing widely in Sri Lanka, importing the spice to European countries during the 16th and 17th centuries [19]. During the Dutch occupation in the 17th century, cinnamon cultivation started in Java, and the East India Company became the main cinnamon exporter to European countries [21]. Although Ceylon cinnamon cultivation diminished, Sri Lanka remains the main source of cinnamon oils, and Ceylon cinnamon oil from Sri Lanka has been broadly used by both pharmaceutical and food industries. Pharmaceutical industries also use Chinese cinnamon oils [10,21].

CULTIVATION OF CINNAMON

The average production rate of cinnamon is about 27,500 to 35,000 tons per year [22]. Cinnamon has mainly been cultivated in Sri Lanka, Seychelles, Madagascar and China [10,22,23]. In addition, it has been cultivated in India and Vietnam on a small scale [10,21]. Cinnamon can easily grow under tropical conditions in different soil types, ranging from the silver sands of the west coast of Sri Lanka to the loamy soils of its south coast. It has however been reported that soil quality and climate changes affect the production and quality of cinnamon. For example, the best cinnamon is produced in sandy soils enriched with humus. The optimum temperature for cinnamon cultivation is between 20–30 °C with an annual rainfall range of 1250–2500 mm. Cinnamon is commonly propagated by seedling and vegetative propagations. The application of fertilizer containing urea, phosphate, and potash is known to increase cinnamon production [23,24].

CHEMICAL COMPOSITION OF CINNAMON

The main compounds isolated and identified in cinnamon (*C. zeylanicum*) belong to two chemical classes: polyphenols and volatile phenols. Among polyphenols, cinnamon contains mainly vanillic, caffeic, gallic, protocatechuic, *p*-coumaric, and ferulic acids (Figure 1) [25]. With regards to volatile components, the chemical composition of cinnamon essential oils depends on the part of the plant from which they are extracted. In bark essential oil, cinnamaldehyde (Figure 2) is the most represented substance, with a content ranging from 90% to 62%–73%, depending on the type of extraction, this being higher for steam distillation than Soxhlet extraction [26]. The other minor volatile compounds are hydrocarbons and oxygenated compounds (*i.e.*, -caryophyllene, benzyl benzoate, linalool, eugenyl acetate, and cinnamyl acetate) (Figure 2). In cinnamon leaf essential oil, the main component is eugenol, which reaches a concentration of more than 80%. In the essential oil obtained from cinnamon fruit and flowers, *(E)*-cinnamyl acetate and caryophyllene are the major components (Figure 2) [27,28,29].

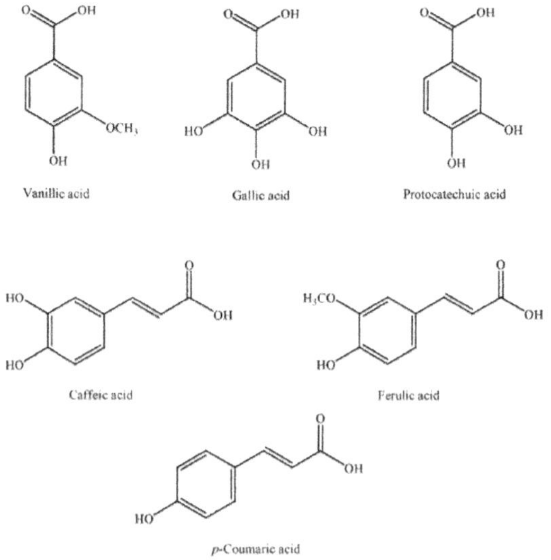

Figure 1: Polyphenolic constituents of cinnamon

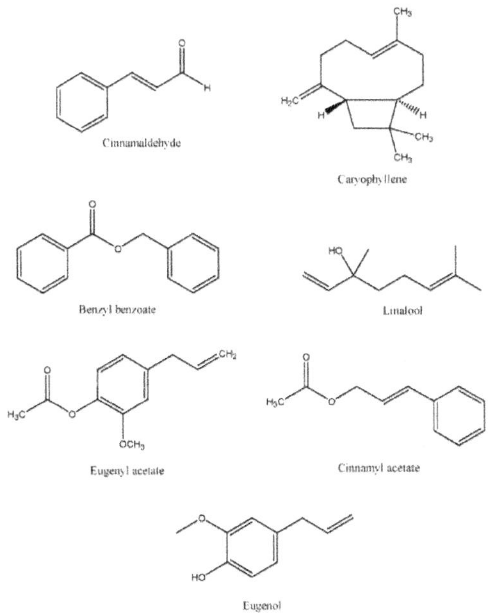

Figure 2: Major and minor chemical compounds of cinnamon essential oil

TRADITIONAL USES

Cinnamon has been known as one of the most common spices and food flavoring additives since ancient times [19]. For instance, it has been used as a flavor in sweets and chewing gum due to the pleasant and refreshing effect that develops in the mouth. It also shows beneficial effects on oral health and is used for toothaches, oral infections, and to remove bad breath [30]. Cinnamon has also been used to treat acne and melisma [31]. Moreover, it has been used for the treatment of gastrointestinal and colonic [32]. Ayurvedic literature shows that cinnamon has potent antiemetic, anti-diarrheal, anti-flatulent, and stimulant activities [33]. Cinnamon has a coagulant effect and therefore it can be used against hemorrhaging [34]. Cinnamon increases the blood flow in the uterus and improves tissue regeneration [35]. Moreover, it possesses potent antibacterial, antifungal, antitermitic, larvicidal, nematicidal, and insecticidal properties [14,36,37,38,39,40,41]. More recently, scientific reports showed that cinnamon has potent neuroprotective, hepatoprotective, cardioprotective and gastroprotective effects due to its potent antioxidant and anti-inflammatory properties [13,42]. Cinnamon essential oil could be also used in aromatherapy, which is the therapeutic use of plant essential oils that can be absorbed into the body via the skin or the olfactory system. A recent research articled showed the benefits deriving from the use of cinnamon oil in massage for alleviating menstrual pain [43].

CLINICAL IMPACTS

As far as the high therapeutic potential of cinnamon is concerned, there are numerous clinical studies on this spice. A search on the Clinical Trials Gov. database with the keyword "cinnamon" showed that there are 28 clinical trials, including 17 completed studies, six recruited studies, and one terminated study [44]. Most of these clinical trials are focused on its anti-diabetes and glucose lowering effects. In this context, some ongoing clinical trials are about the bioavailability of cinnamon and its beneficial effects on gingivitis, polycystic ovary syndrome, body fat, and blood glucose level in diabetic patients. Details of completed clinical trials on cinnamon are summarized in Table 1.

Table 1: Details of our search in http://clinicaltrial.gov website [44] with keyword "cinnamon"

Clinical Trials	Title	Primary Outcome Measures and Treatments	Results
NCT02074423	A Human Clinical Trial Evaluating the Effect of MealShape™ in Blood Glucose Level Following Consumption of Standard Meal	measurements of blood glucose incremental area under the curve between 0 and 120 min, after consumption of a standard meal, compared the consumption of MealShape cinnamon extract (acute administration of 1 g corresponding to 2 capsules of 500 mg)	Cinnamon hydro-alcoholic extract may provide a natural and safe solution for the reduction of postprandial hyperglycemia and therefore help to reduce the risks of developing metabolic disorders.
NCT00846898	Is There a Metabolic Effect of Cinnamon on glycosylated hemoglobin A1c (HbA1c), Blood Pressure and Serum Lipids in Type 2 Diabetes Mellitus? (cinnamon)	measurements of blood profiles of HbA1c levels, after administration of cinnamon capsules (2 g per day for 12 weeks)	No study results posted on ClinicalTrials.gov* [44]
NCT00331279	The Effect of Cinnamon Extract on Insulin Resistance Parameters in Polycystic Ovary Syndrome: A Pilot Study	measurements of fasting glucose, fasting insulin, Homeostasis Model Assessment – Insulin Resistance (HOMA-IR), Quantitative Insulin Sensitivity Check Index (QUICKI), insulin sensitivity index (Matsuda), after administration of 2 cinnamon tablets (500 mg of purified aqueous extract of cinnamon for 8 weeks)	No study results posted on ClinicalTrials.gov* [44]
NCT00951639	Cassia Cinnamon for Glucose Uptake In Young Women	measurements of blood glucose, after the treatment with a cinnamon food supplement (5 g encapsulated ground bark administered once in experimental session)	No study results posted on ClinicalTrials.gov* [44]
NCT00237640	Effect of Cinnamon on Glucose and Lipid Levels in Non-Insulin Dependent Type 2 Diabetes Mellitus	measurements of HbA1c, glucose, total cholesterol, low-density lipoprotein (LDL cholesterol), high-density lipoprotein (HDL cholesterol), and triglycerides levels, after the treatment with cinnamon (500 mg capsule twice daily for 3 months)	Cinnamon taken at a dose of 1 g daily for 3 months produced no significant change in fasting glucose, lipid, A1C, or insulin levels.
NCT00371800	The Effect of Cinnamon on HbA1c Among Adolescents With Type 1 Diabetes	measurements of blood profiles of HbA1c levels, after the treatment with cinnamon (1 gram/day for 90 days).	No study results posted on ClinicalTrials.gov* [44]
NCT01350284	The Effect of Natural Food Flavourings on Gastrointestinal and Cardiovascular Physiological Responses (CinnGastEmpt)	measurements of the effect of 3 g cinnamon on gastric emptying half time	An aliquot of 3 g cinnamon did not alter the postprandial response to a high-fat test meal. No evidence was found to support the use of 3 g cinnamon supplementation for the prevention or treatment of metabolic disease
NCT01027585	The Effects of Cinnamon on Postprandial Blood Glucose, and Insulin in Subjects With Impaired Glucose Tolerance	measurements of postprandial blood glucose, and plasma concentrations of insulin in subjects with impaired glucose tolerance, after the treatment with cinnamon capsules (doses not provided, for 5 months)	No study results posted on ClinicalTrials.gov* [44]
NCT01085019	Impact of Spices and Herbs on Endothelial Function	measurements of circulating level of plasma lipoproteins-lipids, oxidative stress, endothelial activation and inflammatory markers, after daily consumption of spices and herbs, among which cinnamon in capsules (2.8 g/day for 4 weeks)	No study results posted on ClinicalTrials.gov* [44]
NCT00718796	Naturopathic Treatment for the Prevention of Cardiovascular Disease (CVD)	evaluation of metabolic syndrome and general cardiovascular risk profile (Framingham Heart Study), after naturopathic approach with some spices (among which cinnamon) to CVD prevention over the course of 1 year	Naturopathic approach to CVD primary prevention significantly reduced CVD risk over usual care plus biometric screening and reduced costs to society and employers in this multi-worksite-based study.

NCT02193438	Physiologic Effect of Spices Ingestion	determination of resting energy expenditure, calculation of the resting energy expenditure from continuous measurement of oxygen consumption and carbon dioxid production (indirect calorimetry), heart rate variability, power spectral analysis of heart rate variability from continuous measurement of very low, low and high frequency range electrocardiographic signals, after the ingestion of a single dose of cinnamon extract (dose not provided).	No study results posted on ClinicalTrials.gov* [44]
NCT02234206	A Clinical Trial to Study the Safety and Efficacy of Chandrakanthi Choornam in Patients With Low Sperm Count	measurements of sperm concentration, proportion of sperm motility changes in the percentage of total and progressive motility of sperm proportion of sperm morphology changes in the percentage of sperm cells with normal forms, after the treatment with Chandrakanthi Choornam (dose non provided), which is a formulation consisting of 25 ingredients, among which Cinnamomum verum (bark) and Cinnamomum taraula (leaf) for 3 months	No study results posted on ClinicalTrials.gov* [44]
NCT00954902	Effects of Antioxidants on Cardiovascular Risk Measures (Spice Study)	measurements of Interleukin 6 (IL-6) response to psychological stress at time points equal to and greater than 90 min post task, after treatment with a high antioxidant spice blend (14.5 g blend of spice, among which cinnamon, incorporated into a delivery meal	Inclusion of spices may attenuate postprandial lipemia via inhibition of Phospholipase (PL) and Phospholipase A_2 (PLA$_2$).
NCT01752868	Can Fish Oil and Phytochemical Supplements Mimic Anti-Aging Effects of Calorie Restriction?	measurements of carotid-femoral pulse wave velocity, after the treatment with a combination of 10 nutritional supplements, among which cinnamon bark, for 6 months	No study results posted on ClinicalTrials.gov* [44]
NCT01667523	The Effect of Capsaicin and Cinnamaldehyde on Intestinal Permeability	evaluation of the effect of capsaicin and cinnamaldehyde infusion on intestinal permeability, after the administration of cinnamaldehyde (70 mg per intervention administered intraduodenally)	No study results posted on ClinicalTrials.gov* [44]
NCT01895816	Herbal Tonic Fertile Supplement(ZO2C5)	measurements of sperm count variation and semen analysis according World Health Organization methods. after the treatment with mixed herbals drug, in which cinnamon is one of the bioactive components for 6 months (dose not provided)	No study results posted on ClinicalTrials.gov* [44]

* as reported in http://clinicaltrial.gov/ website, the absence of posted results could be due to the facts that: the study may not be subject to U.S. Federal requirements to submit results, or the study has been completed, but the deadline for results submission has not passed, or the results have been submitted but have not yet been posted (for example, pending review by ClinicalTrials.gov) [44].

ANTIBACTERIAL EFFECTS OF CINNAMON ESSENTIAL OIL AND CINNAMON EXTRACTS

One of the most well-established properties of cinnamon extracts, essential oils and their components is the antibacterial activity against Gram-positive and Gram-negative bacteria responsible for human infectious diseases and degradation of food or cosmetics. In the literature, there are a number of studies showing the antibacterial activity of cinnamon essential oils obtained from different botanical parts and extraction methods. A search was conducted on the PubMed database [45], using the keywords *"antibacterial activity of cinnamon"*. The results returned 45 papers from 2010 up to 2015; the most interesting of these were summarized and critically discussed to provide a consistent review.

Antibacterial Activity of Cinnamon against Bacteria Responsible for Human Infectious Diseases

In 2011, the antibacterial activities of several *C. zeylanicum* bark extracts, obtained with different organic solvents, as ethyl acetate, acetone and methanol, were tested *in vitro* against *Klebsiella pneumonia* 13883, *Bacillus megaterium* NRS, *Pseudomonas aeruginosa* ATCC 27859, *Staphylococcus aureus* 6538 P, *Escherichia coli* ATCC 8739, *Enterobacter cloacae* ATCC 13047,*Corynebacterium xerosis* UC 9165, *Streptococcus faecalis* DC 74, by the disk-diffusion method. The results showed that the antibacterial activity, expressed as inhibition zone, ranges from 7 to 18 mm for the application of 30 µL, suggesting a high antibacterial activity [46]. In the same year, Mandal *et al.* showed that the ethanolic extract of stem bark *C. zeylanicum* exerted antibacterial activity against clinical isolates of methicillin resistant *S.*

aureus (MRSA), from Kolkata, India. The antibacterial activity was expressed as both diameters of inhibition and minimum inhibitory concentration (MIC) values at different times of incubation. The cinnamon extract, which showed a diameter of inhibition zone ranging from 22 to 27 mm, resulted to be bactericidal after 6 h of incubation. The authors concluded that *C. zeylanicum* could be considered a valuable support in the treatment of infection and may contribute to the development of potential antimicrobial agents against MRSA bacteria [47].

As part of the studies on the antibacterial activity of cinnamon, the sensibility of two clinical strains of *Moraxella catarrhalis*(an important cause of lower respiratory tract infection, resistant to conventional antimicrobial agents) to the hydro-ethanolic extract of *C. zeylanicum* bark and clove powder, was tested using disk-diffusion and broth dilution methods. The results showed that cinnamon extract is active against both strains and, therefore, it represents an alternative source of natural antimicrobial substances for use in clinical practice for the treatment of cases of *M. cattarhalis* [48]. In 2012, Guerra *et al.* published an investigation on the antibacterial activity of the combination of *C. zeylanicum* essential oil and antibiotics, in which additive and synergistic effects were shown [49]. More recently, Yap *et al.* reached similar results. In fact, the authors showed that the combination of piperacillin and cinnamon bark essential oil induced a considerable reduction in the registered MIC values against a clinical strain of beta-lactamase-producing *E. coli*. The authors concluded that a reduced use of antibiotics could be employed as a treatment strategy to decrease the adverse effects and possibly to reverse the beta-lactam antibiotic resistance [50].

In the same year, cinnamon bark essential oil obtained through hydro-distillation was tested for antibacterial activity (expressed as MIC) against several pathogenic bacterial strains (*Salmonella typhi*, *Salmonella paratyphi A*, *E. coli*, *S. aureus*,*Pseudomonas fluorescens* and *Bacillus licheniformis*) and analyzed with thin layer chromatography (TLC) and gas chromatography coupled with mass spectrometry (GC-MS). The results showed that the tested sample exhibited excellent activity against all the selected strains (MIC values ranged from 2.9 to 4.8 mg/mL). TLC and GC-MS analyses of chemical composition revealed the presence of *t*-cinnamaldehyde (which was the most abundant substance, corresponding to 4.3%), eugenol (0.32%) and minor components such as cuminaldehyde, and γ-terpinene [51].

In 2014, Al-Mariri and Safi studied the antibacterial activity against Gram-negative bacteria (using a microdilution broth susceptibility assay) of cinnamon bark essential oil obtained via hydro-steam distillation. The sample showed good antibacterial activity against the Gram-negative bacteria

(*E. coli* O157:H7, *Yersinia enterocolitica* O9, *Proteus* spp. and *Klebsiella pneumonia*) with very low MIC values (12.5 μL/mL, 6.25 μL/mL, 1.5 μL/mL and 3.125 μL/mL, respectively) [52]. More recently, in 2015 other research groups investigated the antibacterial activity of cinnamon essential oil and extracts and found similar results [53,54]. In this context, another investigation, which is particularly worthy of note, is that from Kim *et al.* They reported that cinnamon bark oil, cinnamaldehyde and eugenol at 0.01% (*v/v*) significantly decreased biofilm formation of enterohemorrhagic *E. coli* O157:H7 (EHEC). Another investigation focusing on Gram-negative bacteria was published by Seukep *et al.* (2013) [55] who studied the *in vitro* antibacterial activity of several Cameroonian dietary plants, including *C. zeylanicum*, against multidrug resistant (MDR) Gram-negative bacteria overexpressing active efflux pumps, which make bacteria resistant to antibiotic treatment. The tested bacteria included both reference (from the American Type Culture Collection, ATCC) and clinical strains of *E. coli*, *Enterobacter aerogenes*, *Providencia stuartii*, *P. aeruginosa*, *Klebsiella pneumoniae*, and *Enterobacter cloacae*. The bacterial efflux pumps can be blocked by various inhibitors, which restore the intracellular concentration and activity of the antibics. In this research the antibacterial activity was evaluated using the liquid microdilution method. Chloramphenicol was used as an antibiotic in the absence and the presence of phenylalanine arginyl ß-naphthylamide (PAßN), a known efflux pump inhibitor, which was used to show the role of efflux pumps in the resistance of the studied bacterial strains. The results revealed that the methanolic bark extract of cinnamon is able to inhibit bacterial growth, with different MIC values, ranging from 64 to 1024 μg/mL, depending on the strains. The authors concluded that the antibacterial activity of that cinnamon methanolic bark extract could be used in the treatment of infectious diseases induced by bacteria expressing MDR phenotypes [56].

Some studies showed that cinnamon extracts and essential oils could be active against oral cavity infections. Chaudhari *et al.* in 2012 [57], showed that cinnamon essential oil was active against *Streptococcus mutans* and concluded that the use of cinnamon essential oils can be a good alternative to other antibacterial compounds against the bacteria responsible for oral infections. More recently, the antibacterial activity of *C. zeylanicum* fresh leaf extract was studied against *Enterococcus faecalis*, one of the main causative factors of pulp and periapical diseases of the oral cavity. *E. faecalis* was grown both on cellulose nitrate membrane and on a tooth model system. The antibacterial activity was determined by the agar diffusion test and microdilution method. The results showed that the obtained inhibition zones vary with increasing concentration (5% to 20%) of cinnamon fresh leaf extract. Moreover, a complete inhibition of bacterial growth was registered after 12 h of contact, using NaOCl as a reference, which suggests that the cinnamon extract is active against both planktonic and biofilm forms; this was also observed *in vivo* [58]. Another piece of recent research has demonstrated that the essential oil obtained from the fresh leaves of *C. zeylanicum* is active against *S. mutans* and *Lactobacillus acidophilus* which are involved in dental plaque formation and caries development. The MIC values obtained from *S. mutans* with the tube dilution bioassay were lower than that of gentamycin (0.31 µL/mL and 0.83 µL/mL, respectively). *L. acidophilus* was less sensitive to this essential oil (1.46 µL/mL). The authors concluded that promising *in vitro* data would require *in vivo* studies to determine the dose to be used in products for oral hygiene, which have no cytotoxicity [59].

The aqueous, hydro-alcoholic and alcoholic dried inner bark extracts of cinnamon obtained using Soxhlet extraction were tested against two acne causing bacteria, *i.e.*, *Propionibacterium acnes* and *Staphylococcus epidermidis*, using the well diffusion method. The results showed that at a concentration of 5 mg/mL the inhibition zone for aqueous and ethanolic dried inner bark extracts against *P. acnes* was 18 ± 1.02 mm and 18 ± 1.6 mm, respectively. The hydro-ethanolic dried inner bark extracts were found to be inactive. The *S. epidermidis* strains were more sensitive towards these extracts, with higher inhibition zones (22 ± 1.7 mm, 22 ± 1.2 mm and 15 ± 1.8 mm for aqueous, hydro-alcoholic and ethanolic extracts, respectively). The authors ascribed the antibacterial activity to the presence of phenolic compounds such as cinnamaldehyde and eugenol, and concluded that these cinnamon extracts could be used to develop new formulations for acne treatment [31].

The antibacterial activity of cinnamon bark essential oil was also tested against 50 clinical strains of *Mycoplasma hominis* isolated from the cervical swabs of randomly selected women. *M. hominis* is responsible for bacterial vaginosis, pelvic inflammation, and pyelonephritis. The essential oil, whose main constituents was cinnamaldehyde (97% *w/w*), showed antibacterial and bactericidal activity, with MIC values ranging from 250 to 1000 µg/mL [60]. Another bacterium involved in sexually transmitted infection is represented by *Haemophilus ducreyi*, a Gram-negative coccobacillus, which is a strict human pathogen responsible for the development of chancroid. Due to the fact that starting from the 1970s, some strains of *H. ducreyi* have shown resistance to penicillin and its derivatives and then to sulfonamides, aminoglycosides, tetracyclines, and chloramphenicol it is of particular concern the search of new compounds given the connection between Human Immunodeficiency Virus 1 (HIV-1) and chancroid. The authors showed the antibacterial activity, expressed as MIC and minimum lethal concentrations, of the essential oil obtained from *C. verum* [61].

Examples of Cinnamon Applications in Food and Cosmetic Industries

Food and cosmetic products can be vectors for many harmful microbial agents that can cause infections. Foodborne pathogens are responsible for infectious diseases that are a growing public health problem worldwide, affecting about 2 million children every year, especially in developing countries. Nevertheless, foodborne diseases are not limited to developing countries, and the research on preservatives able to inhibit bacterial degradation of food and cosmetics is important for the ongoing maintenance and improvement of public health. As reported above, in recent years many investigations have shown the antimicrobial activity of cinnamon essential oil against food poisoning bacteria *in vitro*. Other investigations have studied the protective effects of cinnamon in food matrices, cosmetic products and active packaging and their ability to inhibit pathogen growth without introducing chemical preservatives that consumers could find undesirable. For instance, a recent investigation showed that the essential oil obtained from the bark of *C. cassia* can control the growth of the spoilage microorganism *L. monocytogenes* in meat products contaminated at a concentration of 5 ppm, which did not change the sensorial properties of the products. In particular, cinnamon essential oil reduces the bacterial growth rate significantly in artificially contaminated samples when compared with an untreated control [62]. Similar investigations were performed a few years back by several research groups that studied the antibacterial activity of cinnamon against foodborne pathogens, especially in contaminated meat, such as *Salmonella typhimurium*, *S. aureus* and *E. coli*, *Arcobacter butzeiri* and*Arcobacter skirrowii* [63,64,65]. The following paper is particularly noteworthy because the extract obtained from a cinnamon stick resulted to be active at room temperature (~23 °C) against *L. monocytogenes*, *S. aureus*, and *Salmonella enterica* in a food matrix different from meat and represented by cheese, suggesting that the extract is a potential natural food preservative [66]. Another interesting investigation reports the antibacterial activity of cinnamon bark essential oil and its main constituents, *trans*-cinnamaldehyde and eugenol against *Cronobacter sakazakii* and *C. malonaticus*, which are opportunistic pathogens that cause infection in children and immunocompromised adults. These bacteria are present in many food products; therefore, decreasing the bacterial count would be desirable. The antibacterial activity was assayed in liquid and vapor phases to test the strain susceptibility to both nonvolatile and volatile compounds. The results showed that the MIC values of cinnamon essential oil (ranging from 0.25 to 0.5 mg/mL) in liquid and vapor phase are similar to those registered in the same conditions for *t*-cinnamaldehyde (ranging from 0.128 to 0.3 mg/mL). Eugenol showed higher MIC values (ranging from 0.512 to 1.0 mg/mL), suggesting lower antibacterial activity. Based on these results, the authors concluded that cinnamon essential oil could be incorporated into food packaging materials or used to create an active modified atmosphere to reduce the contamination of *Cronobacter* species [67]. Another study showed that commercial essential oils obtained from the two most common species of cinnamon, *C. cassia* (leaf-branch) and *C. verum* (bark), were tested against *L. monocytogenes*NCTC 11994, *L. monocytogenes* S0580 (isolated from pork meat), *S. typhimurium* ATCC 14028, *S. typhimurium* S0584 (isolated from pig carcass), *E. coli* O157:H7 ATCC 35150 and *E. coli* O157:H7 S0575 (isolated from minced beef), *Brochothrix thermosphacta* ATCC 11509, and *P. fluorescens* ATCC 13525. The antibacterial activity was evaluated using the disk-diffusion methodand both MIC and MBC values were calculated. The essential oils showed high antimicrobial activity against the tested bacteria with MIC values lower than 1 µL/mL. The authors attributed this activity to the main bioactive constituents, especially cinnamaldehyde. They suggested that these essential oils and their main active components could be used as natural alternatives for food preservation to retard or inhibit the bacterial growth of pathogenic and spoilage bacteria and to extend the shelf life of the food products [68].

As far as the cosmetic field is concerned, Herman *et al.* (2013) showed that commercial cinnamon essential oil in a cosmetic emulsion at 2.5% concentration possesses very good antibacterial activity against several contaminants such as *P. aeruginosa*ATCC 27853, *E. coli* ATCC 25922, and *S. aureus* ATCC 29213. The antibacterial activity, evaluated with the disk-diffusion test, was found to be higher than that registered for methylparaben, used as positive control. The diameters of inhibition zones ranged from 24 to 44 mm for the cinnamon essential oil, and from 9 to 8 mm for methylparaben [5].

Another practical application for the antibacterial activity of cinnamon essential oil was reported by Hill *et al.* [69] who tested cinnamon bark extract entrapped in nanoparticles prepared with poly dl-lactide-co-glycolide (PLGA), a biocompatible polymer widely used in the pharmaceutical industry and which could be used in the food industry to deliver antimicrobial compounds to food matrices. The authors tested the antibacterial activity of the nanoparticles loaded with cinnamon extract against *L. monocytogenes* and *S. typhimurium*. The results showed that the nanoparticles exerted antibacterial activity against the tested bacteria. Therefore nanoencapsulation could be a good method to deliver entrapped antibacterial substances to pathogens in food products without a heavy influence on sensorial properties.

Table 2 forms a summary of the antibacterial studies reported in Section 7.

Table 2: List of cinnamon essential oils or extracts active against Gram-negative and Gram-positive bacteria

Type of Sample	Bacteria	References
BARK extracts, obtained with different organic solvents (ethyl acetate, acetone and methanol)	*Klebsiella pneumonia* 13883 *Bacillus megaterium* NRS *Pseudomonas aeroginosa* ATCC 27859 *Staphylococcus aureus* 6538 P *Escherichia coli* ATCC 8739 *Enterobacter cloaca* ATCC 13047 *Corynebacterium xerosis* UC 9165 *Streptococcus faecalis* DC 74	[45]
STEM BARK Ethanolic extract	*Staphylococcus aureus* (MRSA)	[46]
BARK AND CLOVE POWDER Hydroethanolic extract	*Moraxella catarrhalis*	[47]
Combination of piperacillin and cinnamon BARK essential oil	*E. coli* (β-lactamase-producing)	[49]
Essential oil obtained by hydro-distillation of cinnamon BARK	*Salmonella typhi* *Salmonella paratyphi* A *Escherichia coli* *Staphylococcus aureus* *Pseudomonas fluorescens* *Bacillus licheniformis*	[50]
Essential oil obtained by hydro-steam distillation of cinnamon BARK	*Escherichia coli* O157:H7 *Yersinia enterocolitica* O9 *Proteus* spp. *Klebsiella pneumonia*	[51] [52] [53]
Essential oil (BARK and fresh LEAVES)	*Escherichia coli* O157:H7 *Yersinia enterocolitica* O9 *Proteus* spp. *Klebsiella pneumonia* *Streptococcus mutans* *Lactobacillus acidophilus (fresh leaves)* *Mycoplasma hominis* (bark) *Haemophilus ducreyi* (E. O from *C. verum*) *L. monocytogenes* (bark) (E.O from *C. cassia*) *Salmonella typhimurium*	[52] [53] [54] [56] [58] [59] [60] [61] [63]

Methanolic extract of cinnamon BARK	*Escherichia coli*	[55]
	Enterobacter aerogenes	
	Providencia stuartii	
	Pseudomonas aeruginosa	
	Klebsiella pneumoniae	
	Enterobacter cloacae	
Fresh LEAF extract	*Escherichia coli* O157:H7	
	Yersinia enterocolitica O9	[52]
	Proteus spp.	[53]
	Klebsiella pneumonia	[57]
	Enterococcus faecalis	
Aqueous, hydroalcoholic and alcoholic dried inner BARK extracts (Soxhlet)	*Propionibacterium acnes* (hydroethanolic extracts inactive)	[30]
	Staphylococcus epidermidis	
Cinnamon BARK extracts	*Salmonella typhimurium*	
	S. aureus	[62]
	E. coli	
Extract obtained from cinnamon STICK	*L. monocytogenes*	
	S. aureus	[65]
	Salmonella enterica	
BARK essential oil (tested in liquid and vapor phases)	*Cronobacter sakazakii*	[66]
	C. malonaticus	
Commercial essential oils from *C. cassia* (LEAF-BRANCH) and *C. verum* (BARK)	*L. monocytogenes* NCTC 11994	
	L. monocytogenes S0580	
	S. typhimurium ATCC 14028	
	S. typhimurium	
	E. coli O157:H7 ATCC 35150	
	E. coli O157:H7 S0575	[67]
	Brochothrix thermosphacta ATCC 11509	[5]
	P. fluorescens ATCC 13525	[64]
	P. aeruginosa ATCC 27853	
	E. coli ATCC 25922	
	S. aureus ATCC 29213	
	Arcobacter butzeiri	
	Arcobacter skirrowii	
Nanoparticles loaded with cinnamon BARK extract	*L. monocytogenes*	[68]
	S. typhimurium	

TOXICOLOGICAL ASPECTS

Despite the above reported studies that promote the use of cinnamon applications in food and cosmetic products, the oral ingestion or skin applications of cinnamon or its components (*i.e.*, cinnamaldehyde, eugenol, and cinnamic acid) is not always advisable and is recommended only in very small doses. Cinnamon oil should be diluted to less than 2% before oral use [70]. Moreover, it is recommendable to avoid the ingestion of cinnamon bark oil *per os* to patients suffering from liver conditions, in case of alcoholism and when taking paracetamol. This recommendation is related to the capacity of cinnamaldehyde to deplete glutathione [71]. Regarding cinnamon bark, it appears to be safe for most people when taken by mouth in amounts up to six grams daily for six weeks or less [70]. We must also point out that no drug interactions are reported for *C. zeylanicum*. Differently, *C. cassia* bark (at the dose of 2 g in 100 mL) retarded the *in vitro* dissolution of tetracycline. HCl. In fact, only 20% dissolved within 30 min in contrast to 97% when only water was used. Due to the fact that this dose is not uncommon, it is recommended not to use tetracyclines together with cinnamon [72]. Finally, the ingestion of cinnamon oil may cause central nervous system depression, predisposing the patient to aspiration pneumonia [21].

As far as topical cinnamon applications are concerned, it should be remembered that cinnamon is used as a constituent of personal hygiene products, as toilet soaps, mouthwash, toothpaste, as ingredients of beverages and baking products and as flavoring of gums. Therefore, contact dermatitis, perioral dermatitis, stomatitis, gingivitis, glossitis, sub-mucosal inflammation and alteration of the surface epidermis should occur after the contact with products containing cinnamon. Intraoral reaction clinical manifestations consist of pain, erythema, ulcerations, fissures, vesicles, desquamation and hyperpigmentation, and white patches. The main responsible for these manifestations are considered Cinnamaldehyde and cinnamic acid are considered to be mainly responsible for these manifestations because they act as membrane irritants. The severity of the local mucosal reaction depends on the duration of contact and systemic symptoms such as nausea and abdominal pain are rare [33].

CONCLUSIONS AND RECOMMENDATIONS

This paper has reviewed the available references regarding the antibacterial effects of cinnamon and its active constituents, published over the last five years. It has shown that the antibacterial activity of cinnamon is due to bioactive phytochemicals such as cinnamaldehyde and eugenol. Cinnamon use in food products and cosmetics could be a good strategy to reduce or avoid bacterial degradation and thus to reduce the incidence of infection caused by food and cosmetics. In fact, cinnamon is not harmful when used in correct conditions. Regardless, long standing excessive use is not recommended.

Moreover, cinnamon could be used to treat infectious disease. However, there is a lack of clinical trials on the antibacterial effects of cinnamon and its chemical constituents, and therefore its clinical efficacy is not clear. In addition, it is well known that cinnamon essential oil toxicity is one of the most important problems, as some essential oils show the above reported adverse effects on human cells, such as cytotoxicity and cell death. Therefore, the application of natural products in the treatment of infectious diseases may be considered an interesting alternative to common antibiotics, possessing different side effects. In addition, cinnamon can be suggested as an alternative to synthetic antibiotics, especially for the treatment of antibiotic-resistant bacterial infections.

Furthermore, we provide a brief summary of the history, traditional uses, phytochemistry and clinical impacts of cinnamon to provide a better view of this spice and herbal medicine. Finally, we recommend that further studies should be performed on the toxicity of cinnamon prior to its clinical use; studies on the mechanism of the antibacterial effects of its extracts and essential oils; on the separation, purification and identification of the most effective antibacterial constituents of cinnamon and their food- and drug-interactions; and clinical studies to examine the antibacterial effects of extracts and essential oils of cinnamon and its bioactive constituents in different infectious diseases.

ACKNOWLEDGEMENTS

Alessandra Baldi for support in text editing.

AUTHOR CONTRIBUTIONS

Seyed Fazel Nabavi, Eduardo Sobarzo-Sánchez and Maria Daglia designed the paper, Arianna Di Lorenzo collected the literature data, Morteza Izadi analysed the data, and Maria Daglia, Eduardo Sobarzo-Sánchez and Seyed Mohammad Nabavi wrote the paper. All authors participated in the analysis and interpretation of literature data, revised the paper and approved the final manuscript.

REFERENCES

1.   Alinezhad, H.; Azimi, R.; Zare, M.; Ebrahimzadeh, M.A.; Eslami, S.; Nabavi, S.F.; Nabavi, S.M. Antioxidant and Antihemolytic Activities of Ethanolic Extract of Flowers, Leaves, and Stems of Hyssopus officinalis L. Var.*angustifolius*. *Int. J. Food Prop.* 2013, *16*, 1169–1178.

2.   Nabavi, S.F.; Daglia, M.; Moghaddam, A.H.; Habtemariam, S.; Nabavi, S.M. Curcumin and Liver Disease: From Chemistry to Medicine. *Compr. Rev. Food Sci. Saf.* 2014, *13*, 62–77.

3.   Curti, V.; Capelli, E.; Boschi, F.; Nabavi, S.F.; Bongiorno, A.I.; Habtemariam, S.; Nabavi, S.M.; Daglia, M. Modulation of human miR-17–3p expression by methyl 3-*O*-methyl gallate as explanation of its *in vivo* protective activities. *Mol. Nutr. Food Res.* 2014, *58*, 1776–1784.

4.   Nabavi, S.F.; Nabavi, S.M.; Habtemariam, S.; Moghaddam, A.H.; Sureda, A.; Jafari, M.; Latifi, A.M. Hepatoprotective effect of gallic acid isolated from *Peltiphyllum peltatum* against sodium fluoride-induced oxidative stress. *Ind. Crop. Prod.* 2013, *44*, 50–55.

5.   Herman, A.; Herman, A.P.; Domagalska, B.W.; Młynarczyk, A. Essential oils and herbal extracts as antimicrobial agents in cosmetic emulsion. *Indian J. Microbiol.* 2013, *53*, 232–237.

6.   Nabavi, S.M.; Marchese, A.; Izadi, M.; Curti, V.; Daglia, M.; Nabavi, S.F. Plants belonging to the genus *Thymus* as antibacterial agents: From farm to pharmacy. *Food Chem.* 2015, *173*, 339–347.

7.   Simoes, M.; Bennett, R.N.; Rosa, E.A. Understanding antimicrobial activities of phytochemicals against multidrug resistant bacteria and biofilms. *Nat. Prod. Rep.* 2009, *26*, 746–757.

8.   Marchese, A.; Coppo, E.; Sobolev, A.P.; Rossi, D.; Mannina, L.; Daglia, M. Influence of *in vitro* simulated gastroduodenal digestion on the antibacterial activity, metabolic profiling and polyphenols content of green tea (*Camellia sinensis*). *Food Res. Int.* 2014, *63*, 182–191.

9.   Ranasinghe, P.; Jayawardana, R.; Galappaththy, P.; Constantine, G.; de Vas Gunawardana, N.; Katulanda, P. Efficacy and safety of 'true'cinnamon (*Cinnamomum zeylanicum*) as a pharmaceutical agent in diabetes: A systematic review and meta-analysis. *Diabetic Med.* 2012, *29*, 1480–1492.

10.   Ravindran, P.; Shylaja, M.; Nirmal Babu, K.; Krishnamoorthy, B. Botany and crop improvement of cinnamon and cassia. In *Cinnamon and Cassia—The Genus Cinnamomum*; Ravindran, P.N., Babu, K.N., Eds.; CRC Press: Boca Raton, FL, USA, 2004.

11.   Jakhetia, V.; Patel, R.; Khatri, P.; Pahuja, N.; Garg, S.; Pandey, A.; Sharma, S. Cinnamon: A pharmacological review.*JASR* 2010, *1*, 19–23.

12.   Krishnamoorthy, B.; Rema, J. End uses of cinnamon and cassia. In *Cinnamon and Cassia: The Genus Cinnamomum*; Ravindran, P.N., Babu, K.N., Eds.; CRC Press: Boca Raton, FL, USA, 2004.

13.   Khasnavis, S.; Pahan, K. Sodium benzoate, a metabolite of cinnamon and a food additive, upregulates neuroprotective parkinson disease protein DJ-1 in astrocytes and neurons. *J. Neuroimmune Pharmacol.* 2012, *7*, 424–435.

14.   Kim, S.H.; Hyun, S.H.; Choung, S.Y. Anti-diabetic effect of cinnamon extract on blood glucose in db/db mice. *J. Ethnopharmacol.* 2006, *104*,

119–123.

15. Brierley, S.M.; Kelber, O. Use of natural products in gastrointestinal therapies. *Curr. Opin. Pharmacol.* 2011, *11*, 604–611.

16. Al-Jiffri, O.; El-Sayed, Z.; Al-Sharif, F. Urinary tract infection with *Esherichia coli* and antibacterial activity of some plants extracts. *Int. J. Microbiol. Res.* 2011, *2*, 1–7.

17. Leung, E.; Weil, D.E.; Raviglione, M.; Nakatani, H. The WHO policy package to combat antimicrobial resistance. *Bull. World Health Organ.* 2011, *89*, 390–392.

18. Högberg, L.D.; Heddini, A.; Cars, O. The global need for effective antibiotics: Challenges and recent advances. *Trends Pharmacol. Sci.* 2010, *31*, 509–515.

19. Wijesekera, R. Historical overview of the cinnamon industry. *Crit. Rev. Food Sci. Nutr.* 1997, *10*, 1–30.

20. Sulaiman, S.A.B. Extraction of Essential Oil from Cinnamomum Zeylanicum by Various Methods as a Perfume Oil. Bachelor Thesis, University of Malaysia Pahang, Gambang, Pahang, Malaysia, 2013.

21. Barceloux, D.G. Cinnamon (Cinnamomum Species). *Dis.-a-Month* 2009, *55*, 327–335.

22. Madan, M.; Kannan, S. Economics and Marketing of Cinnamon and Cassia–A Global View. In *Cinnamon and Cassia: The Genus Cinnamomum*; Ravindran, P.N., Babu, K.N., Eds.; CRC Press: Boca Raton, FL, USA, 2004.

23. Ranatunga, J.; Senanayake, U.; Wijesekera, R. Cultivation and management of cinnamon. In *Cinnamon and Cassia: The Genus Cinnamomum*; Ravindran, P.N., Babu, K.N., Eds.; CRC Press: Boca Raton, FL, USA, 2004.

24. Thankamani, C.; Sivaraman, K.; Kandiannan, K.; Peter, K. Agronomy of tree spices (clove, nutmeg, cinnamon and allspice)—A review. *J. Spices Aromat. Crops* 1994, *3*, 105–123.

25. Muchuweti, M.; Kativu, E.; Mupure, C.H.; Chidewe, C.; Ndhlala, A.R.; Benhura, M.A.N. Phenolic composition and antioxidant properties of some spices. *Am. J. Food Technol.* 2007, *2*, 414–420.

26. Wong, Y.C.; Ahmad-Mudzaqqirand, M.Y.; Wan-Nurdiyana, W.A. Extraction of Essential Oil from Cinnamon (*Cinnamomum zeylanicum*). *Orient. J. Chem.* 2014, *30*, 37–47.

27. Jayaprakasha, G.K.; Singh, R.P.; Pereira, J.; Sakariah, K.K. Limonoids

from *Citrus reticulata* and their moult inhibiting activity in mosquito *Culex quinquefasciatus* larvae. *Phytochemistry* 1997, *44*, 843–846.

28. Jayaprakasha, G.K.; Jagan Mohan Rao, L.; Sakariah, K.K. Chemical composition of the flower oil of *Cinnamomum zeylanicum* blume. *J. Agric. Food Chem.* 2000, *48*, 4294–4295.

29. Filoche, S.K.; Soma, K.; Sissons, C.H. Antimicrobial effects of essential oils in combination with chlorhexidine digluconate. *Oral Microbiol. Immunol.* 2005, *20*, 221–225.

30. Chaudhary, S.S.; Tariq, M.; Zaman, R.; Imtiyaz, S. The *In vitro* anti-acne activity of two unani drugs. *Anc. Sci. Life* 2013,*33*, 35–38.

31. Vijayan, K.; Thampuran, R.A. Pharmacology and Toxicology of Cinnamon and Cassia. In *Cinnamon and Cassia: The Genus Cinnamomum*; Ravindran, P.N., Babu, K.N., Eds.; CRC Press: Boca Raton, FL, USA, 2004.

32. European Medicines Agency. Assessment Report on *Cinnamomum verum* J.S. Presl, cortex and corticis aetheroleum. EMA/HMPC/246773/2009. Available online: http://www.ema.europa.eu/docs/en_GB/document_library/Herbal_-_HMPC_assessment_report/2011/08/WC500110090.pdf (accessed on 10 May 2011).

33. Hossein, N.; Zahra, Z.; Abolfazl, M.; Mahdi, S.; Ali, K. Effect of *Cinnamon zeylanicum* essence and distillate on the clotting time. *J. Med. Plants Res.* 2013, *7*, 1339–1343.

34. Rao, P.V.; Gan, S.H. Cinnamon: A Multifaceted Medicinal Plant. *Evid. Based Complement. Alternat. Med.* 2014, *2014*, 642942.

35. Reichling, J.; Schnitzler, P.; Suschke, U.; Saller, R. Essential oils of aromatic plants with antibacterial, antifungal, antiviral, and cytotoxic properties–an overview. *Forsch. Komplementärmedizin/Res. Complement. Med.* 2009, *16*, 79–90.

36. Chang, S.-T.; Cheng, S.-S. Antitermitic activity of leaf essential oils and components from *Cinnamomum osmophleum*. *J. Agric. Food Chem.* 2002, *50*, 1389–1392.

37. Mancini-Filho, J.; Van-Koiij, A.; Mancini, D.; Cozzolino, F.; Torres, R. Antioxidant activity of cinnamon (Cinnamomum Zeylanicum, Breyne) extracts. *Boll. Chim. Farm.* 1998, *137*, 443–447.

38. Tung, Y.-T.; Chua, M.-T.; Wang, S.-Y.; Chang, S.-T. Anti-inflammation activities of essential oil and its constituents from indigenous cinnamon (*Cinnamomum osmophloeum*) twigs. *Bioresour. Technol.* 2008, *99*, 3908–3913.

39. Cheng, S.-S.; Liu, J.-Y.; Tsai, K.-H.; Chen, W.-J.; Chang, S.-T. Chemical composition and mosquito larvicidal activity of essential oils from leaves of different *Cinnamomum osmophloeum* provenances. *J. Agric. Food Chem.* 2004, *52*, 4395–4400.

40. Kong, J.-O.; Lee, S.-M.; Moon, Y.-S.; Lee, S.-G.; Ahn, Y.-J. Nematicidal activity of cassia and cinnamon oil compounds and related compounds toward *Bursaphelenchus xylophilus* (Nematoda: Parasitaphelenchidae). *J. Nematol.* 2007, *39*, 31.

41. Moselhy, S.S.; Ali, H.K. Hepatoprotective effect of cinnamon extracts against carbon tetrachloride induced oxidative stress and liver injury in rats. *Biol. Res.* 2009, *42*, 93–98.

42. Alqasoumi, S.; Al-Dosary, M.; Al-Yahya, M.; Al-Mofleh, I. Gastroprotective effect of a popular spice cinnamon *"Cinnamomum zeylanicum"* in rats. *Eur. J. Pharmacol.* 2011, *668*, e42.

43. Hur, M.H.; Lee, M.S.; Seong, K.Y.; Lee, M.K. Aromatherapy massage on the abdomen for alleviating menstrual pain in high school girls: A preliminary controlled clinical study. *Evid.-Based Complement. Altern. Med.* 2012, *2012*, 187163.

44. U.S. National Institutes of Health, ClinicalTrials.gov. Available online: http://clinicaltrial.gov/ (accessed on 1 November 2014).

45. U.S. National Library of Medicine, PubMed database. Available online: http://www.ncbi.nlm.nih.gov/pubmed(accessed on 1 November 2014).

46. Keskin, D.; Toroglu, S. Studies on antimicrobial activities of solvent extracts of different spices. *J. Environ. Biol.* 2011,*32*, 251–256.

47. Mandal, S.; DebMandal, M.; Saha, K.; Pal, N.K. *In vitro* Antibacterial Activity of three Indian Spices against Methicillin-Resistant *Staphylococcus aureus*. *Oman Med. J.* 2011, *26*, 319–323.

48. Rasheed, M.U.; Thajuddin, N. Effect of medicinal plants on Moraxella cattarhalis. *Asian Pac J. Trop. Med.* 2011, *4*, 133–136.

49. Guerra, F.Q.; Mendes, J.M.; Sousa, J.P.; Morais-Braga, M.F.; Santos, B.H.; Melo Coutinho, H.D.; Lima Ede, O. Increasing antibiotic activity against a multidrug-resistant Acinetobacter spp by essential oils of *Citrus limon* and*Cinnamomum zeylanicum*. *Nat. Prod. Res.* 2012, *26*, 2235–2258.

50. Yap, P.S.; Lim, S.H.; Hu, C.P.; Yiap, B.C. Combination of essential oils and antibiotics reduce antibiotic resistance in plasmid-conferred multidrug resistant bacteria. *Phytomedicine* 2013, *20*, 710–713.

51. Naveed, R.; Hussain, I.; Tawab, A.; Tariq, M.; Rahman, M.; Hameed, S.; Mahmood, M.S.; Siddique, A.B.; Iqbal, M. Antimicrobial activity of the bioactive components of essential oils from Pakistani spices against *Salmonella* and other multi-drug resistant bacteria. *BMC Complement. Altern. Med.* 2013, *13*, 265–275.

52. Al-Mariri, A.; Safi, M. *In vitro* Antibacterial Activity of Several Plant Extracts and Oils against Some Gram-Negative Bacteria. *Iran J. Med. Sci.* 2014, *39*, 36–43.

53. Bardají, D.K.; Reis, E.B.; Medeiros, T.C.; Lucarini, R.; Crotti, A.E.; Martins, C.H. Antibacterial activity of commercially available plant-derived essential oils against oral pathogenic bacteria. *Nat. Prod. Res.* 2015. in press.

54. Saleem, M.; Bhatti, H.N.; Jilani, M.I.; Hanif, M.A. Bioanalytical evaluation of *Cinnamomum zeylanicum* essential oil. *Nat. Prod. Res.* 2015. in press.

55. Kim, Y.G.; Lee, J.H.; Kim, S.I.; Baek, K.H.; Lee, J. Cinnamon bark oil and its components inhibit biofilm formation and toxin production. *Int. J. Food Microbiol.* 2015, *195*, 30–39.

56. Seukep, J.A.; Fankam, A.G.; Djeussi, D.E.; Voukeng, I.K.; Tankeo, S.B.; Noumdem, J.A.K.; Kuete, A.H.L.N.; Kuete, V. Antibacterial activities of the methanol extracts of seven Cameroonian dietary plants against bacteria expressing MDR phenotypes. *SpringerPlus* 2013, *2*, 363.

57. Chaudhari, L.K.; Jawale, B.A.; Sharma, S.; Sharma, H.; Kumar, C.D.; Kulkarni, P.A. Antimicrobial activity of commercially available essential oils against *Streptococcus mutans*. *J. Contemp. Dent. Pract.* 2012, *13*, 71–74.

58. Gupta, A.; Duhan, J.; Tewari, S.; Sangwan, P.; Yadav, A.; Singh, G.; Juneja, R.; Saini, H. Comparative evaluation of antimicrobial efficacy of *Syzygium aromaticum*, *Ocimum sanctum* and *Cinnamomum zeylanicum* plant extracts against*Enterococcus faecalis*: A preliminary study. *Int. Endod. J.* 2013, *46*, 775–783.

59. Miller, A.B.; Cates, R.G.; Lawrence, M.; Soria, J.A.; Espinoza, L.V.; Martinez, J.V.; Arbizú, D.A. The antibacterial and antifungal activity of essential oils. *Pharm. Biol.* 2015, *53*, 548–554.

60. Sleha, R.; Mosio, P.; Vydrzalova, M.; Jantovska, A.; Bostikova, V.; Mazurova, J. *In vitro* antimicrobial activities of cinnamon bark oil, anethole, carvacrol, eugenol and guaiazulene against *Mycoplasma hominis* clinical isolates. *Biomed. Pap. Med. Fac. Univ. Palacky Olomouc Czech Repub.* 2014, *158*, 208–211.

61. Lindeman, Z.; Waggoner, M.; Batdorff, A.; Humphreys, T.L. Assessing the antibiotic potential of essential oils against *Haemophilus ducreyi*. *BMC Complement. Altern. Med.* 2014, *14*, 172.

62. Dussault, D.; Vu, K.D.; Lacroix, M. *In vitro* evaluation of antimicrobial activities of various commercial essential oils, oleoresin and pure compounds against food pathogens and application in ham. *Meat Sci.* 2014, *96*, 514–520.

63. Tayel, A.A.; El-Tras, W.F.; Moussa, S.H.; El-Sabbagh, S.M. Surface decontamination and quality enhancement in meat steaks using plant extracts as natural biopreservatives. *Foodborne Pathog. Dis.* 2012, *9*, 755–761.

64. Chen, C.H.; Ravishankar, S.; Marchello, J.; Friedman, M. Antimicrobial activity of plant compounds against *Salmonella Typhimurium* DT104 in ground pork and the influence of heat and storage on the antimicrobial activity. *J. Food Prot.* 2013, *6*, 1264–1269.

65. Irkin, R.; Abay, S.; Aydin, F. Inhibitory effects of some plant essential oils against *Arcobacter butzleri* and potential for rosemary oil as a natural food preservative. *J. Med. Food* 2011, *14*, 291–296.

66. Shan, B.; Cai, Y.Z.; Brooks, J.D.; Corke, H. Potential application of spice and herb extracts as natural preservatives in cheese. *J. Med. Food* 2011, *14*, 284–290.

67. Frankova, A.; Marounek, M.; Mozrova, V.; Weber, J.; Kloucek, P.; Lukesova, D. Antibacterial Activities of Plant-Derived Compounds and Essential Oils toward *Cronobacter sakazakii* and *Cronobacter malonaticus*. *Foodborne Pathog. Dis.* 2014, *11*, 795–797.

68. Mith, H.; Dure´, R.; Delcenserie, V.; Zhiri, A.; Daube, G.; Clinquart, A. Antimicrobial activities of commercial essential oils and their components against food-borne pathogens and food spoilage bacteria. *Food Sci. Nutr.* 2014, *2*, 403–416.

69. Hill, L.E.; Taylor, T.M.; Gomes, C. Antimicrobial efficacy of poly (dl-lactide-co-glycolide) (PLGA) nanoparticles with entrapped cinnamon bark extract against Listeria monocytogenes and *Salmonella typhimurium*. *J. Food Sci.* 2013, *78*, 626–632.

70. National Institutes of Health; U.S. Department of Health and Human Services. Herbs at a glance. Cinnamon. Available online: http:// cinnamonvogue.com/DOWNLOADS/Cinnamon%20Side%20Effects. pdf (accessed on 7 September 2015).

71. Price, S.; Price, L. *Aromatherapy for Health Professionals*, 3rd ed.; Churchill Livingstone Elsevier: London, UK, 2007.

72. Keller, K.; Hänsel, R.; Chandler, R.F. *Adverse Effects of Herbal Drugs: Cinnamomum Species*; De Smet, P.A.G.M., Ed.; Springer Verlag: Heidelberg, Germany, 1992; Volume 1.

Chapter 6

BIOTECHNOLOGICAL PRODUCTION OF OLIGOSACCHARIDES — APPLICATIONS IN THE FOOD INDUSTRY

Tathiana Souza Martins Meyer[1], Ângelo Samir Melim Miguel[1], Daniel Ernesto Rodríguez Fernández[1] and Gisela Maria Dellamora Ortiz[1]

[1]School of Pharmacy, Federal University of Rio de Janeiro, Rio de Janeiro, Brazil

ABSTRACT

Oligosaccharides are carbohydrates, composed of up to twenty monosaccharides linked by glycosydic bonds, widely used in food and pharmaceutical industries. These compounds can be obtained by extraction from natural sources (milk, vegetables, fruits), and by chemical or biotechnological processes. In the last case, chemical structures and composition of the generated oligosaccharides depend on the type and source of enzymes, and on process conditions, including the initial concentration of substrate. Among the various functions of nondigestible oligosaccharides, one that has attracted attention is its prebiotic potential. The intestinal benefits of prebiotics, such as fructooligosaccharides and inulin as well as their symbiotic association with probiotic bacteria, encompass prevention and treatment of infectious diseases, including viral or bacterial diarrhea, and chronic inflammatory diseases such as ulcerative colitis. Other benefits attributed to prebiotics and probiotics include treatment of inflammatory intestinal and irritable bowel syndrome, prevention of cancer, and modulation of the immune system, mineral absorption and lipid metabolism. Fructooligosaccharides (FOS), galactooligosaccharides (GOS) and chitooligosaccharides (COS) have been widely studied for their prebiotic properties. Moreover, novel oligosaccharides with potential prebiotic activity are currently under investigation. This review will focus mainly on the biotechnological production, health benefits and applications of non-natural oligosaccharides in the food industry.

INTRODUCTION

Consumers all around the world are increasingly aware and concerned about safety and the quality of food. Besides the push towards replacement of chemical additives by those obtained from natural sources, this awareness has led to a rising demand for enrichment of foods with bioactive compounds that have beneficial effects on human health [1]. Therefore, nowadays, a variety of gluten free and products enriched with dietary fiber, or containing probiotics and/or prebiotic and functional oligosaccharides are available in the market [2].

Oligosaccharides are carbohydrates, composed of up to twenty monosaccharides linked by glycosydic bonds, widely used in food and pharmaceutical industries. These compounds are obtained from natural sources and through chemical or biotechnological processes [3,4].

Among the various functions of non-digestible oligosaccharides, one that has attracted attention is its prebiotic potential. A prebiotic can be defined as "selectively fermented ingredients that allow specific changes, both in the composition and/or activity in the gastrointestinal microbiota that confers benefits upon host well-being and health" [5]. An oligosaccharide to be regarded as prebiotic must not be hydrolyzed or absorbed in the upper part of the gastrointestinal tract; and must be assimilated selectively by one or by a limited number of beneficial microorganisms in the colon, promoting benefic luminal or systemic effects. To improve colonic function, live microorganisms can be administered in adequate amounts, being known as probiotics; and to be used in food, these organisms must be able to survive passage through the gut; to proliferate and to colonize the digestive tract; and must be safe and effective [6,7].

The intestinal benefits of prebiotics, such as fructooligosaccharides and inulin as well as their symbiotic association with probiotic bacteria, encompass prevention and treatment of infectious diseases, including viral or bacterial diarrhea, and chronic inflammatory diseases such as ulcerative colitis [8]. The mechanisms of action of probiotics against gastrointestinal pathogens consist mainly on competition for nutrients and sites of access, production of antimicrobial metabolites, changes in environmental conditions, and modulation of the immune response of the host. Other benefits attributed to prebiotics and probiotics include treatment of inflammatory intestinal and irritable bowel syndrome, prevention of cancer, and modulation of the immune system, mineral absorption and lipid metabolism [8,9].

Oligosaccharides can be obtained by extraction from natural sources (milk, vegetables, fruits), and by chemical or biotechnological processes [10,11]. Mixtures of oligosaccharides with different degrees of polymerization and glycosidic linkages are usually formed in the enzymatic processes. Chemical structures and composition of these mixtures depend on the type and source of enzymes, and on process conditions, including the initial concentration of substrate [11,12]. Depending on the initial substrate, production of oligosaccharides can involve different steps: hydrolysis of glycosidic bonds giving rise to monomers, followed by generation of disaccharides and other oligomers through the action of transferases [13,14].

FRUCTOOLIGOSACCHARIDES

Fructans are carbohydrates in which one or more fructosylfructose links constitute the majority of glycosidic bonds [15]. These carbohydrates can be of the inulin-type with β-(2,1)-D-fructofuranosyl units, found in plants and synthesized by fungi. Additionally, there are the levan-type fructans with β-(6,2)-D-fructofuranosyl units, found in plants and synthesized by bacteria [16].

Levan is a polymer with very high molecular weight that can reach 10^7 Da [17]. In contrast to levan, inulin from chicory consists of a mixture of oligomers and polymers with a degree of polymerization (DP) that varies from two to approximately sixty units (Figure 1; Table 1) [18]. Around 10% of the fructan chains in native chicory inulin have a DP in the range between two and five [5].

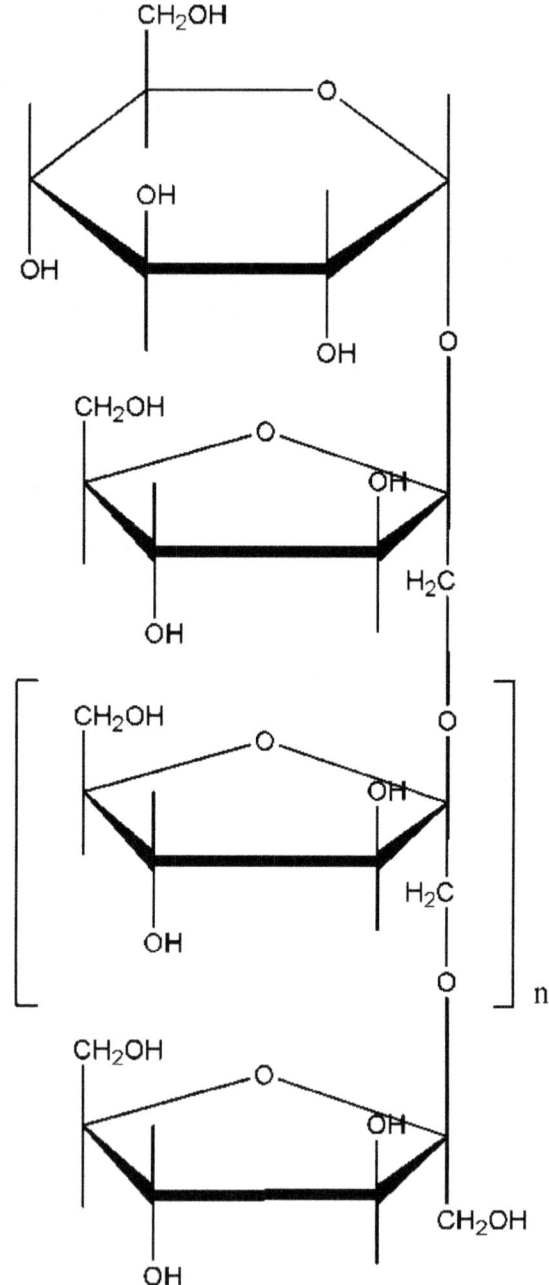

Figure 1: Structure of inulin, a linear fructosyl polymer linked by β-(2,1) bonds (n=3-65), attached to a terminal glucosyl residue by an α-(1,2) bond.

Table 1: Structure and biological activity of prebiotics

Prebiotics	Chemical structure	Properties	Applicability / Health benefits	Reference
FOS	Fructosyl units linked by β-(2,1) bonds, attached to a terminal glucosyl residue by α-(1,2) bond (Variants: Inulin type β-1,2 and Levan type β-2,6 linkages between fructosyl units in the main chain) DP=3-10.	Soluble fibers; Gel formation; Sugar replacement; Moderate sweetness; Stable (depending on matrix).	Prevention of intestinal infections and extra intestinal infections (e.g. respiratory tract; Inhibition of pathogens, ordering intestinal flora; Regulation of intestinal immune system; Enhancement of immune response; Stimulation of probiotic growth of Lactobacilli and Bifidobacteria species; Optimization of colonic function and metabolism; Production of short chain fatty acids; Increase of mineral absorption; Reduction of food intake and obesity management and control of diabetes type 2; Prevention of cancer.	[8,24,25, 61,67,69, 72,170, 176-186]
Inulin	Mixture of linear fructosyl polymers and oligomers (DP = 3-65) linked by β-(2,1) bonds, attached to a terminal glucosyl residue by α(1−2) bond.	Soluble fiber; Water adsorption; Gel formation; Modifier of viscosity, texture, colour and sensory aspects of food formulations; Replacement for fat and sugar; Low calorimetric value; Moderate sweetness.	Stimulation of probiotic growth; Lowering effect on cholesterol LDL and triglycerides levels; Influence on inflammatory markers and development of gut associated lymphoid tissue (GALT); Regulation of intestinal immune system; Enhancement of immune response; Increase of mineral absorption (Calcium, Iron and Magnesium); Prevention of cancer.	[8,19,67, 176,187-195]
GOS	Mixture of galactopyranosyl oligomers (DP= 3-8) linked mostly by β-(1,4) or β-(1,6) bonds, although low proportions of β (1,2) or β-(1,3) linkages may also be present. Terminal glucosyl residues	Stable in acidic conditions and in higher temperatures; Soluble; Cryoprotector activity; Low ability to crystalyze; Incorporated in various functional foods.	Stimuli of probiotic growth; Reorder intestinal flora; Regulation of intestinal immune system; Reinforcement of intestinal barrier; Inhibition of adhesion of pathogens; Mimic molecular receptors, inhibit microbial adherence;	[8,105, 196-206]

	are linked by β-(1,4) bonds to galactosyl units.		Prevent infections (e.g. *Clostridium difficile* diarrhea); Prevention of cancer; Enhance mineral absorption; Reduce food intake, helping obesity management. Use in diabetic foods, free from carbohydrates that increase the level of postprandial glucose; Use in specialized foods for individuals intolerant to lactose.	
Lactulose	Galactosyl β-(1,4) fructose.	Sweetener, sugar replacement;	Induces growth of *Bifidobacterium* both *in vitro* and *in vivo*; use as laxative in the treatment of constipation; Optimization of colonic function and metabolism, reducing colon pH and ammonia concentration; Increased mineral absorption; Treatment of portal systemic encephalophathy and chronic constipation; Uses in diabetic and dairy foods.	[205, 207-211]
COS	Chitin: β-(1,4) linked N-acetyl-D-glucosamine residues; Chitosan: β-(1,4) linked D-glucosamine polymer. DP=2-8	Antimicrobial activity of chitosan depends on degree of polymerization, amino groups content and degree of acetylation; Chelation of metal trace elements and essential nutrients; Flocculation and adsorption capacity mainly because of the cationic macromolecular	Antimicrobial and antioxidant activity; Use as food preservative; Use as dietary supplements in functional foods; Prebiotic activity; Hypocholesterolemic;	[149, 212, 213]
XOS	Xylose oligomers connected by β-(1,4) linkages (DP=3-6).	Stable in a large range of pH values (2,5-8,0); Thermal stability (up to 100°C); Antioxidant effects; Antifreezing activity; Low cariogenicity; Low calorimetric value; Low glycemic index.	Inhibition of pathogens growth, reordering intestinal flora; Stimulation of probiotic growth; Reinforcement of intestinal barrier; Optimization of colonic function and metabolism; Obesity management, reduction of food intake and weight.	[159, 162, 179, 214-216]
IMO	Glucosyl residues linked to maltose or isomaltose by α-(1,6) glycosidic bonds.	Low sweetness; Low viscosity; Bulking properties; Humectant; Prevention of sucrose crystallization.	Optimization of colonic function and metabolism, reduces nitrogenated products; Increase caecum weight; Antidiabetic effects; Improve lipid metabolism and obesity management.	[2, 217-219]
SOS	Oligomers composed by galactosyl units linked to sucrose by α-(1,6) bonds. Most abundant are raffinose and stachyose.	Stabilizer properties; Cryoprotectant effect.	Prevention of pathogen proliferation.	[2, 156, 213, 220, 221]

FOS: Fructooligosaccharides; GOS: Galactooligosaccharides; COS: Chitooligosaccharides;

XOS: Xylooligosaccharides; IMO: Isomaltooligosaccharides; SOS: Soybean oligosaccharides

- FOS: Fructooligosaccharides; GOS: Galactooligosaccharides; COS: Chitooligosaccharides;
- XOS: Xylooligosaccharides; IMO: Isomaltooligosaccharides; SOS: Soybean oligosaccharides

Inulin-type fructooligosaccharides are made up of two or more fructosyl moieties linked by β-(2,1) bonds and united at the non-reducing end to a terminal glucose residue by an α-(1,2) glycosidic bond (Table 1) [19]. The term fructooligosaccharides (FOS) is mainly used for fructose oligomers that contain one glucose unit and from two to four fructose units bound together by β-(2,1) glycosidic linkages [20,21]. Nevertheless, oligofructose and FOS may be regarded as synonyms for the mixture of small inulin oligomers with DP<10 [6,22]; while short chain FOS (sc-FOS) are fructose oligomers mainly composed of 1-kestose (GF_2), nystose (GF_3), and ^{1}F-fructofuranosylnystose (GF_4) (Figure 2) [23-25].

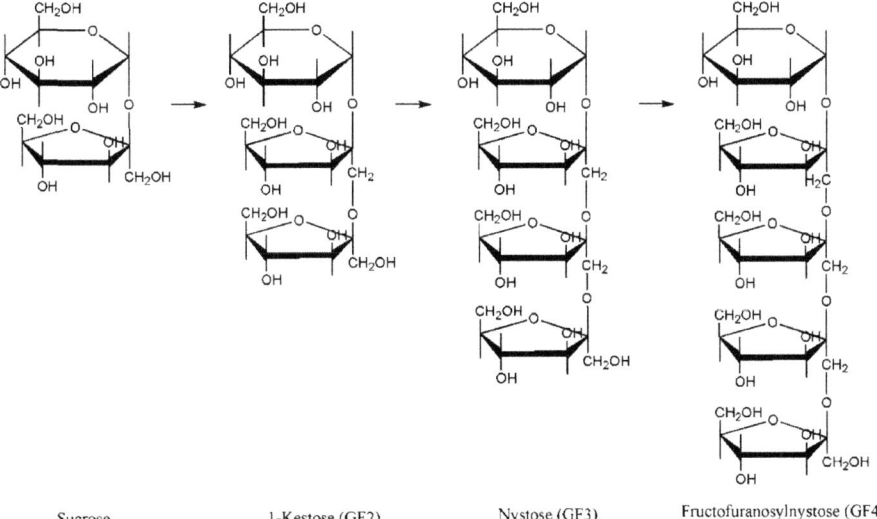

Sucrose	1-Kestose (GF2)	Nystose (GF3)	Fructofuranosylnystose (GF4)

Figure 2: Structures of typical fructooligosaccharides (FOS), derived from sucrose. FOS consist of a glucosyl residue α-(1,2) linked to two or more β-(1,2) fructosyl units. Synthesis of these FOS is catalyzed by fructosyltransferases, requiring a second sucrose molecule as a fructosyl residue donor.

Fructans have storage and protective functions in many commonly consumed plants, being a typical part of the diet. Some food sources are richer in high molecular weight fructans, such as inulin, while others have higher levels of sc-FOS [26].

FOS are found in low levels in natural sources such as asparagus, sugar beet, garlic, chicory, onion, Jerusalem artichoke, wheat, honey, banana, barley, tomato, and rye [27-29]. Apart from usually occurring in low concentrations, seasonal conditions also limit their large-scale production from these sources [30].

For this reason, enzymatic processes are used for the industrial production of FOS. One route involves the controlled hydrolysis of long chain fructans (Table 2) [31,32], which results in a large amount of FOS mostly without glucose in their structures. The other route is the synthesis from sucrose, which leads to sc-FOS that contain a molecule of glucose in their structures [11,33]. The present review will focus on the synthesis of FOS from sucrose.

FOS are produced from sucrose by the action of microbial enzymes with high transfuctosylating activity: β-D-fructosyltransferase (FTase, EC 2.4.1.9) and β-fructofuranosidase (FFase, EC 3.2.1.26) (Table 2) [34]. Since FTase possesses almost only the transfructosylating activity, it is able to cleave the β-1,2 linkage of sucrose, transferring the fructosyl group to an acceptor molecule, with the resulting formation of fructooligosaccharides and release of glucose [35]. This enzyme shows little affinity towards water as an acceptor, therefore the hydrolase activity of FTase is very low [36].

FFase can catalyze both hydrolytic and transfructosylating reactions, nevertheless, transfructosylation only takes place when sucrose concentrations are higher than 500 g L^{-1} [27,34,36-38]. The production of FOS by the action of FFase on sucrose can occur either by reverse hydrolysis or by transfructosylation [36].

TABLE 2: Obtention and industrial production of prebiotics

Type of prebiotics	Obtention source	Enzyme processing	Microbial producer	Industrial product and manufacturer	References
FOS	Enzymatic reactions: fructosyltransferases using sucrose as a substrate or from inulin using microbial endoinulinases.	Fructosyltransferases or β-fructofuranosidases; Levansucrases; Endoinulinases.	*B. macerans* *Z. mobilis* *L. reutri* *A. niger* *A. japonicus* *A. foetidus* *A. sydowi* *A. pullans* *C. purpurea* *F. oxysporum* *P. citrinum* *P. frequentans* *P. spinulosum* *P. rigulosum* *P. parasitica* *S. brevicaulis* *S. cerevisiae* *K. marxianus*	Neosugar Actilight NutraFlora P-95 - GTC Nutrition Raftilose P95 - Orafti Group	[21,29, 170, 176, 222]
Inulin	Natural product, extraction from plants	Not applicable	Not apllicable	Inulin-S – SigmaAldrich Fibruline - Trades S.A. Fibrex - Danisco Sugar Frutafit CLR DP8, Fruta- fit HD DP10, Frutafit TEX DP5, Inulin TEX – Sensus Inulin GR, HP, HP-gel, HPX, LS, ST, Raftilin ST, Raftilose P95, Raftiline HP - Orafti Group	[189, 194, 222]
GOS	Enzymatic transgalactosylation reactions, using lactose as substrate; Fermentation process.	β-Galactosidases	*Aspergillus sp.* *Bacillus sp.* *B. circulans* *Kluyveromyces sp.* *B. bifidum*	Vivinal GOS Syrup - Bolculo Domo or Friesland Foods Domo	[105,120, 198, 202-205, 223-227]

			S. singularis	Purimune - GTC	
			S. thermophilus	Nutrition	
			C. laurentii	Oligomate 55NP -	
				Yakult	
				Pharmaceutical Inc.	
				Cup Oligo H-70®	
				Kowa Company	
				BiMuno - Clasado	
				Ltd.	
Lactulose	Thermal-alkaline isomerisation of lactose; Enzymatic transgalactosylation of fructose.	β-Galactosidases β-Glycosidases	A. oryzae S. fragilis K. lactis P. furiosus S. solfataricus	Sigma Aldrich Discovery Fine Chemicals Solvay	[209, 210, 228-230]
XOS	Enzymatic degradation of xylans	Endo-β-1,4-xylanases, exo-β-1,4-xylosidases, α-glucuronosidases, α-L-arabinofuranosidases, acetylxylan esterases, ferulic acid esterases and p-coumaric acid esterases.	T. reesei T. harzianu T. viride T. kmingii T. longibrachiatum P. chyrosporium G. trabeum A. oryzae	Xylooligo™- Suntory Ltd. YOGHURINA - Suntory Ltd. MARUSHIGE GENKISU - Marushige Ueda Co. L-ONE - Enzamin Laboratory Inc. SUKKIRI KAICHO Lotte Co.	[152, 159, 232]
COS	Enzymatic or chemical depolimerization and deacetylation of chitin and chitosan	Chitosanases and other non-specific enzymes (papain, and lysozyme)	S. coelicolor B. pumilus Bacillus sp. S. kurssanocii	Qingdao BZ-Oligo Co, Ltd. BioCHOS. AMSBIO	[213 232-235]
SOS	Directly extracted from soybean	Not applicable	Not applicable	Soya-oligo - The Calpis Food Industry Co.	[152]
IMO	Enzymatic hydrolysis of starch	α-Amylases or pullulanases, β-amylases and α-glucosidases in sequence. Pullulanases	A. niger Bacillus spp. B.subtilis B. stearothermophilus	Isomalto-900 - Showa Sangyo	[12, 152, 159, 236]
			and T. maritima A. carbonarious L. mesenteroides		

FOS are produced at industrial scale from concentrated sucrose solutions using fungal transfructosylating enzymes mainly from strains of *Aspergillus niger, Aspergillus oryzae* and*Aureobasidium pullulans* [27,29,30]. Moreover, production of FTase from bacteria (*Lactobacillus*) and yeasts (*Rhodotorula, Candida, Cryptococcus* sp) has been reported [39,40]. The main enzymes used for industrial production of FOS generally give rise to a mixture of molecules with the inulin-type structure, [1]F-FOS, whereas those from yeasts usually form levan-type FOS ([6]F-FOS) or neoFOS ([6]G-FOS) [41].

The enzymes from *Aureobasidium pullulans* and from *Aspergillus niger* are highly regiospecific in the fructosyl transfer reaction, transferring one fructosyl moiety from sucrose to the 1-OH of the furanoside of another fructose molecule or fructooligosaccharide, with high selectivity [27]. This synthesis is a complex process in which several reactions occur simultaneously, both in parallel and in series, because sc-FOS are also potential substrates of FTase [42].

Catalytic and physicochemical properties of the producing enzymes, as well as production conditions and composition of FOS are different, depending on the microbial strain. For instance, fungal FTases have molecular masses ranging between 180,000 and 600,000, and are homopolymers with two to six monomer units [43].

Fructosyltransferase from *Aureobasidium pullulans* was submitted to preparative scale chromatographic separation on a weak anion-exchanger [42]. The molecular weight of the enzyme determined by size-exclusion chromatography was 570,000. Analysis of the action of FTase on a FOS substrate (Actilight 950P) showed that sucrose was the only donor of fructosyl moiety used in the transfer reaction catalyzed by the enzyme, while the acceptor could be another molecule of fructose or FOS [42].

A transferase isolated and purified from *Aspergillus aculeatus* exhibited pH and temperature optima of 6.0 and 60°C, respectively, remaining stable with no decrease in activity after 5 h under such conditions [44]. The enzyme was monomeric with a molecular mass of 85 kDa. On the other hand, FFase I from *A. pullulans* DSM2404 had a molecular weight of 430,000 [45]. The biocatalyst from *A. aculeatus* showed both transfructosylation and hydrolytic activity, and the transfructosylation ratio increased to 88% at 600 mg mL^{-1} of sucrose [44]. Conditions such as sucrose concentration (400 mg mL^{-1}), temperature (60°C) and pH (5.6) favored synthesis of high levels of GF$_3$ and GF$_4$. The major products were GF$_2$ after 4 h and GF$_4$ after 8 h of reaction. Prolonged incubation for 16 h resulted in the conversion of GF$_4$ into GF$_2$ due to hydrolase activity.

The theoretical yield of FOS from sucrose is 75% if 1-kestose is the only FOS produced [46]. However, production yields of FOS are typically low (55–60%) due to the hydrolytic activity which gives rise to glucose and fructose as reaction byproducts [27] and/or to the fact that glucose acts as an inhibitor of the enzymes, reducing the reaction efficiency [36,47,48]. To improve FOS production yields, glucose oxidase has been used to remove glucose via transformation to gluconic acid [49] and glucose isomerase has been used to interconvert glucose to fructose [46]. Nevertheless, it is necessary to seek for strains among the microbial diversity with high transfructosylating activity, able to produce high yields of oligosaccharides and low yields of monomeric sugars [35].

In addition, the supply of sc-FOS is limited compared to their increasing demand in the food industry, because enzymes such as fructosyltransferases are not widely commercially available [50]. For this reason, the production of FOS is usually carried out in a two-stage process, in which the first stage consists of the microbial production of the enzyme with transfructosylation activity, while the second involves the reaction of the produced enzyme with sucrose (substrate) to generate FOS [29].

A commercial pectinase preparation from *Aspergillus aculeatus*, Pectinase Ultra SP-L, contains FTase [51,52] besides being composed of different pectinolytic and cellulolytic enzymes. The preparation, used in the food industry to reduce the viscosity of fruit juices [42,53], was the only commercially available source of FTase according to [42].

Enzymes from *Aspergillus japonicus*, *Aspergillus aculeatus* (Pectinex Ultra SP-L) and *Aureobasidium pullulans* were used to determine the reaction conditions required to obtain high yields of sc-FOS [34,51,54]. High concentrations of sucrose (600-850 g L^{-1}), pH (4.5–6.5), temperature (50–60°C), reaction time (3–5 h) and high ratios of transferase and hydrolase activities of the enzyme favored transfructosylation over hydrolysis reaction [44,53].

In a recent study, twenty-five commercial enzyme preparations used in the food industry were screened for transfructosylation activity. Three preparations showed high transfructosylation activity from sucrose, high ratios of transferase over hydrolase activity, selectivity for the synthesis of sc-FOS and did not hydrolyze the produced sc-FOS after a 12 h reaction time [55]. Among these enzymes, a cellulolytic enzyme preparation, Rohapect CM, catalyzed the synthesis of sc-FOS with relatively high production yield (63.8%), under cost-effective conditions of temperature (50°C), sucrose concentration (2.103 M) and enzyme concentration (6.6 TU/mL), which could provide a process with potential application at industrial scale [50].

The synthesis of FOS from sucrose is economically advantageous because sucrose is less expensive than inulin; however, the use of enzymes as catalysts for industrial processes is expensive. Furthermore, the recovery of soluble enzymes for reuse is not economically feasible. In contrast, enzyme immobilization usually confers high storage and long-term operational stability, facilitates the recovery and reuse of the biocatalyst, allowing a cost-efficient use of the enzyme in continuous operation, among other advantages [56,57].

In this context, the commercial enzyme preparation from *Aspergillus aculeatus* (Pectinex Ultra SP-L) has been studied for production of FOS in free and immobilized form. Immobilization of the enzyme onto Eupergit C led to retention of enzyme activity for 20 days of batch operation, and both free and immobilized enzyme produced FOS from sucrose with a yield around 57% [58]. Similarly, production of FOS using the enzyme preparation immobilized onto epoxy-activated Sepabeads EC (Sepabeads EC-EP5) reached a yield of 61% after 36 h of reaction [59].

Synthesis of FOS by dried alginate entrapped enzymes (DALGEEs) was recently reported [60]. FTase from *Aspergillus aculeatus*, contained in Pectinex Ultra SP-L, was entrapped in alginate gel beads, which were then submitted to dehydration. The dried alginate biocatalysts were evaluated for the synthesis of FOS from sucrose in a continuous fixed-bed reactor. A 40-fold enhancement of the space-time-yield of the fixed-bed bioreactor was observed when using DALGEEs compared with conventional gel beads. The fixed-bed reactor packed with DALGEEs presented excellent operational stability since the composition of the outlet was nearly constant during at least 700 h, with an average FOS concentration of 275 g/L.

A partially purified β-fructofuranosidase from the commercial enzyme preparation Viscozyme L was covalently immobilized on glutaraldehyde-activated chitosan particles [61]. Thermal stability of the immobilized biocatalyst was around 100-fold higher at 60°C when compared to the free enzyme. The biocatalyst also showed a high operational stability, which allowed its reuse for at least 50 cycles without significant loss of activity. The average yield of FOS production from sucrose was 55%.

An alternative to the enzymatic production of FOS is the use of either free or immobilized whole cells in bioreactors [62]. Production of these oligosaccharides via fermentation processes has the advantage of obviating purification of FOS-producing enzymes from the cell extracts [29,63,64].

An integrated one-stage method for production of FOS via sucrose fermentation by *Aureobasidium pullulans* was developed and optimized with experimental design tools. To maximize production of FOS, temperature and agitation speed were optimized. A production yield of FOS from sucrose of 64% was obtained in 48 h of fermentation under the optimum conditions (32°C and 385 rpm) [62].

Two filamentous fungi, *Cladosporium cladosporioides* and *Penicillium sizovae*, with mycelium-bound transfructosylating activity were recently isolated. *C. cladosporioides* and *P. sizovae* provided maximum FOS yields of 56% and 31%, respectively. *C. cladosporioides* synthesized a mixture of FOS ([1]F-FOS, [6]F-FOS and [6]G-FOS, including a non-conventional disaccharide (blastose)) with different glycosidic linkages, which could afford certain benefits regarding their bioactivity [41].

Two food companies in Japan and Korea use different commercial processes for the continuous production of FOS with immobilized cells of *Aspergillus niger* and *Aureobasidium pullulans*, respectively, both entrapped in calcium alginate gel [27,63]. Calcium alginate has also been employed to immobilize mycelia of *A. japonicus* aiming to establish FOS-producing processes [65,66].

Immobilization of whole cells of *Aspergillus japonicus* ATCC 20236 onto different lignocellulosic materials was also undertaken to produce fructooligosaccharides. Cells immobilized in the different support materials showed FOS production and FFase activity ranging from 128.35 to 138.73 g/L and from 26.83 to 44.81 U/mL, respectively. Corncobs were the best support for immobilization, providing the highest results of microorganism immobilization, FOS and FFase production. In addition, use of immobilized cells led to higher FOS productivity and yield, as well as higher transfructosylation over hydrolysis ratio of FFase than free cells [64].

Several important health benefits are associated with the consumption of FOS as food ingredients. These include modulation of colonic microflora; improvement of the gastrointestinal physiology; activation of the immune system; enhancement of the bioavailability of minerals; reduction of the levels of serum cholesterol, triglycerides and phospholipids; and prevention of colonic carcinogenesis [34,44,67,68].

Among the different FOS, 1-kestose is considered to have better therapeutic properties than those with higher degree of polymerization [69]. The chain length is an important factor influencing the physiological effect of the oligomer in the host [69] and fermentation by bifidobacteria and lactobacilli species [70].

In this context, fermentation of oligosaccharides was evaluated using pure FOS mixtures containing three FOS species (GF_2, GF_3 and GF_4). Only two oligosaccharides (GF_2 and GF_3) were consumed by *Lactobacillus* strains. Moreover, none of the investigated strains metabolized the GF_4 species, suggesting an intracellular metabolism after the FOS transport [70]. This transfer apparently involves an ATP-dependent transport system with specificity for a limited scope of substrates [71].

Moreover, β-fructofuranosidase activity enables bifidobacteria to degrade FOS. Nevertheless, this property is strain-dependent. Some strains consume both fructose and oligofructose, with different preferences and degradation rates [72].

FOS can be used as calorie-free and non-carcinogenic sweeteners. 1-Kestose has enhanced sweetening power compared to other sc-FOS, and 1-kestose-rich sc-FOS syrups can be used as sugar for diabetics [27,73].

Other types of FOS, such as the levan-type and the neo-FOS, have very promising properties; however, they are not yet commercially available [53,74,75].

GALACTOOLIGOSACCHARIDES

Lactose is a disaccharide formed by the condensation of glucose and galactose molecules, and is the most important component of mammalian milk, present in a concentration range from 2.0% to 10%. Lactose can be obtained at industrial scale from whey during cheese production, with dry weight around 80-85%, using crystallization techniques [76-78]. In the past, whey was considered a waste, although, nowadays, it is used to produce whey powders products, improving economic and environmental aspects of the by-products [79].

Lactose presents a great importance for food and pharmaceutical industries, being used in various food products such as chocolate, confectionary and other processed products, as well as carrier of medicines in dry powder inhalation preparations, excipient of tablets [80]. In humans, lactose can cause abdominal discomfort due to its maldigestion, which reaches approximately 70% of the world's adult population [81]. β-Galactosidase (β-D-galactoside galactohydrolase, E.C. 3.2.1.23) plays an important role in human health because it is able to catalyze the hydrolysis of lactose in glucose and galactose, and because of that, it is often referred to as lactase. In addition, the transglycosylation reaction can also occur, in which galactooligosaccharides (GOS) are produced, and their structures can differ in regiochemistry of glycosidic linkage and degree of polymerization (Figures 3-5; Tables 1 and 2) [82,83].

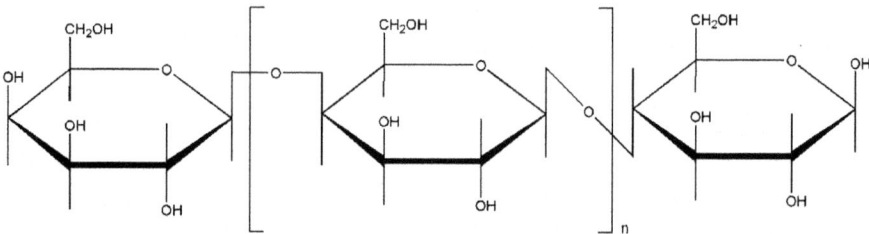

Figure 3: Structure of a galactooligosaccharide (GOS) derived from lactose, a β-(1,4) linked galactosyl oligomer (n=1-4), attached to a terminal glucosyl residue by a β-(1,4) bond. GOS are synthesized by the reverse action of β-galactosidases on lactose in higher concentrations.

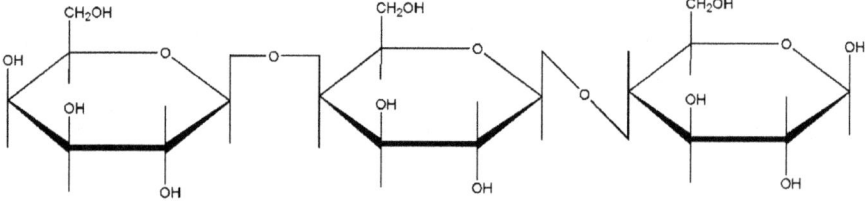

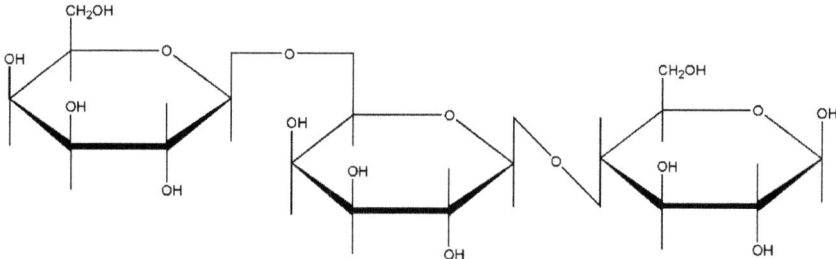

Figure 4: Examples of structures of galactooligosaccharides: 4-galactosyl lactose (top) and 6-galactosyl lactose (bottom) are represented, showing usual regiochemistry differences in galactosyl linkages

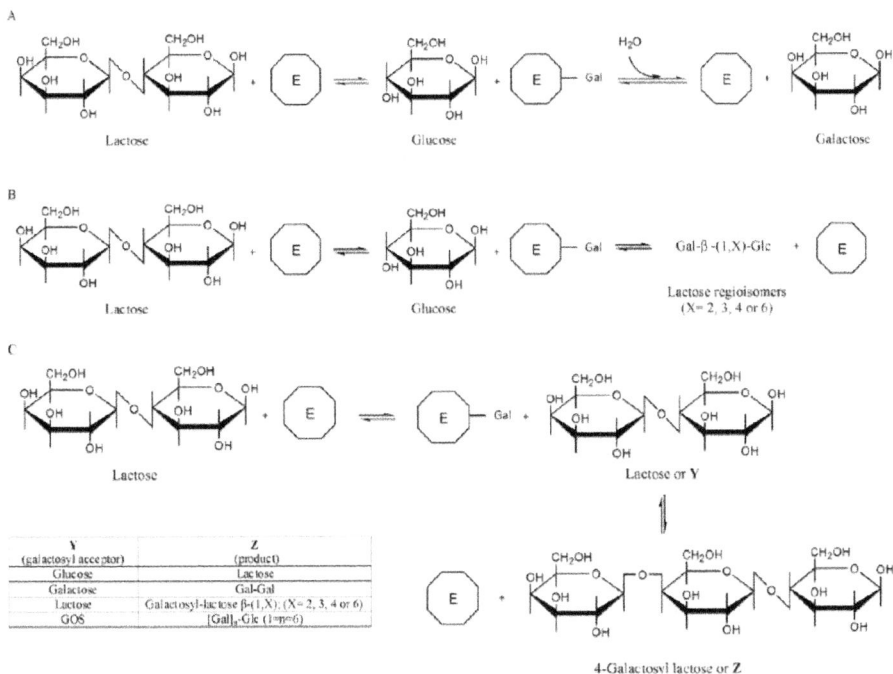

Figure 5: Enzymatic synthesis of GOS by transgalactosylation reactions. Transgalactosylation is the transfer of the galactosyl residue, after the cleavage of lactose, to an acceptor molecule containing a hydroxyl group. When the acceptor is water (A), a galactose is formed by lactose hydrolysis, whereas if the acceptor is a sugar, a disaccharide or a GOS may be formed. In intramolecular transgalactosylation (B), galactosyl donor and acceptor are the same (glucose), only linkage position changes. In intermolecular transgalactosylation (C), there is an enzymatic transfer to another nucleophilic acceptor (Y), which can be all the sugars present in the reaction media, resulting in GOS mixtures.

Despite the fact that enzymes such as β-glycosidases and β-glucosidases, which also hydrolyze carbohydrates, are able to catalyze transglycosylation reactions, β-galactosidase is the most used enzyme in dairy industry to produce GOS. β-galactosidases from *Kluyveromyces* sp. and *Aspergillus* sp. are the most used in industry because products from those microorganisms are considered as GRAS [84].

Galactooligosaccharides can be defined as a mixture of substances produced from lactose, with two to eight saccharide units, in which one of the units is a terminal glucose and the remaining units are galactose and disaccharides comprising two units of galactose [85]. Several of these GOS are recognized as prebiotics, because they are non-digestible saccharides and can be used selectively by bifidobacteria and lactobacilli in human intestine, and thus improve host health [86, 87].

Conversion of lactose into GOS is catalyzed by β-galactosidases in a kinetically controlled reaction that involves competition between hydrolysis and transgalactosylation. The thermodynamically favored hydrolysis of lactose, which generates D-galactose and D-glucose, competes with the transferase activity that produces a complex mixture of galactose-based di- and oligosaccharides. Transgalactosylation involves direct galactosyl transfer (intramolecular reaction) to D-glucose yielding regio-isomers of lactose, and indirect transgalactosylation (intermolecular) giving rise to disaccharides, trisaccharides, and tetrasaccharides, and eventually longer GOS. The interaction in the active site of the enzyme differs with the acceptor. When the acceptor is water, glucose and galactose are formed; whereas if the acceptor is a sugar, reaction results in GOS [86, 87].

Therefore, high lactose concentrations and low water contents are favorable for GOS synthesis, being the initial lactose concentrations the most important factor, independently of the enzyme source. In general, higher lactose concentrations than 30% are necessary to favor synthesis over hydrolysis [87]. However, at the same lactose concentrations, different yields of GOS can be obtained, because β-galactosidases from different sources, with different structures and/or mechanisms, exhibit different selectivity for water and saccharides. Moreover, GOS yields depend on process conditions, such as temperature, reaction time, pH and enzyme/substrate ratio [88]. However, GOS production can be affected by glucose and/or galactose that are recognized as inhibitors of hydrolysis for many β-galactosidases [89,90].

The reaction time and initial concentration of lactose are considerably important to favor GOS production, since they are simultaneously synthesized and hydrolyzed by β-galactosidase, being regulated by the kinetics of synthesis and hydrolysis. Additionally, lactose concentration can increase formation of GOS due to increased availability of galactosyl and decreased availability of water [82,91]. Additionally, reverse micelle systems, in which the enzyme is entrapped in an aqueous micelle surrounded by organic solvent, provide decrease of the thermodynamic activity of water [92,93]. Chen et al. 2003 [93] reported that the transgalactosylation capability of low concentrations of β-galactosidase and lactose, operating in reverse micelles system, was similar to high concentrations of enzyme and substrate in an aqueous system. Authors also showed that GOS production decreases with the increase in water content.

Production of GOS can be improved increasing the reaction temperature. Lactose has relatively low solubility at room temperature, which increases with increasing temperature. Therefore high temperatures are desirable since they allow the increase of lactose concentration [94,95]. Besides the possibility to increase the solubility of subtracts and products, high temperature is advantageous due to the reduced risk of microbial contamination, lower viscosity and improved transfer rates [96]. However, this is not a general rule, Boon et al. (1998) [97] reported that the increase of initial lactose concentration achieved at high temperature does not influence GOS yield using β-galactosidase from*Pyrococcus furiosus*. Another problem of carrying out GOS synthesis at high temperature is the occurrence of Maillard reaction and enzyme inactivation. Bruins et al. (2003) [95] noted that in addition to enzyme inactivation with the increase of temperature (80°C or above), Maillard reactions almost doubled the rate of enzyme inactivation. Therefore, the development of new thermostable enzymes, through recombinant DNA technology, has been undertaken in order to improve the GOS yield [98-102]. Hansson et al (2001) [103] verified an increase of GOS yield due to an increase of transgalactosylation/hydrolysis ratio by changing a phenylalanine residue to tyrosine in β-glucosidase from *Pyrococcus furiosus*, using site directed mutagenesis.

Another strategy to decrease water activity, and carry out catalysis with both high lactose concentration and temperature, demonstrated by Maugard *et al.* (2003) [104], is the use of microwave irradiation. GOS was produced using immobilized β-galactosidase from *Kluyveromyces lactis* along with organic solvents. In these conditions the selectivity for GOS synthesis was increased 217-fold, compared to a reaction carried out under conventional heating.

Similarly to temperature, pH value can affect the GOS yield, possibly through the control of synthesis and degradation [105] According to Huber *et al.* (1976) [106], that studied β-galactosidase from*Escherichia coli* K-12, higher pH values than 7.8 increased transgalactosylation/hydrolysis ratio, which decreased at lower pH values than 6.0. In contrast, Hsu *et al.* (2006) [107] observed that β-galactosidase from *Bifidobacterium longum* CCRC 15708 exhibits its maximum activity at pH 7.0. This enzyme was stable between pH 6.5-7.0, and after three hours in these conditions, 20% of its activity was lost.

In general, oligosaccharides, including galactooligosaccharides, are produced using sucrose or starch, whey, among other substrates with high quality and low cost. The process designed to convert raw material into oligosaccharides must be inexpensive and focused on increasing the productivity and stability of enzymes. In this context, immobilization of biocatalysts can reduce the process costs due to some advantages; such as possibility to reuse the biocatalyst, applying a series of batchwise or continuous reactions; the biocatalyst can exhibit more stability than the native counterpart; besides this, immobilization can reduce costs of downstream, since separation of the biocatalyst from the product can be minimized [108-110]. Recently, several authors have employed immobilized β-galactosidase to produce GOS, applying different strategies with promising results [111-114]. Urrutia et al (2013) [111] immobilized *Bacillus circulans* β-galactosidase in glyoxyl agarose. The enzyme did not lose the synthetic capacity, and retained 92% of its activity along 10 reaction batches, producing 1956 g GOS/g protein at the end of 10 batches. Palai et al (2014) [112] immobilized β-galactosidase in hydrophobic polyvinylidene fluoride and the reaction for GOS production was carried out with partial recirculation loop. Both GOS concentration and selectivity for GOS production increased with increasing initial lactose concentrations, with maximum GOS production of 30% at 50°C, and feed flow rate of 0.5 mL/min. A novel economic and efficient method to produce GOS through cellulose-binding fusion β-galactosidase was developed by Lu et al (2012) [113]. A fusion protein, formed by β-galactosidase from*Lactobacillus bulgaricus* L3 and a cellulose binding domain were employed for immobilization by adsorption onto microcrystalline cellulose. The immobilization was conducted with efficiency of 61% and the maximum GOS yield was 49% (w/w). Moreover, enzymatic activity of 85% and yield over 40% (w/w) were maintained after twenty batches. Warmerdam *et al.* (2014) [114] carried out GOS production in a packed-bed reactor using commercial β-galactosidase (Biolacta N5) immobilized on Eupergit C250L. GOS productivity was six-fold higher in one run in the packed-bed reactor than observed in one run in a batch reactor.

Smart polymers have been studied to develop GOS production processes. Poly-N-isopropyl acrylamide is a thermo-responsive poly-N-isopropyl acrylamide (PNIPAAm), which presents good solubility in water and distinct phase transition at its lower critical solution temperature (LCST). It is applied in different areas, such as medicine, biotechnology, and engineering [115,116]. Based on these advantages, Palai et al (2014) [117] developed a useful bioconjugate between PNIPAAm and β-galactosidase. The constructed PNIPAAm-β-galactosidase (PNbG) can be used in catalysis and, after that; it can be easily separated from the solution by heating at a temperature above its LCST. Further on, Palai et al (2015) [118] continued the GOS production research using this bioconjugate. A maximum GOS yield of 35 % was obtained at pH 6 and 40°C. An increase in GOS yield was observed when the temperature was risen from 30 to 40°C. At 45°C or above, after prolonged time, enzyme deactivation occurred. Moreover, bioconjugates could be reutilized at least ten times; and the separation was done by simple decantation after addition of 0.05 M NaCl and heating at 40°C.

The use of resting or living cells for GOS production appears to be interesting due to its low cost when compared to the use of purified enzyme. Despite the complexity of biocatalysis processes involving whole cells, glucose and galactose can be consumed by them. The consumption of the monosaccharides is interesting because their presence in foods is undesirable, since they do not exhibit prebiotic effect, increase caloric value of food, and can inhibit the activity of certain β-galactosidases [119].

Nevertheless, the use of whole cells can be exploited in order to selectively improve GOS production [120]. Beta-galactosidase form *Aspergillus oryzae* was used to produce GOS from lactose, followed by fermentation with *Kluyveromyces marxianus* cells, that consumed mono and disaccharides. GOS with 95% purity containing mostly tri- and tetrasaccharides were obtained [120]. Association of β-galactosidase and cells can be applied to develop GOS enriched food products. During yogurt manufacturing, GOS was produced by addition of a commercial β-galactosidase, since starter and probiotic culture were not able to provide it. Thus, this yogurt with low lactose content can be useful for lactose intolerant people. Moreover, GOS was stable during storage, probably because it was not metabolized by microbial culture and enzyme was inactivated by yogurt pH [121].

Products containing GOS were launched for the first time in Japan in the 1980s. Due to their various and important health benefits, applications of GOS gradually increased worldwide. These oligosaccharides can be found in diverse products such as yogurt, bakery products, beverages, snack bars among others [122]. GOS are able to stimulate the growth of bifidobacteria and lactobacilli in the lumen despite other members of the microbiota that were considered potentially harmful. These oligosaccharides can prevent bacterial adherence due to their properties of mimicking host cell receptors in which bacterial adhesion occurs [123]. GOS can hinder the development of colon cancer, effect which can be attributed to their capacity of delaying fermentation processes, and reducing the activity of genotoxic bacterial enzymes associated with this disease [124]. Mineral absorption can be stimulated by GOS administration, and their effect on calcium absorption was verified. GOS can be used to alleviate constipation, which is relatively common in elderly people and pregnant women. It occurs due to increased bacterial growth and fecal weight; besides this, short fatty acids stimulate intestinal peristalsis and increase osmotic pressure of fecal weight. Moreover, GOS have been reported as indirectly acting on mucosal and systemic immune activity, and also as having protective effects against allergic manifestations [125].

CHITOOLIGOSACCHARIDES

In the last years, studies of production and application of chitooligosaccharides (COS) have increased due to their biodegradability, biorenewability, biocompatibility, physiological inertness and hydrophilicity, properties that serve as a basis for the use of COS as functional food or to preserve food from degradation.

Chitin is one of the most abundant natural compounds on earth and its production is mainly based on the extraction from marine species (shrimps, crabs, lobsters, krills, etc.) [126]. Chitin is a copolymer of N-acetyl-D-glucosamine and D-glucosamine units linked by β-(1,4) glycosidic bonds, where N-acetyl-D-glucosamine units are predominant in the polymeric chain as shown in Figure 6A [127]. Chitin obtained from natural sources has a complex composition, containing several minerals, proteins, lipids, pigments and other compounds. Chitosan, an important derivative from chitin, is the deacetylated form of chitin, where N-acetyl groups are removed by chemical methods (Figure 6B).

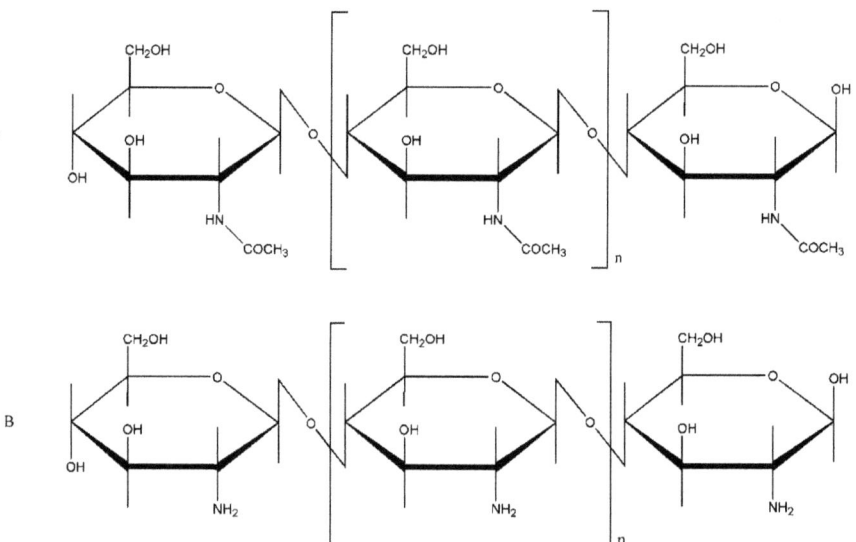

Figure 6: Structures of chitooligosaccharides (n= 3-7), A – Chitin (β-1,4-linked N-acetyl-D-glucosamine residues); B – Chitosan (β-1,4-linked D-glucosamine polymer).

A considerable amount of residues from processing of fish and crustaceans, rich in chitin and chitosan, are considered hazardous wastes and at the same time have high potential commercial value as raw material [128]. It is possible to obtain chitooligosaccharides from those residues, after prior demineralization and deproteinization by acid and alkali treatments [129].

Chitooligosaccharides are produced by chemical methods or by enzymatic methods from chitosan, produced by alkaline N-deacetylation. At industrial scale, the chemical route is used to produce chitooligosaccharides; however, this methodology presents several disadvantages such as high cost, low yield due to indiscriminate breaks of the polymer chain, production of toxic compounds due to modification on the chitin structure, as well as, corrosion and environmental hazards [130].

The enzymatic process is an attractive solution to overcome the above-mentioned disadvantages, due to their specific action on the substrate, despite the economic costs. Enzymatic hydrolysis of chitin or chitosan involves several enzymes: chitinase, chitosanase, lysozyme and cellulase [131]

According to Mourya *et al.* (2014) [132], various specific enzymes as chitosanases, chitinases and other nonspecific enzymes can hydrolyze chitin and chitosan. Action of chitinases and chitosanases are related to the degree of acetylation of the biopolymers. A novel flow chart for COS production from chitin employing chitinases and chitosanases has been reported (Figure 7) [130].

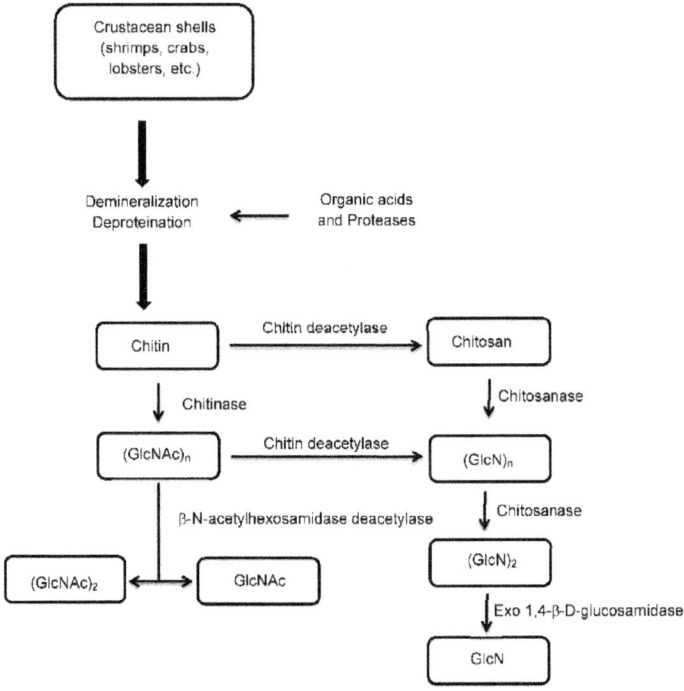

Figure 7: Global flow chart for production of COS by enzymatic hydrolysis of chitin and chitosan (adapted from Jung and Park 2014 [130]).

Chitnases are chitinolytic enzymes hydrolyzing the glycoside bonds between the sugars, which have the unique capacity to hydrolyze the GlcNAc-GlcNAc (2-acetamido-2-deoxy-β-D-glucose) links. Pre-treatment with acid solution is necessary to break down the crystalline structure of chitin and increase the availability of substrate to the action of enzymes. Chitosanases are enzymes that hydrolyze chitosan, classified according to the substrate specificity towards chitosan, which act specifically on the deacetylated (D–D) bonds [133].

In recent years, many scientific papers reported the application of chitinolytic enzymes, from different microorganisms, for the hydrolysis of chitin and chitosan. Enzymes for hydrolysis can be free or immobilized in non-toxic and inert supports.

Fernandes de Assis et al. 2010 [134] reported that COS yields of 54% were obtained after 10 minutes of hydrolysis reaction. Initial concentration of chitosan was 1% and the final oligomers concentration was 5.43 mg/mL. Production yields decreased when hydrolysis reaction time exceeded 10 minutes.

Gao et al. 2012 [135] determined that the optimal enzyme/chitosan ratio was 7.3 U/mg chitosan at 55°C to produce COS from chitosan employing chitinases from Bacillus cereus, achieving a hydrolysis yield of 76%. The yields of COS (GlcN)2, (GlcN)3 and (GlcN)4 were 13.2; 32.6 and 30.2%, respectively.

Ming et al. 2006 [136] producing chitooligosaccharides, reported pH range 4.5-6 as the optimal for chitinase activity, reaching 20 g/L of chitooligossacharides from an initial concentration of 50g/L of chitosan, which means a system with 40% of yield in the conversion of chitosan into chitooligosaccharides. Also, employing free and immobilized chitosanase from Bacillus pumilus, Kuriowa et al. 2009 [137], produced chitooligosaccharides in batch and continuous systems. In a system with free enzyme at batch conditions a concentration of 2.8 g/L was achieved, from an initial concentration of 5g/L after 40 minutes of treatment. Another system used was a membrane reactor with cutoff 2000Da. Enzyme concentration of 940 U/L, 40 minutes of residence time and 35°C were reported as the optimal conditions to attain 2.6 g/L of chitooligosaccharides. The membrane bioreactor with the free enzyme was able to maintain a constant rate of chitooligosaccharides production for 96 hours, after that time concentration decreased due to inactivation of enzymes. In order to extend the period of operation, the use of immobilized enzymes was evaluated in the membrane bioreactor. The maximum total concentration of chitooligosaccharides was 2.3 g/L with 620 U/L of immobilized chitinase during 1 month, however it is important to point out that yield was 46%, lower when compared to free enzyme tests.

COS can be applied as food preservatives due to their antimicrobial activity and as functional food, mainly in prebiotics and to help the absorption of important minerals, as calcium. Antimicrobial activity of COS depends on the degree of polymerization (DP) and the degree of deacetylation (DD) as summarized in Table 1.

Inhibitory effects of COS were tested on both Gram (-) and Gram (+) bacteria, including *Escherichia coli, Pseudomonas fluorescens, Salmonella typhimurium, Vibrio parahaemolyticus, Listeria monocytogenes, Bacillus megaterium, Bacillus cereus, Staphylococcus aureus, Lactobacillus plantarum, Lactobacillus brevis* and *Lactobacillus bulgaricus* [138]. Solutions containing 1% (w/v) COS with different molecular weights inhibited bacterial growth by 1-5 log cycles. For Gram (-) bacteria the antimicrobial activity was inversely proportional to the molecular size of oligomers, which means higher antibacterial activity was found with lower molecular weight of oligomers (1 kDa). This phenomenon was not observed for Gram (+) bacteria.

The proposed mechanism of antibacterial activity for COS with DP>12 was cellular lysis [139]. This would be due to the cationic charges of COS that could link to the negative charges present in the cell walls, leading to the formation of large bacterial clusters, which might block the nutrition transport across the bacterial cell and result in death of the bacteria. Highly deacetylated COS were shown to be more effective at inhibiting the growth of *Staphylococcus aureus, Escherichia coli, Pseudomonas aeruginosa, Streptococcus fecalis* and *Samonella typhimurium* than COS with low degree of deacetylation [140].

It has been suggested that COS are able to pass through the bacterial cell wall and be incorporated in the cytoplasm of Gram (+) bacteria [141]. Those low molecular weight compounds can have importance in gene expression related to regulation of stress, autolysis and energy metabolism.

Chitooligosaccharides with DP 4 were demonstrated to have higher antimicrobial effect on four bacteria species (*Escherichia coli, Staphylococcus aureus, Streptococcus lactis, Bacillus subtilis*) and six fungi (*Saccharomyces cerevisiae, Rhodotorula bacarum, Mucor circinelloides, Rhizopus apiculatus,Penicillium charlesii, Aspergillus niger*) [142]. At the same time, degrees of deacetylation over 90% were shown to be more efficient in the inhibition of microbial growth. In addition, chitooligosaccharides with low molecular weight were able to cross the cell wall and interact with DNA in the cytoplasm suppressing the growth of microorganisms. Highly deacetylated

COS have many free amines, which can bond to negatively charged residues at the cell wall, leading to formation of aggregates of microorganisms. Those aggregates precipitate, resulting in death of the microorganism.

Chitooligosaccharides can be employed as preservatives due to their antioxidative properties. Antioxidant activity of chitooligosaccharides depends on their degree of deacetylation and molecular weights [143]. It was shown that 90% deacetylated medium molecular weight COS have the highest free radical scavenging activity for DPPH, hydroxyl, superoxide and carbon centered radicals [144]. Antioxidant properties are closely related to the amino and hydroxyl groups, which can react with unstable free radicals to form stable macromolecule radicals [145,146].

According to Halden *et al.* 2013 [147] COS could be applied as feed additives or hypocholesterolemic agents. Based on their study, hypercolesterol concentration in blood is directly related to the generation of reactive oxygen species. Thus, chitooligosaccharides can be used to scavenge the free radicals on the body, triggering the enhanced synthesis of catalase and superoxide dismutase and decreasing lipid peroxidation.

COS were conjugated with phenolic acid (PAC-COS) to improve the antioxidant properties of the oligosaccharides in the presence of reactive oxygen species (2,2-diphenyl-1-picrylhydrazyl (DPPH), hydroxyl (OH) and nitric oxide (NO)) [148]. The increase on the antioxidant activity is associated to the structure of phenolic acids and the substitutions on the aromatic ring of the side chain.

Chitooligosacharides can be considered as prebiotics because they are non-digestible food ingredients with beneficial effects on probiotic bacteria (*Lactobacillus* and *Bifidobacterium*) present in the gastrointestinal tract [5]. In fact, prebiotic activities of COS preparations (0.1 to 0.5%) with varying degree of polymerization (2 to 8) were reported [149]. Assays were conducted with three strains of probiotic bacteria, *Bifidobacterium bifidum KCTC 3440, Bifidobacterium infantis KCTC 3249* and*Lactobacillus casei KCTC 3109.*

However, an opposite effect was shown on the population of *Lactobacillus* and *Bifidobacterium* when chitooligosaccharides were tested as prebiotic agents in healthy rats [150]. Chitooligosaccharides have been demonstrated to have a weaker prebiotic effect over *Lactobacillus* and *Bifidobacterium* when compared with other oligosaccharides as fructooligosaccharides, mannanoligosaccharides; and galactooligosaccharides [151].

Chitooligosaccharides from marine species, mainly shrimps and crabs, are produced and commercialized by several companies (Table 2), such as:

Qingdao BZ-Oligo Co., Ltd: Monomers of chitosan oligosaccharides are obtained by enzymatic hydrolysis, chemical derivatization and column chromatography. The degree of polymerization is from 2 to 10.

- BioCHOS: Preparation of chitooligosaccharides (CHOS) made by controlled enzymatic degradation of chitosan.

- AMSBIO: Preparation of a series of chitosan-oligosaccharides from dimer to hexamer by hydrolysis of chitosan from crab shells. All oligomers are chromatographically pure, not less than 98%, confirmed by high performance liquid chromatography.

NOVEL OLIGOSACCHARIDES

Typical oligosaccharides like FOS and GOS in particular have been widely studied for their prebiotic effects. However, a number of other non-digestible oligosaccharides (NDOs), to which less rigorous study has been so far applied, have at least indications of prebiotic potential. Those with the most accumulated evidence to date are isomalto-oligosaccharides (IMO), soybean oligosaccharides (SOS), xylo-oligosaccharides (XOS) and lactosucrose. Together with FOS, GOS, and lactulose, all of these oligosaccharides are recognized in the Japanese functional food regulation system as ingredients with beneficial health effects [152].

A great interest resides on the identification, evaluation and commercialization of new products with improved functional properties and benefic health effects such as higher ability to modulate microbiota. Arabinoxylo-oligosaccharides (AXOS), levan-type FOS, gentio-oligosaccharides (GenOS) and pectin-derived oligosaccharides (POS) are examples of these new potential products.

Isomalto-Oligosaccharides

Isomalto-oligosaccharides are usually found as a mixture of oligosaccharides with predominantly α-(1,6)-linked glucose residues with a degree of polymerization (DP) ranging from 2–6, and oligosaccharides with a mixture of α-(1,6) and occasionally α-(1,4) glycosidic bonds such as panose (Figure 8; Table 1) [152].

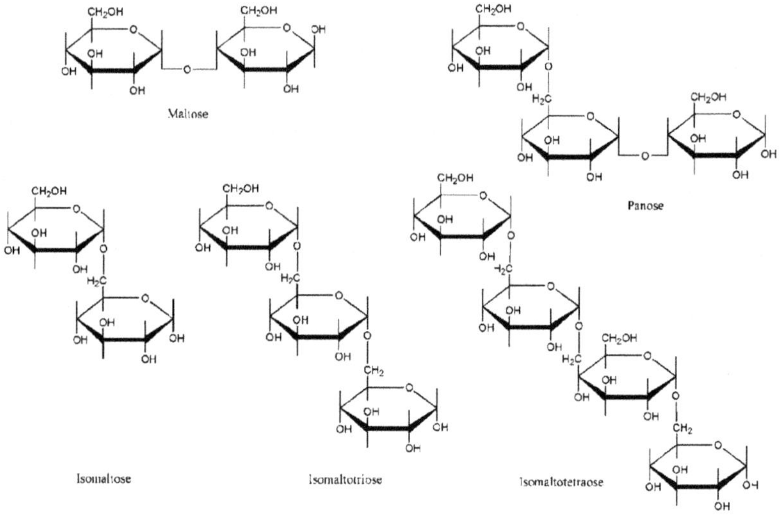

Figure 8: Examples of structures of isomalto-oligosaccharides. Glucosyl residues are linked to maltose or isomaltose by α-(1,6) glycosidic bonds.

Isomalto-oligosaccharides, like malto-oligosaccharides, are produced using starch as the raw material. Isomalto-900, a commercial product, is produced from cornstarch and consists of isomaltose, isomaltotriose and panose. Starch dextrans are easily converted to IMO, which are the market leaders in the dietary carbohydrate sector of functional foods in Japan. However, unlike malto-oligosaccharides, there is evidence to suggest that isomalto-oligosaccharides induce a bifidogenic response [11].

IMO occur naturally in various fermented foods and sugars such as sake, soybean sauce and honey. They are a product of an enzymatic transfer reaction, using a combination of immobilized enzymes. Initially, starch is liquefied using α-amylase (EC 3.2.1.1) and pullulanase (EC 3.2.1.41), and, in a second stage, the intermediary product is processed by both β-amylase (EC 3.2.1.2) and α-glucosidase (EC 3.2.1.20). Beta-amylase first hydrolyzes the liquefied starch to maltose. The transglucosidase activity of α-glucosidase then produces isomalto-oligosaccharides mixtures which contain oligosaccharides with both α-(1,6)- and α-(1,4)-linked glucose residues (Table 2) [153].

In recent years, much research has been focused on improvement of the efficiency of IMO production by screening for new and better enzymes for high yield IMO synthesis. Efforts also have been made to develop novel processes such as synthesis of IMO from sucrose using free or immobilized dextransucrase and dextranase, and efficient conversion of maltose into IMO using immobilized transglucosidase, or using an enzyme membrane reactor

[153,154].

IMO are mild in taste and relatively inexpensive to produce. These oligosaccharides have desirable physicochemical characteristics such as relatively low sweetness, low viscosity and bulking properties. IMOs have been developed to prevent dental caries, as substitute sugars for diabetics [155], or to improve the intestinal flora [152].

Several companies currently manufacture isomaltooligosaccharides, of which Showa Sangyo (Japan) is the major producer. Of the emerging prebiotic oligosaccharides, IMO are used in the largest quantities for food applications. In Japan, the volume of IMOs manufactured is estimated to be three times greater than for either FOS or GOS [152]. Among other oligosaccharides, which are widely used as food ingredients or additives [156] based on their nutritional and health benefits [157], IMO are interesting due to availability, high stability and low cost [154].

Unlike other prebiotic oligosaccharides, considerable digestion of IMO occurs during intestinal transit. A large portion of this ingredient reaches the colon and intestinal enzymes degrade the remainder, leading to a rise in blood glucose levels [154]. Thus, a part of the IMO survives gastric transit to be fermented by the intestinal microbiota [152]. *In vitro* fermentation studies have shown that IMO promote the selective proliferation of bifidobacteria in the fecal microbiota [158]. However, further controlled human feeding studies employing culture and molecular techniques are required to determine the impact of IMO on the intestinal microbiota.

Beneficial effects of IMO consumption have been reported in a few human feeding studies investigating health parameters in specific populations. IMOs stimulate bowel movement and help to decrease total cholesterol levels with an intake of 10 g/d in elderly people [158].The limited data for physiological effects showed only improved defecation pattern (frequency and stool bulk via increases in microbial biomass) and lowering of total cholesterol levels [158,159]. In conclusion, the data for the bifidogenic effects of isomalto-oligosaccharides are less consistent than for other typical oligosaccharides like inulin or oligofructose [155].

Soybean Oligosaccharides

Unlike other oligosaccharides, soybean oligosaccharides are extracted directly from the raw material and do not require enzymatic manufacturing processes. These α-galactooligosaccharides include and consist of galactosyl residues linked to the glucose moiety of sucrose by α-(1,6) bonds (Figure 9, Table 1) [2].

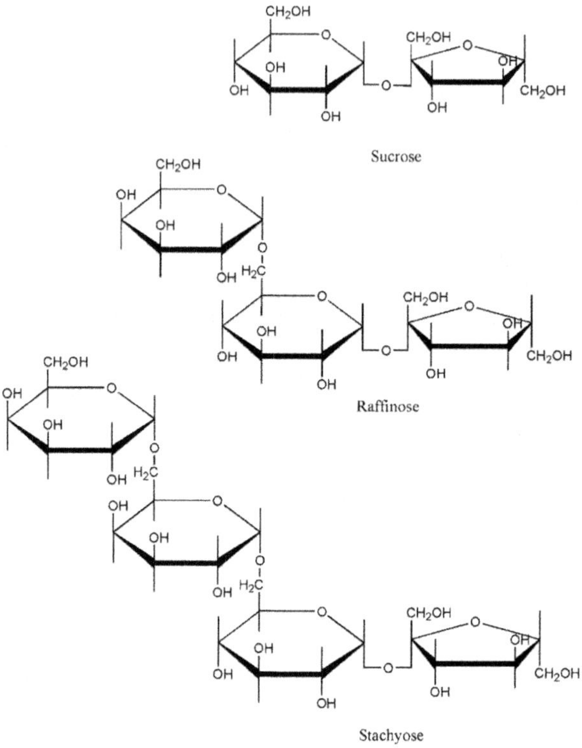

Figure 9: Examples of the main soybean oligosaccharides, raffinose and stachyose, derived from sucrose, showing galactosyl residues linked to sucrose by α-(1,6) bonds.

Soybean whey, a by-product from the production of soy protein isolates and concentrates, is composed mainly of raffinose (DP 3), stachyose (DP 4) and verbascose (DP 5), as well as sucrose, glucose and fructose. The most abundant sugars are extracted from the soybean whey and concentrated to produce soybean oligosaccharide syrup (Table 2), rather than being commercially synthesized using enzymatic processes [158].

Raffinose and stachyose are resistant to digestion, since α-galactosidase activity (required to hydrolyze these carbohydrates) is not present among human digestive enzymes and, therefore, reach the colon intact, where they act as prebiotics, stimulating the growth of bifidobacteria. Apart from being acknowledged as non-digestible, human studies on the effects of these oligosaccharides are scarce. Their physiological actions appear to be similar to the other galactooligosaccharides; they are bifidogenic and promote other effects expected from this change in colon microbiota. Calpis Co. (formerly known as Calpis Food Industry Co.) produces soybean oligosaccharides in Japan [11].

Xylo-Oligosaccharides

Xylo-oligosaccharides (XOS) are sugar oligomers of xylose units linked by β-(1,4) linkages (Figure 10,Table 1). The number of xylose residues can vary from 2 to 10, but mainly consist of xylobiose, xylotriose and xylo-tetraose [152], which are found naturally in bamboo shoots, fruits, vegetables, milk and honey [160]. In addition to xylose residues, xylans are usually found in combination with arabinofuranosyl, glucopyranosyl uronic acid or its 4-O-methyl derivative (2- or 3-acetyl or phenolic substituents), resulting in branched XOS with diverse biological properties [153].

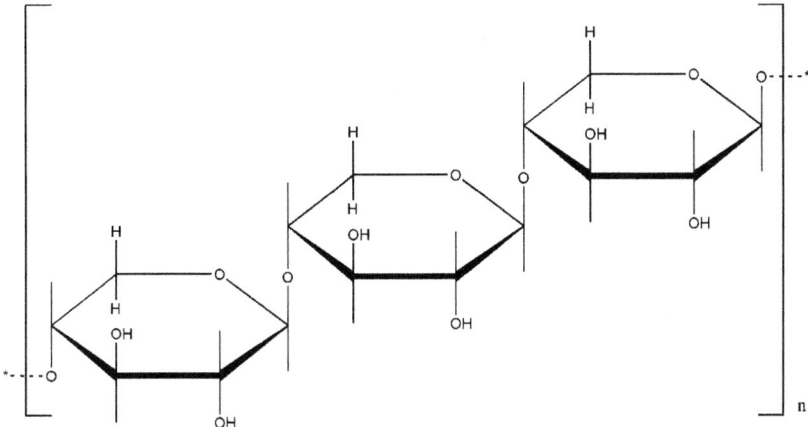

Figure 10: Partial structure of xylo-oligosaccharides (n=3-6) produced by enzymatic hydrolysis of xylan hemicelluloses, catalyzed by β-xylanases.

Their production at an industrial scale is carried out from lignocellulosic materials (LCMs). XOS can be used for several purposes, among them, food-related applications. The LCM for XOS production comes from a variety of feedstocks (from forestry, agriculture, industry or urban solid wastes) that show similarities in composition. The raw material for xylo-oligosaccharide synthesis is the polysaccharide xylan, which is extracted mainly from corncobs besides hardwoods, straws, bagasses, hulls, malt cakes and bran [160].

XOS production from LCM is not simple or economical because it depends on two treatment steps. The first step is the xylan extraction from LCM, which includes a chemical pretreatment. Although there are multiple treatments for xylan extraction (alkaline hydrolysis using NaOH, KOH, $Ca(OH)_2$, ammonia or a mixture of bases, oxidizing agents, salts or alcohols to remove lignin or pectic substances), there is no favorite consensus among them. Once the extracted xylan is in a soluble form, the second step includes the xylanase enzymatic reaction or the hydrolytic degradation of xylan by steam, water or dilute solutions of mineral acids [160]. For the enzymatic production of XOS, xylan is enzymatically hydrolysed to xylo-oligosaccharides by endo-β-1,4-xylanases (EC 3.2.1.8) (Table 2). Enzyme complexes with controlled exo-xylanase and/or β-xylosidase activity are required, to avoid the production of xylose, which may cause inhibition effects in XOS production. For food related applications, a DP of 2–4 is the most desirable [160]. Therefore, development of efficient and economical xylanase based bioprocesses for use in XOS production is necessary. Many microorganisms well known as producers of xylanolytic enzymes may be promising for novel production processes [161].

The process yields predominantly linear β-(1,4)-linked XOS (mainly xylobiose, xylotriose and xylotetraose) as well as some oligosaccharides with branched arabinose residues. For the production of food-grade XOS, a refining step is necessary. Vacuum evaporation increases the XOS concentration and removes volatile compounds such as acetic acid and the flavours of their precursors. In order to obtain higher-purity oligosaccharide products, the monosaccharide xylose and high molecular mass carbohydrates, as well as non-saccharide components can be removed from the oligosaccharides using membrane filtration techniques, organic solvent extraction, adsorption in different materials and chromatographic separation techniques used for XOS purification. Chromatographic methods, however, are not suitable for economic reasons for large-scale production of XOS intended to be used in the food industry [153].

XOS can be metabolized by bifidobacteria and lactobacilli in pure culture. In relation to human health, XOS selectively enhanced the growth of bifidobacteria thus promoting a favorable intestinal environment [152]. XOS is a promising oligosaccharide class that stimulates increased levels of bifidobacteria to a greater extent than do FOS or other oligosaccharides [161]. However, to date well-controlled animal and human feeding studies to confirm the prebiotic activity of XOS are still scarce. While they show promise, more research is required before XOS can conclusively be claimed as prebiotics. Besides the potential prebiotic effect, immunostimulating effects, antioxidant activity, anti-allergy, anti-infection and anti-inflammatory properties were reported for XOS [162-164].

In addition to the beneficial health effects, XOS have interesting physicochemical properties; they are only moderately sweet, have an acceptable odor, are noncariogenic and low caloric, stable over a wide range of pH values (2.5–8.0), even the relatively low pH value of gastric juice, and temperatures up to 100°C. Most oligosaccharides can be hydrolyzed, resulting in the loss of nutritional and physicochemical properties at acidic pH values, when treated at high temperatures for short periods, or when submitted to prolonged storage under room conditions. These properties make them suitable for incorporation into many food products such as in combination with soymilk, soft drinks, dairy products, sweets and confectionaries [158].

XOS show a remarkable potential for practical utilization in many fields, including pharmaceuticals, feed formulations and agricultural applications. Nevertheless, their most important market developments correspond to food-related applications, however, their comparatively high production costs impair market development of these oligosaccharides, and further improvements in process technology are necessary [11].

Arabinoxylo-Oligosaccharides

Arabinoxylo-oligosaccharides (AXOS) are an example of a novel prebiotic dietary fiber. They can be isolated from wheat bran and consist of xylan chains with a variable substitution of arabinose side chains (Table 3) [158]. On an industrial scale, AXOS are generated through the enzymatic cleavage of AX with endoxylanases, resulting in various molecules differing in DP (between 3 and 67) and degree of substitution of arabinosyl residues [165].

Table 3: Novel oligosaccharides with prebiotic activities.

Novel oligosaccharides	Chemical structure	References
Lactosucrose	4-galactosyl sucrose	[2, 156]
Arabinogalactooligosaccharides	Galactan oligomers β-(1,3) or (1,6) attached to arabinofuranose residues.	[19]
Arabinoxylooligosaccharides	Xylan randomly attached to arabinofuranose residues by α-(1,3) or α-(1,2) linkages.	[165, 237]
Arabinooligosaccharides	Arabinosyl units linked by α-(1,5) bonds.	[2]
Pectic oligosaccharides	Linear backbone of α-(1,4) linked D-galacturonic acid units randomly acetylated and/or methylated.	[171, 172, 238, 239]
Galacturonan	Linear chain of α-(1,4) linked D-galacturonic acids	
Rhamnogalacturonan	α-(1,4) linked galacturonic acid and α-(1,2) linked rhamnose units	
Mannan oligosaccharides	Mannose α-(1,6) linked backbone and α-(1,2) and α-(1,3) linked branches.	[176]

Oligodextrans	Glucosyl units linked by α-(1,4) bonds.	[176, 240]
Gentiooligosac-charides	Glucosyl units linked by β-(1,6) bonds.	
Beta-glucan oli-gosaccharides	Glucosyl units linked by β-(1,3/1,4) or β-(1,3/1,6) bonds.	
Cyclodextrins	α-(1,4) linked cyclic – glucosyl units.	

The fiber properties include an improvement of bowel habit and positive change of the fermentation in the colon, whereas they were also shown to possess bifidogenic properties [158]. There are indications that AXOS have an effect against type II diabetes. AXOS decrease postprandial glucose levels and insulin response, and increase postprandial ghrelin in healthy humans [156,166].

This bifidogenic effect is strongly influenced by the complexity of the AXOS molecules and decreases with increasing average DP and degree of substitution [72,166]. Genome sequence analysis reveals that several bifidobacterial strains contain genes possibly coding for enzymes involved in the debranching of side groups and in the cleavage of the xylose backbones of AXOS [72]. This kind of specialization together with the potential to degrade xylose backbones intracellularly could explain the selective growth stimulation of bifidobacteria by AXOS.

Novel Fructooligosaccharides

There is an increasing interest in novel molecules with prebiotic and physiological effects. Some fungi are able to synthesize levan-type FOS containing fructosyl units linked by β-(2,6) linkages (6-kestose being first in the series) (Table 1), or neolevan type FOS containing a fructosyl unit also linked by this type of linkage to a glucose (neokestose, neonystose, or neofructofuranosylnystose). Such FOS have been metabolized by different bifidobacteria strains when supplied as the sole carbon source [167].

Levan-type FOS were synthesized by acid hydrolysis of β-(2,6)-linked polymers containing a glucose at one terminus (levans), these have been produced by several microorganisms growing in sucrose-based medium [168]. The discovery of novel enzymes that synthesize β-(2,6)-linked FOS from sucrose may, however, provide a non-pollutant alternative to acid hydrolysis of levans. Because there is an existing process to produce inulin-type FOS, an enzymatic method involving the hydrolysis of levan to produce levan-FOS may be possible. However, with the lack of an available plant source of levan, as there is for inulin, it is possible to derive an enzymatic process to produce levan-type FOS from microbial levan, using levansucrase (Table 2) and endolevanases [169].

Marx *et al.* 2000 [170] observed that levan-type FOS obtained via the acid hydrolysis of levans were metabolized by different bifidobacteria strains, thus further demonstrating their prebiotic potential. Nevertheless, the levan-type FOS prebiotic properties have not been fully characterized, possibly due to their limited availability.

The production of levan-type FOS has not reached industrial levels [171], despite several reports demonstrating their potential applications as food and feed additives in agriculture as well as their pharmaceutical applications.

Pectic Oligosaccharides

Pectic oligosaccharides (POS) (Table 3) are obtained by pectin depolymerization. Pectins are ramified heteropolymers made up of a linear backbone of α-(1,4)-linked D-galacturonic acid units (which can be randomly acetylated and/or methylated).

POS have been proposed as a new class of prebiotics capable of exerting a number of health-promoting effects. Among these are protection of colonic cells against pathogenic microorganisms [172], stimulation of apoptosis of human colonic adenocarcinoma cells [173] and *in vivo* synergistic empowerment of immunomodulation caused by galactooligosaccharides (GalOS) and fructooligosaccharides (FOS). Other benefits include potential for cardiovascular protection *in vivo*, reduction of damage by heavy metals, antiobesity effects, dermatological applications and antitoxic, antiinfection, antibacterial and antioxidant properties. Additionally, *in vivo* and *in vitro* studies have confirmed that acidic POS are not cytotoxic or mutagenic, being suitable for use in foods for children and babies [173].

Gentio-Oligosaccharides

Gentio-oligosaccharides (GenOS) consist of 2–5 glucose residues linked by β-(1,6) glycosidic linkages (Table 3). These oligosaccharides are not hydrolysed in the stomach or small intestine and therefore reach the colon intact, thus fulfilling a criterion of a prebiotic [11]. GenOS were further reported to possess bifidogenic activity [153]. GenOS are usually produced from glucose syrup by enzymatic transglucosylation or by biocatalytic glycosylation with cultured cells. Despite the prebiotic potential of GenOS, research on the novel production of GenOS is sparse. Gentio-oligosaccharides are produced in Japan by Nihon Shokuhin Kako [11].

PERSPECTIVES

Function and application of chitooligosacharides frequently depend on their size, and, therefore, the degrees of polymerization and acetylation. Substrate-enzyme synergisms determine the molecular weight of the generated COS. Gutierrez-Román *et al.* 2014 [174] tested three chitolytic enzymes ChiA, ChiB and ChiC, alone and in combination. In addition, three chitanases were tested in synergism with a chitobiase and a non-catalytic binding protein. When evaluated individually, ChiA was unable to hydrolyze chitin while ChiB and ChiC were able to degrade chitin and generate chitin monomers and dimers. When enzymes were tested pairwise (ChiA-ChiB, ChiA-ChiC, and ChiB-ChiC) the production of dimers and trimers was much higher, and monomers significantly lower than those seen with ChiB or ChiC individually. However, higher concentrations of COS were obtained when the authors tested the four enzymes in combination with non-catalytic binding protein acting on chitin.

Further studies must be focused on the action of the enzymes on substrates with different degrees of polymerization and acetylation and N-acetylation pattern to improve the comprehension of that synergism. In addition, researches involving synergism of non-catalytic binding proteins and hydrolytic enzymes should be developed in order to increase the understanding of oligomers syntheses [127]. Consequently, to produce size-specific chitooligosaccharides by enzymatic hydrolysis, further studies on genetic modification are necessary to overproduce enzymes and non-catalytic binding proteins, which will have a great impact on the quality of oligomers obtained and on the productivity of industrial processes.

Another important challenge in the development of biotechnological processes that employ agro-food industry residues as raw material is the direct fermentation of those raw materials. Obviously, direct fermentation of raw materials is closely related with the aforementioned aspects, since fermentative processes involve microbial growth and enzymatic hydrolysis, and process conditions that in many cases are different from physiological conditions. Moreover, it is important to give attention to screening of new enzymes from extremophile microorganisms, which usually catalyze reactions under non-physiological conditions such as high salinity, high temperature and low water activity [175].

As important part of the biotechnological process, bioreactors and enzyme (free or immobilized) are essential and need special attention to improve yields and productivities. Free enzymes in batch systems are the most conventional technology employed in the production of oligosaccharides by enzymatic hydrolysis. However, it has several important drawbacks, because enzymes are unstable, can be employed once and accumulation of products usually reduces their activity. These drawbacks are related directly to the quality of the product and the yield of the process. Development of novel technologies in order to solve those snags employing immobilized enzymes in column reactor and membrane systems have been studied. Column reactor packing with immobilized enzymes allows continuous production of oligosaccharides and has important advantages, such as increased operational stability of the enzyme and reduced accumulation which otherwise could lead to enzyme inhibition. The poorer affinity of immobilized enzymes is the main disadvantage of the application of column reactors at industrial scale. Studies should be directed towards the improvement of enzyme-support affinity. Membrane reactors are considered a new and attractive technology to produce oligosaccharides, in which enzymes are confined in the reaction side and continuously reused, with obvious implications for the efficiency and economy of the process. Low-cost and low-energy consumption are other important advantages to increase its utilization. The main limitation for industrial application of membrane reactors are fouling and polarization phenomena, which decrease considerably permeate flux, containing the produced oligosaccharides [176]. The main challenge to be studied in order to implement this technology advantageously in the industry is how to reduce the effect of these problems without affecting the stability of enzymes.

REFERENCES

1.   Miguel, A S M, Martins-Meyer, T S, Figueiredo, E V C, Lobo, B W P, Dellamora-Ortiz, G M. Enzymes in Bakery: Current and Future Trends. In: Muzzalupo, I, editor. Food Industry. 1ed. Rijeka: InTech; 2013. p. 287-321. DOI: 10.5772/53168.ch14

2.   Patel S, Goyal A. Functional oligosaccharides: production, properties and applications. World Journal of Microbiology and Biotechnology. 2011;27:1119-1128. DOI: 10.1007/s11274-010-0558-5

3.   Villares, J M M. Prebiotics in Infant Formulas: Risks and Benefits. In: Watson, R R, Preedy, V R, editors. Bioactive Foods in Promoting Health. London: Academic Press; 2010. p. 117-129. ch8

4.   Pinelo M, Jonsson G, Meyer AS. Membrane technology for purification of enzymatically produced oligosaccharides: Molecular and operational features affecting performance. Separation and Purification Technology. 2009;70:1-11. DOI:10.1016/j.seppur.2009.08.010

5.   Gibson, G R, Probert, H M, Loo, J V, Rastall, R A, Roberfroid, M B. Dietary modulation of the human colonic microbiota: updating the concept of prebiotics. Nutrition Research Reviews. 2004;17:259-275.

DOI: 10.1079/NRR200479

6.   Roberfroid, M B. Prebiotics: Concept, Definition, Criteria, Methodologies, and Products. In: Gibson, G R, Roberfroid, M B, editors. Handbook of Prebiotics. Boca Raton: CRC Press; 2008. p. 39-68. Ch. 4.

7.   Rioux, K P, Madsen, K L, Fedorak, R N. The Role of Enteric Microflora in Inflammatory Bowel Disease: Human and Animal Studies with Probiotics and Prebiotics. Gastroenterology Clinics of North America. 2005;34:465-482. DOI: 10.1016/j.gtc.2005.05.005

8.   Saad N, Delattre C, Urdaci M, Schmitter J M, Bressollier P. An overview of the last advances in probiotic and prebiotic field. LWT - Food Science and Technology. 2013;50:1-16. DOI: 10.1016/j.lwt.2012.05.014

9.   Scheid, M M A, Moreno, Y M F, Marostica-Junior, M R, Pastore, G M. Effect of prebiotics on the health of the elderly. Food Research International. 2013;53:426-432. DOI: 10.1016/j.foodres.2013.04.003

10.   Bruins, M E. Oligosaccharide Producion with Thermophilic Enzymes [thesis]. Wageningen: Wageningen University; 2003.

11.   Crittenden, R G, Playne, M J. Production, properties and applications of food-grade oligosaccharides. Trends in Food Science and Technology. 1996;7:353-361. DOI: 10.1016/S0924-2244(96)10038-8

12.   Mussatto, S I, Mancilha, I M. Non-digestible oligosaccharides: A review. Carbohydrate Polymers. 2007;68:587-597. DOI: 10.1016/j.carbpol.2006.12.011

13.   Courtois, J. Oligosaccharides from land plants and algae: production and applications in therapeutics and biotechnology. Current Opinion in Microbiology. 2009;12:261-273. DOI 10.1016/j.mib.2009.04.007

14.   Grout, D H G, Gabin, V. Glycosidases and glycosyl transferases in glycoside and oligosaccharide synthesis. Current Opinion in Chemical Biology. 1998;2:98-111. DOI: 10.1016/S1367-5931(98)80041-0

15.   Casci, T, Rastall, R A. Manufacture of Prebiotic Oligosaccharides. In: Gibson, G R, Rastall, R A, editors. Prebiotics: Development and Application. Chichester: John Wiley & Sons, Ltd.; 2006. p. 29-55. Ch 2.

16.   Patel, V, Saunders, G, Bucke, C. Production of fructooligosaccharides by *Fusarium oxysporum*. Biotechnology Letters. 1994;16:1139-1144. DOI: 10.1007/BF01020840

17.   Viikari, L, Gisler, R. By-products in the fermentation of sucrose by different *Zymomonas*-strains. Applied Microbiology and Biotechnology. 1986;23:240-244. DOI: 10.1007/BF00261922

18.   Van Loo, J, Coussement, P, De Leenheer, L, Hoebregs, H, Smits, G. On the

presence of inulin and oligofructose as natural ingredients in the Western diet. Critical Reviews in Food Science and Nutrition. 1995;35:525-552. DOI: 10.1080/10408399509527714

19. Barreteau, H, Delattre, C, Michaud, P. Production of oligosaccharides as promising new food additive generation. Food Technology and Biotechnology. 2006;44:323-333.

20. Madlova, A, Antosova, M, Barathova, M, Polakovic, M, Stefuca, V, Bales, V. Biotransformation of sucrose to fructooligosaccharides: the choice of microorganisms and optimization of process conditions. Progress in Biotechnology. 2000;17:151-155. DOI:10.1016/S0921-0423(00)80061-1

21. Silva, M F, Rigo, D, Mossi, V, Golunski, S, Kuhn, G O, Di Luccio, M, Dallago, R, Oliveira, D, Oliveira, J V, Treichel, H. Enzymatic synthesis of fructooligosaccharides by inulinases from *Aspergillus niger* and *Kluyveromyces marxianus* NRRL Y-7571 in aqueous–organic medium. Food Chemistry. 2013;138:148-153. DOI: 10.1016/j.foodchem.2012.09.118

22. Roberfroid, M B. Functional foods: Concepts and application to inulin and oligofructose. British Journal of Nutrition. 2002;87:S139-S143. DOI: 10.1079/BJN/2002529

23. L'homme, C, Puigserver, A, Biagini, A. Effect of food-processing on the degradation of fructooligosaccharides in fruit. Food Chemistry. 2003;82:533-537. DOI: 10.1016/S0308-8146(03)00003-7

24. Sabater-Molina, M, Larqué, E, Torrella, F, Zamora, S. Dietary fructooligosaccharides and potential benefits on health. Journal of Physiology and Biochemistry. 2009;65:315-328. DOI: 10.1007/BF03180584

25. Vega, R, Zuniga-Hansen, M E. The effect of processing conditions on the stability of fructooligosaccharides in acidic food products. Food Chemistry. 2015;173:784-789. DOI: 10.1016/j.foodchem.2014.10.119

26. Birkett, A M, Francis, C C. Short-Chain Fructo-Oligosaccharide. A Low Molecular Weight Fructan. In: Cho, S S, Finocchiaro, E T, editors. Handbook of Prebiotics and Probiotics Ingredients: Health Benefits and Food Applications. Boca Raton: CRC Press; 2010. p. 13-42. Ch. 2.

27. Yun, J W. Fructooligosaccharides – Occurrence, preparation and application. Enzyme and Microbial Technology. 1996;19:107-117. DOI: 10.1016/0141-0229(95)00188-3

28. Ziemer, C J, Gibson, G R. An overview of probiotics, prebiotics and synbiotics in the functional food concept: perspectives and future

strategies. International Dairy Journal. 1998;8:473-479. DOI: 10.1016/S0958-6946(98)00071-5

29. Sangeetha, P T, Ramesh, M N, Prapulla, S G. Recent trends in the microbial production, analysis, and application of fructooligosaccharides. Trends in Food Science and Technology. 2005;16:442-457. DOI: 10.1016/j.tifs.2005.05.003

30. Sangeetha, P T, Ramesh, M N, Prapulla, S G. Maximization of fructooligosaccharide production by two stage continuous process and its scale up. Journal of Food Engineering. 2005;68:57-64. DOI: 10.1016/j.jfoodeng.2004.05.022

31. Roberfroid, M B. Inulin-type fructans: Functional food ingredients. Journal of Nutrition. 2007;137:2493S-2502S.

32. Silva, M F, Rigo, D, Mossi, V, Golunski, S, Kuhn, G O, Di Luccio, M, Dallago, R, Oliveira, D, Oliveira, J V, Treichel, H. Enzymatic synthesis of fructooligosaccharides by inulinases from *Aspergillus niger* and *Kluyveromyces marxianus* NRRL Y-7571 in aqueous–organic medium. Food Chemistry. 2013;138:148-153. DOI: 10.1016/j.foodchem.2012.09.118

33. Arrizon, J, Urias-Silvas, J E, Sandoval, G, Mancilla-Margalli, N A, Gschaedler, A C, Morel, S, Monsan, P. Production and Bioactivity of Fructan-Type Oligosaccharides. In: Moreno, F J, Sanz, M L, editors. Food Oligosaccharides: Production, Analysis and Bioactivity. Chichester: John Wiley & Sons, Ltd.; 2014. p. 184-199. Ch. 11.

34. Chen, W-C, Liu, C-H. Production of β-fructofuranosidase by *Aspergillus japonicus*. Enzyme and Microbial Technology. 1996;18:153-160. DOI: 10.1016/0141-0229(95)00099-2

35. Ganaie, M A, Gupta, U S, Kango, N. Screening of biocatalysts for transformation of sucrose to fructooligosaccharides. Journal of Molecular Catalysis B: Enzymatic. 2013;97:12-17. DOI: 10.1016/j.molcatb.2013.07.008

36. Antosova, M, Polakovic, M. Fructosyltransferases: The enzymes catalyzing production of fructooligosaccharides. Chemical Papers. 2001;55:350-358.

37. Kim M-H, In M-J, Cha H J, Yoo Y J. An empirical rate equation for the fructooligosaccharide-producing reaction catalyzed by β-fructofuranosidase. Journal of Fermentation and Bioengineering. 1996;82:458-463. DOI: 10.1016/S0922-338X(97)86983-8

38. Fernandez, R C, Ottoni, C A, Silva, E S, Matsubara, R M S, Carter, J M, Magossi, L R, Wada, M A A, Rodrigues, M F A, Maresma, B G, Maiorano,

A E. Screening of β-fructofuranosidase-producing microorganisms and effect of pH and temperature on enzymatic rate. Applied Microbiology and Biotechnology. 2007;75:87-93. DOI: 10.1007/s00253-006-0803-x

39. Perez, E R, Trujillo, L E, Arrieta, J G, Pérez, H, Brizuela, M A, Trujillo, G, Hernández, L. A pH shift-based procedure to screen fructooligosaccharides fermenting yeast or bacterial strains. Biotecnología Aplicada. 2010;27:216-220.

40. Maugeri, F, Hernalsteens, S. Screening of yeast strains for transfructosylating activity. Journal of Molecular Catalysis B: Enzymatic. 2007;49:43-49. DOI: 10.1016/j.molcatb.2007.08.001

41. Zambelli, P, Fernandez-Arrojo, L, Romano, D, Santos-Moriano, P, Gimeno-Perez, M, Poveda, A, Gandolfi, R, Fernández-Lobato, M, Molinari, F, Plou, F J. Production of fructooligosaccharides by mycelium-bound transfructosylation activity present in *Cladosporium cladosporioides* and *Penicillium sizovae*. Process Biochemistry. 2014;49:2174-2180. DOI: 10.1016/j.procbio.2014.09.021

42. Antosova, M, Illeova, V, Vandakova, M, Druzkovska, A, Polakovic, M. Chromatographic separation and kinetic properties of fructosyltransferase from *Aureobasidium pullulans*. Journal of Biotechnology. 2008;135:58-63. DOI: 10.1016/j.jbiotec.2008.02.016

43. Maiorano, A E, Piccoli, R M, Silva, E S, Rodrigues, M F A. Microbial production of fructosyltransferases for synthesis of pre-biotics. Biotechnol Letters. 2008;30:1867-1877. DOI: 10.1007/s10529-008-9793-3

44. Nemukula, A, Mutanda, T, Wilhelmi, B S, Whiteley, C G. Response surface methodology: Synthesis of short chain fructooligosaccharides with a fructosyltransferase from *Aspergillus aculeatus*. Bioresource Technology. 2009;100:2040-2045. DOI: 10.1016/j.biortech.2008.10.022

45. Yoshikawa, J, Amachi, S, Shinoyama, H, Fujii, T. Purification and some properties of β-fructofuranosidase I formed by *Aureobasidium pullulans* DSM 2404. Journal of Bioscience and Bioengineering. 2007;103:491-493. DOI: 10.1263/jbb.103.491

46. Yoshikawa, J, Amachi, S, Shinoyama, H, Fujii, T. Production of fructooligosaccharides by crude enzyme preparations of β-fructofuranosidase from *Aureobasidium pullulans*. Biotechnology Letters. 2008;30:535-539. DOI: 10.1007/s10529-007-9568-2

47. Jung, K H, Yun, J W, Kang, K R, Lim, J Y, Lee, J H. Mathematical model for enzymatic production of fructo-oligosaccharides from sucrose. Enzyme and Microbial Technology. 1989;11:491-494. DOI: 10.1016/0141-0229(89)90029-X

48. Duan, K J, Chen, J S, Sheu, D C. Kinetic studies and mathematical model for enzymatic production of fructooligosaccharides from sucrose. Enzyme and Microbial Technology. 1994;16:334-339. DOI: 10.1016/0141-0229(94)90176-7

49. Sheu, D C, Lio, P J, Chen, S T, Lin, C T., Duan, K J. Production of fructooligosaccharides in high yield using a mixed enzyme system of β-fructofuranosidase and glucose oxidase. Biotechnology Letters. 2001;23:1499-1503. DOI: 10.1023/A:1011689531625

50. Vega, R, Zuniga-Hansen, M E. Enzymatic synthesis of fructooligosaccharides with high 1-kestose concentrations using response surface methodology. Bioresource Technology. 2011;102:10180-10186. DOI:10.1016/j.biortech.2011.09.025

51. Hang, Y D, Woodams, E E. Fructosyltransferase activity of commercial enzyme preparations used in fruit juice processing. Biotechnology Letters. 1995;17:741-745. DOI: 10.1007/BF00130361

52. Tanriseven, A, Gokmen, F. Novel method for the production of a mixture containing fructooligosaccharides and isomaltooligosaccharides. Biotechnology Techniques. 1999;13:207-210. DOI: 10.1023/A:1008961016065

53. Ghazi, I, Fernandez-Arrojo, L, Gomez De Segura, A, Alcalde, M, Plou, F J, Ballesteros, A. Beet sugar syrup and molasses as low-cost feedstock for the enzymatic production of fructo-oligosaccharides. Journal of Agricultural and Food Chemistry. 2006;54:2964-2968. DOI: 10.1021/jf053023b

54. Madlova, A, Antosova, M, Barathova, M, Polakovic, M, Stefuca, V, Bales, V. Screening of microorganisms for transfructosylating activity and optimization of biotransformation of sucrose to fructooligosacharides. Chemical Papers. 1999;53:366-369.

55. Vega-Paulino, R J, Zuniga-Hansen, M E. Potential application of commercial enzyme preparations for industrial production of short-chain fructooligosaccharides. Journal of Molecular Catalysis B: Enzymatic. 2012;76:44-51. DOI: 10.1016/j.molcatb.2011.12.007

56. Guisan, J M. Immobilization of Enzymes as the 21st Century Begins. An Already Solved Problem or Still an Exciting Challenge? In: Guisan, J M, editor. Immobilization of Enzymes and Cells. Second Edition. Totowa: Humana Press; 2006. p. 1-13. ch1.

57. Sheldon, R A, van Pelt, S. Enzyme immobilisation in biocatalysis: why, what and how. Chemical Society Reviews. 2013;42:6223-6235. DOI: 10.1039/c3cs60075k

58. Tanriseven, A, Aslan, Y. Immobilization of Pectinex Ultra SP-L to produce fructooligosaccharides. Enzyme and Microbial Technology. 2005;36:550-554. DOI: 10.1016/j.enzmictec.2004.12.001

59. Ghazi, I, De Segura, A G, Fernandez-Arrojo, L, Alcalde, M, Yates, M, Rojas-Cervantes, M L, Plou, F J, Ballesteros, A. Immobilisation of fructosyltransferase from *Aspergillus aculeatus* on epoxy-activated Sepabeads EC for the synthesis of fructo-oligosaccharides. Journal of Molecular Catalysis B: Enzymatic. 2005;35:19-27. DOI: 10.1016/j. molcatb.2005.04.013

60. Fernandez-Arrojo, L, Rodriguez-Colinas, B, Gutierrez-Alonso, P, Fernandez-Lobato, M, Alcalde, M, Ballesteros, A O, Plou, F J. Dried alginate-entrapped enzymes (DALGEEs) and their application to the production of fructooligosaccharides. Process Biochemistry. 2013;48:677-682. DOI: 10.1016/j.procbio.2013.02.015

61. Lorenzoni, A S G, Aydos, L F, Klein, M P, Rodrigues, R C, Hertz, P F. Fructooligosaccharides synthesis by highly stable immobilized β-fructofuranosidase from *Aspergillus aculeatus*. Carbohydrate Polymers. 2014;103:193-197. DOI: 10.1016/j.carbpol.2013.12.038

62. Dominguez, A, Nobre, C, Rodrigues, L R, Peres, A M, Torres, D, Rocha, I, Lima, N, Teixeira, J. New improved method for fructooligosaccharides production by *Aureobasidium pullulans*. Carbohydrate Polymers. 2012;89:1174-1179. DOI: 10.1016/j.carbpol.2012.03.091

63. Chien, C-S, Lee, W-C, Lin, T-J. Immobilization of *Aspergillus japonicus* by entrapping cells in gluten for production of fructooligosaccharides. Enzyme and Microbial Technology. 2001;29:252-257. DOI: 10.1016/ S0141-0229(01)00384-2

64. Mussatto, S I, Aguilar, C N, Rodrigues, L R, Teixeira, J A. Fructooligosaccharides and β-fructofuranosidase production by *Aspergillus japonicus* immobilized on lignocellulosic materials. Journal of Molecular Catalysis B: Enzymatic. 2009;59:76-81. DOI: 10.1016/j.molcatb.2009.01.005

65. Cheng, C Y, Duan, K J, Sheu, D C, Lin, C T, Li, S Y. Production of fructooligosaccharides by immobilized mycelium of *Aspergillus japonicus*. Journal of Chemical Technology and Biotechnology. 1996;66:135-138. DOI: 10.1002/(SICI)1097-4660(199606)66:2<135::AID-JCTB479>3.0.CO;2-S

66. Cruz, R, Cruz, V D, Belini, M Z, Belote, J G, Vieira, C R. Production of fructooligosaccharides by the mycelia of *Aspergillus japonicus* immobilized in calcium alginate. Bioresource Technology.

1998;65:139-143. DOI: 10.1016/S0960-8524(98)00005-4

67. Charalampopoulos D, Rastall R A. Prebiotics in foods. Current Opinion in Biotechnology. 2012;23:187-191. DOI; 10.1016/j.copbio.2011.12.028

68. Lopez, H W, Coudray, C, Levrat-Verny, M-A, Feillet-Coudray, C, Demigne, C, Remesy, C. Fructooligosaccharides enhance mineral apparent absorption and counteract the deleterious effects of phytic acid on mineral homeostasis in rats. Journal of Nutrition and Biochemistry. 2000;11:500-508. DOI: 10.1016/S0955-2863(00)00109-1

69. Biedrzycka, E, Bielecka, M. Prebiotic effectiveness of fructans of different degrees of polymerization. Trends in Food Science and Technology. 2004;15:170-175. DOI: 10.1016/j.tifs.2003.09.014

70. Kaplan, H, Hutkins, R W. Fermentation of fructooligosaccharides by lactic acid bacteria and bifidobacteria. Applied and Environmental Microbiology. 2000;66:2682-2684.

71. Kaplan, H, Hutkins, R W. Metabolism of fructooligosaccharides by *Lactobacillus paracasei* 1195. Applied and Environmental Microbiology. 2003;69:2217-2222. DOI: 10.1128/AEM.69.4.2217-2222.2003

72. De Vuyst, L, Moens, F, Selak, M, Riviere, A, Leroy, F. Summer Meeting 2013: growth and physiology of bifidobacteria. Journal of Applied Microbiology. 2013;116:477-491. DOI: 10.1111/jam.12415

73. Mabel, M J, Sangeetha, P T, Platel, K, Srinivasan, K, Prapulla, S G. Physicochemical characterization of fructooligosaccharides and evaluation of their suitability as a potential sweetener for diabetics. Carbohydrate Research. 2008;343:56-66. DOI: 10.1016/j.carres.2007.10.012

74. Marx, S P, Winkler, S, Hartmeier, W. Metabolization of beta-(2,6)-linked fructose-oligosaccharides by different bifidobacteria. FEMS Microbiology Letters 2000;182:163-169. DOI: 10.1111/j.1574-6968.2000.tb08891.x

75. Grizard, D, Barthomeuf, C. Enzymatic synthesis and structure determination of NEO-FOS. Food Biotechnology. 1999;13:93-105. DOI: 10.1080/08905439609549963

76. Holsinger, V H. Lactose. In: Wong, N P, Jenness, R, Keeney, M, Marth, E H, editors. Fundamentals of Dairy Chemistry. 3rd ed. New York: Van Nostrand Reinhold Co.; 1988. p. 279-342.

77. Yang, S T, Silva, E M. Novel products and new technologies for use of a familiar carbohydrate, milk lactose. Journal of Dairy Science. 1995;78:2541-2562.

78. Paterson, A H J. Production and uses of lactose. In: McSweeney, P L H, Fox, P F, editors. Lactose, Water, Salts and Minor Constituents. 3rd ed. New York: Springer; 2009. p 105-120.

79. Fox, P F. Lactose: Chemistry and Properties. In: McSweeney, P L H, Fox, P F, editors. Lactose, Water, Salts and Minor Constituents. 3rd ed. New York: Springer; 2009. p.1–15. DOI: 10.1007/978-0-387-84865-5Fox 2009

80. Schaafsma, G. Lactose and lactose derivatives as bioactive ingredients in human nutrition. International Dairy Journal. 2008;18:458-465. DOI: 10.1016/j.idairyj.2007.11.013Schaafsma 2008

81. Paige, D M. Lactose Intolerance. In: Caballero, B, Allen, L, Prentice, A, editors. Encyclopedia of Human Nutrition. 2nd ed. Oxford: Elsevier Ltd.; 2005. p. 113-120.

82. Mahoney, R R. Galatosyl-oligosaccharide formation during lactose hydrolysis: a review. Food Chemistry. 1998;63:147-154. DOI: 10.1016/S0308-8146(98)00020-X

83. Wallenfels, K, Malhotra, O P. Beta-galactosidase. In: Boyer, P D, editor. The Enzymes. 2nd ed. New York: Academic Press Inc.; 1960. p. 409-430.

84. Lomer, M C E, Parkes, G C, Sanderson, J D. Review article: Lactose intolerance in clinical practice – Myths and realities. Alimentary Pharmacology and Therapeutics. 2008;27:93-103. DOI: 10.1111/j.1365-2036.2007.03557.x

85. Tzortzis, G, Vulevic, J. Galacto-oligosaccharide Prebiotics. In: Charalampopoulos, D, Rastall, R A, editors. Prebiotics and Probiotics Science and Technology. New York: Springer; 2009. p. 207-244.

86. Villamiel, M, Montilla, A, Olano, A, Corzo, N. Production and Bioactivity of Oligosaccharides Derived from Lactose. In: Moreno, F J, Sanz, M L, editors. Food Oligosaccharides: Production, Analysis and Bioactivity. John Wiley & Sons. 2014, p. 137.

87. Oliveira, C, Guimarães, P M R, Domingues, L. Recombinant microbial systems for improved β-galactosidase production and biotechnological applications. Biotechnology Advances. 2011;29:600-609. DOI: 10.1016/j.biotechadv.2011.03.008

88. Torres, D P M, Gonçalves, M F, Teixeira, J A, Rodrigues, L R. Galactooligosaccharides: production, properties, applications, and significance as prebiotics. Comprehensive Reviews in Food Science and Food Safety. 2010;9:438-454. DOI: 10.1111/j.1541-4337.2010.00119.x

89. Martinez-Villaluenga, C, Cardelle-Cobas, A, Corzo, N, Olano, A, Villamiel,

M. Optimization of conditions for galactooligosaccharides synthesis during lactose hydrolysis by β-galactosidase from *Kluyveromyces lactis* (Lactozym 3000 L HP G). Food Chemistry. 2008;107:258-264. DOI: 10.1016/j.foodchem.2007.08011

90. Hatzinikolaou, D G, Katsifas, E, Mamma, D, Karagouni, A D, Christakopoulos, P, Kekos, D. Modeling of the simultaneous hydrolysis–ultrafiltration of whey permeate by a thermostable β-galactosidase from *Aspergillus niger*. Biochemical Engineering Journal. 2005;24:161-172. DOI: 10.1016/j.bej.2005.02.011

91. Jurado, E, Camacho, F, Luzón, G, Vicaria, J M. Kinetic models of activity for β-galactosidases: influence of pH, ionic concentration and temperature. Enzyme and Microbial Technology. 2004;34:33-40. DOI: 10.1016/j.enzmictec.2003.07.004

92. Buchholz, K, Kasche, V, Bornscheuer, U T. Equilibrium and Kinetically Controlled Reactions Catalysed by Enzymes. In: Biocatalysts and Enzyme Technology. Weinheim: Wiley-VCH Verlag GmbH& Co.; 2005. p. 42-46.

93. Chen, S-X, Wei, D-Z, Hu, Z-H. Synthesis of galacto-oligosaccharides in AOT/isooctane reverse micelles by beta-galactosidase. Journal of Molecular Catalysis B: Enzymatic. 2001;16:109-114. DOI: 10.1016/S1381-1177(01)00051-0

94. Chen, C W, Ou-Yang, C C, Yeh, C W. Synthesis of galactooligosaccharides and transgalactosylation modeling in reverse micelles. Enzyme and Microbial Technology. 2003;33:497-507. DOI: 10.1016/S0141-0229(03)00155-8

95. Roos, Y H. Solid and Liquid States of Lactose. In: McSweeney, P L H, Fox, P F, editors. Lactose, Water, Salts and Minor Constituents. 3rd ed. New York: Springer; 2009. p. 17-33.

96. Bruins, M E, van Hellemond, E W, Janssen, A E M, Boom, R M. Maillard reactions and increased enzyme inactivation during oligosaccharide synthesis by a hyperthermophilic glycosidase. Biotechnology and Bioengineering. 2003;81:546-552. DOI: 10.1002/bit.10498

97. Bruins, M E, Janssen, A E M, Boom, R M. Thermozymes and their applications—a review of recent literature and patents. Applied Biochemistry and Biotechnology. 2001;90:155-186. DOI: 10.1385/ABAB:90:2:155

98. Boon, M A, van der Oost, J, de Vos, A E M, van't Riet, K. Synthesis of oligosaccharides catalyzed by thermostable β-glucosidase from *Pyrococcus furiosus*. Applied Biochemistry and Biotechnology.

1998;75: 269-278. DOI: 10.1007/BF02787780

99. Chen, W, Chen, H, Xia, Y, Yang, J, Zhao, J, Tian, F, Zhang, H P, Zhang, H. Immobilization of recombinant thermostable β-galactosidase from*Bacillus stearothermophilus* for lactose hydrolysis in milk Journal of Dairy Science. 2009;92:491-498. DOI: 10.3168/jds.2008-1618

100. Petzelbauer, I, Reiter, A, Splechtna, B, Kosma, P, Nidetzky, B. Transgalactosylation by thermostable β-glycosidases from *Pyrococcus furiosus* and*Sulfolobus solfataricus*. Binding interactions of nucleophiles with the galactosylated enzyme intermediate makes major contributions to the formation of new beta-glycosides during lactose conversion. European Journal of Biochemistry. 2000;267:5055-5066. DOI: 10.1046/j.1432-1327.2000.01562.x

101. Petzelbauer I, Zeleny R, Reiter A, Kulbe D, Nidetzky B. Development of an ultra-high-temperature process for the enzymatic hydrolysis of lactose: II. Oligosaccharide formation by two thermostable β-glycosidases. Biotechnology and Bioengineering. 2000;69:140-149. DOI: 10.1002/(SICI)1097-0290(20000720)692<140::AID-BIT3>3.0.CO;2-R

102. Reuter, S, Rusborg, N A, Zimmermann, W. β-Galactooligosaccharide synthesis with β-galactosidases from *Sulfolobus solfataricus*, *Aspergillus oryzae*, and *Escherichia coli*. Enzyme and Microbial Technology. 1999;25:509-516. DOI: 10.1016/S0141-0229(99)00074-5

103. Chen, W, Chen, H, Xia, Y, Zhao, J, Tian, F, Zhang, H. Production, purification, and characterization of a potential thermostable galactosidase for milk lactose hydrolysis from *Bacillus stearothermophilus*. Journal of Dairy Science. 2008;91:1751-1758. DOI: 10.3168/jds.2007/617

104. Hansson, T, Kaper, T, van der Oost, J, de Vos, W M, Adlercreutz, P. Improved oligosaccharide synthesis by protein engineering of beta-glucosidase CelB from hyperthermophilic *Pyrococcus furiosus*. Biotechnology and Bioengineering. 2001;73:203-210. DOI: 10.1002/bit.1052

105. Maugard T, Gaunt D, Legoy MD, Besson T. Microwave-assisted synthesis of galacto-oligosaccharides from lactose with immobilized β-galactosidase from *Kluyveromyces lactis*. Biotechnology Letters.2003;25:623-629. DOI: 10.1023/A:1023060030558

106. Gosling, A, Stevens, G W, Barber, A R, Kentish, S E, Gras, S L. Recent advances refining galactooligosaccharide production from lactose. Food Chemistry. 2010;121:307-318. DOI: 10.1016/j.foodchem.2009.12.063

107. Huber RE, Kurz G, Wallenfels K. A quantitation of the factors which affect the hydrolase and transgalactosylase activities of beta-galactosidase (*E.*

coli) on lactose. Biochemistry.1976;15:1994-2001.

108. Hsu, C-A, Yu, R-C, Chou, C-C. Purification and characterization of a sodium-stimulated β-galactosidase from *Bifidobacterium longum* CCRC 15708. World Journal of Microbiology and Biotechnology. 2006;22:355-361. DOI 10.1007/s11274-005-9041-0

109. Seibela, J, Buchholz, K. Tools in Oligosaccharide Synthesis: Current Research and Application. Advances in Carbohydrate Chemistry and Biochemistry. 2010;63:101-138. DOI: 10.1016/S0065-2318(10)63004-1

110. Kunst, T. Protein Modification to Optimize Functionality: Protein Hydrolysates Galactosidase. In: Whitaker, J R, Voragen, A G J, Wong, D W S, editors. Handbook of Food Enzymology. New York: M. Dekker; 2003, p. 221-236.

111. Kawamoto, T, Tanaka, A. Entrapment of Biocatalysts by Prepolymer Methods. In: Whitaker, J R, Voragen, A G J, Wong, D W S, editors. Handbook of Food Enzymology. New York: M. Dekker; 2003. p. 331-342.

112. Urrutia, P, Mateo, C, Guisan, J M, Wilson, L, Illanes, A. Immobilization of *Bacillus circulans* β-galactosidase and its application in the synthesis of galacto-oligosaccharides under repeated-batch operation. Biochemical Engineering Journal. 2013;77:41-48. DOI: 10.1016/j.bej.2013.04.015

113. Palai, T, Singh, A K, Bhattacharya, P K. Enzyme β-galactosidase immobilized on membrane surface for galacto-oligosaccharides formation from lactose: Kinetic study with feed flow under recirculation loop. Biochemical Engineering Journal. 2014;88:68-76. DOI: 10.1016/j.bej.2014.03.017

114. Lu. L, Xu, S, Zhao, R, Zhang, D, Li, Z, Li, Y, Xiao, M. Synthesis of galactooligosaccharides by CBD fusion β-galactosidase immobilized on cellulose. Bioresource Technology. 2012;116:327-333. DOI: 10.1016/j.biortech.2012.03.108

115. Warmerdam, A, Benjamins, E, de Leeuw, T F, Broekhuis, T A, Boom, R M, Janssen, A E M. Galacto-oligosaccharide production with immobilized β-galactosidase in a packed-bed reactor vs. free β-galactosidase in a batch reactor. Food and Bioproducts Processing. 2014;92:383-392. DOI: 10.1016/j.fbp.2013.08.014

116. Ivanov, A E, Edink, E, Kumar, A, Galaev, I Y, Arendsen, A F, Bruggink, A, Mattiasson, B. Conjugation of penicillin acylase with the reactive copolymer of N-isopropylacrylamide: a step toward a thermosensitive industrial biocatalyst. Biotechnology Progress. 2003;19:1167-1175. DOI: 10.1021/bp0201455

117. Ward MA, Georgiu TK. Thermoresponsive polymers for biomedical applications. Polymers. 2011;3:1215-1242 DOI: 10.3390/polym3031215

118. Palai T, Kumar A, Bhattacharya PK. Synthesis and characterization of thermo-responsive poly-N-isopropylacrylamide bioconjugates for application in the formation of galacto-oligosaccharides. Enzyme and Microbial Technology. 2014;55:40-49. DOI:10.1016/j.enzmitec.2013.12.003

119. Palai T, Kumar A, Bhattacharya PK. Kinetic studies and model development for the formation ofgalacto-oligosaccharides from lactose using synthesized thermo-responsive bioconjugate. Enzyme and Microbial Technology. 2015;70:42-49. DOI: 10.1016/j.enzmictec.2014.12.010

120. Torres DPM, Gonçalves MPF, Teixeira JA, Rodrigues L R. Galacto-Oligosaccharides: Production, Properties, Applications, and Significance as Prebiotics. Comprehensive Reviews in Food Science and Food Safety. 2010;9:438-454. DOI: 10.1111/j.1541-4337.2010.00119.x

121. Guerrero C, Vera C, Novoa C, Dumont J, Acevedo F, Illanes A. Purification of highly concentrated galacto-oligosaccharide preparations by selective fermentation with yeasts. International Dairy Journal.2014;39:78-88. DOI: 10.1016/j.idairyj.2014.05.011

122. . Venica CI, Bergamini C V, Rebechi S R, Perotti M C Galacto-oligosaccharides formation during manufacture of different varieties of yogurt. Stability through storage LWT - Food Science and Technology. DOI: 10.1016/j.lwt.2015.02.032

123. Nauta A, Bakker-Zierikzee AM, Schoterman MHC. Galacto-Oligosaccharides. In: Cho, SS, Finocchiaro, ET, editors. Handbook of prebiotics and probiotics ingredients: health benefits and food applications. Boca Raton: Taylor and Francis Group;2010. p. 75-93.

124. Shoaf K, Mulvey GL, Armstrong GD, Hutkins RW. Prebiotic galactooligosaccharides reduce adherence of enteropathogenic Escherichia coli to tissue culture cells, Infection and. Immunity. 2006;74:6920-6928. DOI: 10.1128/IAI.01030-06

125. Rowland IR, Tanaka R. The effects of transgalactosylated oligosaccharides on gut flora metabolism in rats associated with a human fecal microflora, Journal of Applied Bacteriology. 1993;74:667-674. DOI: 10.1111/j.1365-2672.1993.tb05201.x

126. Rivero-Urgell M, Santamaria-Orleans A. Oligosaccharides: application in infant food. Early Human Development. 2001;65:S43-S52. DOI: 10.1016/S0378-3782(01)00202-X

127. Kim SK, Mendis E. Bioactive compounds from marine processing

byproducts – A review. Food Research International. 2006;39:383-393. DOI: 10.1016/j.foodres.2005.10.010

128. Adrangi S, Faramarsi MA. From bacteria to human: A journey into the world of chitinases. Biotechnology Advances. 2013;31:1786-1795. DOI: 10.1016/j.biotechadv.2013.09.012

129. Ordonez-Del Pazo T, Antelo LT, Franco-Uria A, Perez-Martin RI, Sotelo C.G, Alonso AA. Fish discards management in selected Spanish and Portuguese metiers: Identification and potential valorization. Trends in Food Science & Technology. 2014;36:29-43. DOI: 10.1016/j.tifs.2013.12.006

130. Newton R, Telfer T, Little D. Perspectives on the utilization of aquaculture coproduct in Europe and Asia: Prospects for value addition and improved resource efficiency. Critical Reviews in Food Science and Nutrition. 2014;54:495-510. DOI:10.1080/10408398.2011.588349

131. Jung W.-J.; Park R.-D. Bioproduction of Chitooligosaccharides: Present and Perspectives. Marine Drugs. 2014;12:5328-5356. DOI: 10.3390/md12105328

132. Yang Y, Yu B. Recent advances in the synthesis of chitooligosaccharides and congeners. Tetrahedron. 2014;70:1023-1046. DOI: 10.1016/j.tet.2013.11.064

133. Mourya VK, Inamdar NN, Choudhari YM. Chitooligosaccharides: Synthesis, Characterization and Applications. Polymer Science, Serie A. 2011;53:583-612. DOI: 10.1134/S0965545X11070066

134. Cheng, CY, Chang CH, Wu YJ, Li YK. Exploration of Glycosyl Hydrolase Family 75, a Chitosanase from *Aspergillus fumigatus*. The Journal of Biological Chemistry. 2006;281:3137-3144. DOI: 10.1074/jbc.M512506200

135. Fernandes de Assis, C, Araujo, N K, Pagnoncelli, M G B, da Silva Pedrini, M R, Ribeiro de Macedo, G, dos Santos, E S. Chitooligosaccharides enzymatic production by *Metarhizium anisopliae*. Bioprocess and Biosystem Engineering 2010;33:893-899. DOI: 10.1007/s00449-010-0412-z

136. Gao XA, Zhang YF, Park RD, Huang X, Zhao XY, Xie J, Jin RD. Preparation of chitooligosaccharides from chitosan using crude enzyme of Bacillus cereus D-11. Journal of Applied Biological Chemistry. 2012;55:13-17. DOI: 10.3839/jabc.2011.053

137. Ming M, Kuroiwa T, Ichikawa S, Sato S, Mukataka S. Production of chitosan-oligosaccharides at high concentration by immobilized chitosanase. Food Science and Technology Research. 2006;12:85-90.

138. Kuroiwa T, Izuta H, Nabetani H, Nakajima M, Sato S, Mukataka S, Ichikawa S. Selective and stable production of physiologically active chitosan oligosaccharides using an enzymatic membrane bioreactor. Process Biochemistry. 2009;44:283-287. DOI: 10.1016/j.procbio.2008.10.020

139. No HK, Young PN, Ho LS, Meyers SP. Antibacterial activity of chitosans and chitosan oligomers with different molecular weights. International Journal of Food Microbiology. 2002;74:65-72. DOI: International Journal of Food Microbiology

140. Li K; Xing R, Liu S, Qin Y, Yu H, Li P. Size and pH effects of chitooligomers on antibacterial activity against *Staphylococcus aureus*. International Journal of Biological Macromolecules. 2014;64:302-305. DOI: 10.1016/j.ijbiomac.2013.11.037

141. Chung YCh, Su YP, Chen, CC, Jia G, Wang HL, Gaston Wu JC, Lin JG. Relationship between antibacterial activity of chitosan and surface characteristics of cell wall. Acta Pharmacologica Sinica. 2004;25:932-936.

142. Benhabiles MS, Salah R, Lounici H, Drouiche N, Goosen MFA, Mameri N. Antibacterial activity of chitin, chitosan and its oligomers prepared from shrimp shell waste. Food Hydrocolloids. 2012;29:48-56. DOI: 10.1016/j.foodhyd.2012.02.013

143. Wang, Y.; Zhou, P.; Yu, J.; Pan, X.; Wang, P.; Lan, W.; Tao, Sh. Antimicrobial effect of chitooligosaccharides produced by chitosanase from Pseudomonas CUY8. Asia Pacific Journal of Clinical Nutrition. 2007;16:174-177.

144. Ngo DH, Wijesekara I, Vo TS.; Tan QV, Kim SK. Marine food-derived functional ingredients as potential antioxidants in the food industry: An overview. Food Research International. 2011;44:523-529. DOI: 10.1016/j.foodres.2010.12.030

145. Je JY, Park PJ, Kim SK. Free radical scavenging properties of heterochitooligosaccharides using an ESR spectroscopy. Food and Chemical Toxicology. 2004;42:381-387. DOI: 10.1016/j.fct.2003.10.001

146. Fernandes JC, Eaton P, Nascimento H, Gião MS, Ramos ÓS, Belo L, Santos-Silva A, Pintado ME, Malcata FX. Antioxidant activity of chitooligosaccharides upon two biological systems: Erythrocytes and bacteriophages. Carbohydrate Polymers. 2010;79:1101-1106. DOI: 10.1016/j.carbpol.2009.10.050

147. Kim SK, Rajapaksea Ni. Enzymatic production and biological activities of chitosan oligosaccharides (COS): A review. Carbohydrate Polymers.

2005;62:357-368. DOI: 10.1016/j.carbpol.2005.08.012

148. Halder SK, Adak A, Maity C, Jana A, Das A, Paul T, Ghosh K, Mohapatra PKD, Pati BR, Mondal KC. Exploitation of fermented shrimp-shells hydrolysate as functional food: Assessment of antioxidant, hypocholesterolemic and prebiotic activities. Indian Journal of Experimental Biology. 2013;51:924-934.

149. Eom TK, Senevirathnea M, Kim SK. Synthesis of phenolic acid conjugated chitooligosaccharides and evaluation of their antioxidant activity. environmental toxicology and pharmacology. 2012;34:519-527. DOI: 10.1016/j.etap.2012.05.004

150. Lee HW, Park YS, Jung JS, Shin WS. Chitosan oligosaccharides, dp 2–8, have prebiotic effect on the *Bifidobacterium bifidium* and *Lactobacillus*sp. Anaerobe. 2002;8:319-324. DOI: 10.1016/S1075-9964(03)00030-1

151. Koppová I, Bureš M, Šimůnek J. Intestinal bacterial population of healthy rats during the administration of chitosan and chitooligosaccharides. Folia Microbiologica. 2012;57:295-299. DOI: 10.1007/s12223-012-0129-2

152. Pan X, Chen F, Wu T, Tang H, Zhao Z. Prebiotic oligosaccharides change the concentrations of short-chain fatty acids and the microbial population of mouse bowel. Journal of Zhejiang University SCIENCE B. 2009;10:258-263.

153. Gibson, G R, Rastall, R A. Prebiotics: Development & Application. Chichester: John Wiley & Sons Ltd; 2006.

154. Nguyen, T H, Haltrich, D. Microbial production of prebiotic oligosaccharides. In: McNeil, B, Archer, D, Giavasis, I, Harvey, L. Microbial Production of Food Ingredients, Enzymes and Nutraceuticals. Woodhead Publishing; 2013. p. 494-530. DOI: 10.1533/9780857093547.2.494. ch18.

155. Zhang, L, Su, Y, Zheng, Y, Jiang, Z, Shi, J, Zhu, Y, Jiang, Y. Sandwich structured enzyme membrane reactor for efficient conversion of maltose into isomaltooligosaccharides. Bioresource Technology. 2010;101:9144-9149. DOI: 10.1016/j.biortech.2010.07.001

156. Bharti, S K, Kumar, A, Krishnan, S, Gupta, A K, Kumar, A. Mechanism-based antidiabetic activity of Fructo- and isomalto-oligosaccharides: Validation by *in vivo*, *in silico* and *in vitro* interaction potential. Process Biochemistry. 2015;50:317-327. DOI: 10.1016/j.procbio.2014.10.014

157. Qiang, X, YongLie, C, QianBing, W. Health benefit application of functional oligosaccharides. Carbohydrate Polymers. 2009;77:435-441. DOI: 10.1016/j.carbpol.2009.03.016

158. Meyer, D. Chapter Two - Health Benefits of Prebiotic Fibers. Advances in Food and Nutrition Research. 2015;74:47-91.DOI: 10.1016/bs.afnr.2014.11.002

159. Candela, M, Maccaferri, S, Turroni, S, Carnevali, P, Brigidi, P. Functional intestinal microbiome, new frontiers in prebiotic design. International Journal of Food Microbiology. 2010;140:93-101. DOI: 10.1016/j.ijfoodmicro.2010.04.017

160. Vazquez, M J, Alonso, J L, Domınguez, H, Parajo, J C. Xylooligosaccharides: manufacture and applications. Trends in Food Science and Technology. 2000;11:387-393.

161. Carvalho, A F A, Oliva Neto, P, Silva, D F, Pastore, G M. Xylo-oligosaccharides from lignocellulosic materials: Chemical structure, health benefits and production by chemical and enzymatic hydrolysis. Food Research International. 2013;51:75-85. DOI: 10.1016/j.foodres.2012.11.021

162. Chung, Y C, Hsu, C K, Ko, C Y, Chan, Y C. Dietary intake of xylooligosaccharides improves the intestinal microbiota, fecal moisture, and pH value in the elderly. Nutrition Research. 2007;27:756-761. DOI: 10.1016/j.nutres.2007.09.014

163. Moura, P, Cabanas, S, Lourenço, P, Girio, F, Loureiro-Dias, M C, Esteves, M P. *In vitro* fermentation of selected xylo-oligosaccharides by piglet intestinal microbiota. LWT - Food Science and Technology. 2008;41:1952-1961. DOI: 10.1016/j.lwt.2007.11.007

164. Veenashri, B R, Muralikrishna, G. *In vitro* anti-oxidant activity of xylo-oligosaccharides derived from cereal and millet brans – A comparative study. Food Chemistry. 2011;126:1475-1481. DOI: 10.1016/j.foodchem.2010.11.163

165. Broekaert, W F, Courtin, C M, Verbeke, K, Van De Wiele, T, Verstraete, W, Delcour, J A. Prebiotic and other health-related effects of cereal-derived arabinoxylans, arabinoxylan-oligosaccharides, and xylooligosaccharides. Critical Reviews in Food Science and Nutrition. 2011;51:178-194. DOI: 10.1080/10408390903044768

166. Grootaert, C, Delcour, J A, Courtin, C M, Broekaert, W F, Verstraete, W, Wiele, T V. Microbial metabolism and prebiotic potency of arabinoxylan oligosaccharides in the human intestine. Trends in Food Science and Technology. 2007;18:64-71. DOI: 10.1016/j.tifs.2006.08.004

167. Fialho, M B, Simões, K, Barros, C A, Bom Pessoni, R A, Braga, M R, Figueiredo-Ribeiro, R C L. Production of 6-kestose by the filamentous fungus*Gliocladium virens* as affected by sucrose concentration.

Mycoscience. 2013;54:198-205. DOI: 10.1016/j.myc.2012.09.012

168. Bekers, M, Laukevics, J, Upite, D, Kaminska, E, Vigants, A, Viesturs, U, Pankova, L, Danilevics, A. Fructooligosaccharide and levan producing activity of *Zymomonas mobilis* extracellular levansucrase. Process Biochemistry. 2002;38:701-706. DOI: 10.1016/S0032-9592(02)00189-9

169. Álvaro-Benito, M, Abreu, M, Fernández-Arrojo, L, Plou, F J, Jimenez-Barbero, J, Ballesteros, A, Polaina, J, Fernández-Lobato, M. Characterization of a β-fructofuranosidase from *Schwanniomyces occidentalis* with transfructosylating activity yielding the prebiotic 6-kestose. Journal of Biotechnology. 2007;132:75-81. DOI: 10.1016/j.jbiotec.2007.07.939

170. Marx, S, Winkler, S, Hartmeier, W. Metabolization of beta-(2,6)-linked fructose-oligosaccharides by different bifidobacteria. FEMS Microbiology Letters. 2000;182:163-169. DOI: 10.1111/j.1574-6968.2000.tb09376.x

171. Porras-Domínguez, J R, Ávila-Fernández, A, Rodríguez-Alegría, M E, Miranda-Molina, A, Escalante, A, González-Cervantes, R, Olvera, C, Munguía, A L. Levan-type FOS production using a *Bacillus licheniformis* endolevanase. Process Biochemistry. 2014;49:783-790. DOI: 10.1016/j.procbio.2014.02.005

172. Chen, J, Liang, R H, Liu, W, Li, T, Liu, C M, Wu, S S, Wang, Z J. Pectic-oligosaccharides prepared by dynamic high-pressure microfluidization and their *in vitro* fermentation properties. Carbohydrate Polymers. 2013;91:175-182. DOI: 10.1016/j.carbpol.2012.08.021

173. Gullon, B, Gomez, B, Martinez-Sabajanes, M, Yanez, R, Parajo, J C, Alonso, J L. Pectic oligosaccharides: Manufacture and functional properties. Trends in Food Science and Technology. 2013;30:153-161. DOI: 10.1016/j.tifs.2013.01.006

174. Gutiérrez-Román MI, Dunn MF, Tinoco-Valencia R, Holguín-Meléndez F, Huerta-Palacios G, Guillén-Navarro K. Potentiation of the synergistic activities of chitinases ChiA, ChiB and ChiC from Serratia marcescens CFFSUR-B2 by chitobiase (Chb) and chitin binding protein (CBP). World Journal of Microbiology Biotechnology. 2014;30:33-42. DOI: 10.1007/s11274-013-1421-2

175. Karan R, Cape MD, DasSarma S. Function and biotechnology of extremophilic enzymes in low water activity. Aquatic Biosystems. 2012;8:4.

176. Pinelo M, Jonsson G, Meyer AS. Membrane technology for purification of enzymatically produced oligosaccharides: Molecular and operational features affecting performance. Separation and Purification Technology.

2009;70:1-11. DOI:10.1016/j.seppur.2009.08.010

177. Cui, S W. Food Carbohydrates: Chemistry, Physical Properties, and Applications. CRC Press - Taylor & Francis Group. 2005.

178. Jakob, F, Pfaff, A, Novoa-Carballal, R, Rübsamc, H, Becker, T, Vogel, R F. Structural analysis of fructans produced by acetic acid bacteria reveals a relation to hydrocolloid function. Carbohydrate Polymers. 2013;92:1234-1242. DOI: 10.1016/j.carbpol.2012.10.054

179. Bruzzese, E, Volpicelli, M, Squeglia, V, Bruzzese, D, Salvini, F, Bisceglia, M, Lionetti, P, Cinquetti, M, Iacono, G, Amarri, S, Guarino, A. A formula containing galacto- and fructo-oligosaccharides prevents intestinal and extra-intestinal infections: An observational study. Clinical Nutrition. 2009;28:156-161. DOI: 10.1016/j.clnu.2009.01.008

180. Gomes, A M P, Malcata, F X. *Bifidobacterium* spp. and *Lactobacillus acidophilus*: biological, biochemical, technological and therapeutical properties relevant for use as probiotics. Trends in Food Science and Technology. 1999;10:139-157. DOI: 10.1016/S0924-2244(99)00033-3

181. Yen, C H, Kuo, Y W, Tseng, Y H, Lee, M C, Chen, H L. Beneficial effects of fructo-oligosaccharides supplementation on fecal bifidobacteria and index of peroxidation status in constipated nursing-home residents: A placebo-controlled, diet-controlled trial. Nutrition. 2011;27:323-328. DOI: 10.1016/j.nut.2010.02.009

182. Jakobsdottir, G, Nyman, M, Fak, F. Designing future prebiotic fiber to target metabolic syndrome. Nutrition. 2014;30:497-502. DOI: 10.1016/j. nut.2013.08.013

183. Choque Delgado, G T, Tamashiro, W M S C, Pastore, G M. Immunomodulatory effects of fructans. Food Research International. 2010;43:1231-1236. DOI: 10.1016/j.foodres.2010.04.023

184. Juskiewicz, J, Semaskaite, A, Zdunczyk, Z, Wroblewska, M, Gruzauskas, R, Juskiewicz, M. Minor effect of the dietary combination of probiotic*Pediococcus acidilactic*i with fructooligosaccharides or polysaccharidases on beneficial changes in the cecum of rats. Nutrition Research. 2007;27:133-139. DOI: 10.1016/j.nutres.2007.01.005

185. Licht, T R, Ebersbach, T, Frøkiær, H. Prebiotics for prevention of gut infections. Trends in Food Science and Technology. 2012;23:70-82. DOI: 10.1016/j.tifs.2011.08.011

186. Van den Heuvel, E G H M, Muijs, T, Brouns, F, Hendriks, H F J. Short-chain fructo-oligosaccharides improve magnesium absorption in adolescent girls with a low calcium intake. Nutrition Research. 2009;29:229-237. DOI: 10.1016/j.nutres.2009.03.005

187. Losada, M A, Olleros, T. Towards a healthier diet for the colon: the influence of fructooligosaccharides and lactobacilli on intestinal health. Nutrition Research. 2002;22:71-84. DOI: 10.1016/S0271-5317(01)00395-5

188. Niness, K R. Inulin and Oligofructose: What Are They? The Journal of Nutrition. 1999; Supplement:1402S-1406S.

189. Glibowski, P, Pikus, S, Jurek, J, Kotowoda, M. Factors affecting inulin crystallization after its complete dissolution. Carbohydrate Polymers. 2014;110:107-112. DOI: 10.1016/j.carbpol.2014.03.080

190. Morris C, Morris GA. The effect of inulin and fructo-oligosaccharide supplementation on the textural, rheological and sensory properties of bread and their role in weight management: A review. Food Chemistry. 2012;133:237-248. DOI: 10.1016/j.foodchem.2012.01.027

191. Beserra, B T S, Fernandes, R, Rosario, V A, Mocellin, M C, Kuntz, M G F, Trindade, E B S M. A systematic review and meta-analysis of the prebiotics and synbiotics effects on glycaemia, insulin concentrations and lipid parameters in adult patients with overweight or obesity. Clinical Nutrition. DOI: 10.1016/ j.clnu.2014.10.004

192. Karimi, R, Azizi, M H, Ghasemlou, M, Vaziri, M. Application of inulin in cheese as prebiotic, fat replacer and texturizer: A review. Carbohydrate Polymers. DOI: 10.1016/j.carbpol.2014.11.029

193. Oliveira, R P S, Perego, P, Oliveira, M N, Converti, A. Growth, organic acids profile and sugar metabolism of *Bifidobacterium lactis* in co-culture with *Streptococcus thermophilus*: The inulin effect. Food Research International. 2012;48:21-27. DOI: 10.1016/j.foodres.2012.02.012

194. Dehghan, P, Gargari, B P, Jafar-abadi, M A. Oligofructose-enriched inulin improves some inflammatory markers and metabolic endotoxemia in women with type 2 diabetes mellitus: A randomized controlled clinical trial. Nutrition. 2014;30:418-423. DOI: 10.1016/j.nut.2013.09.005

195. Apolinário, A C, Damasceno, B P G L, Beltrão, N E M, Pessoa, A, Converti, A, Silva, J A. Inulin-type fructans: A review on different aspects of biochemical and pharmaceutical technology. Carbohydrate Polymers. 2014;101:368-378. DOI: 10.1016/j.carbpol.2013.09.081

196. Adebola, O, Corcoran, O, Morgan, W A. Protective effects of prebiotics inulin and lactulose from cytotoxicity and genotoxicity in human colon adenocarcinoma cells. Food Research International. 2013;52:269-274. DOI: 10.1016/j.foodres.2013.03.024

197. Martínez-Villaluenga, C, Cardelle-Cobas, A, Corzo, N, Olano, A. Study of galactooligosaccharide composition in commercial fermented milks. Journal of Food Composition and Analysis. 2008;21:540-544.

DOI:10.1016/j.jfca.2008.05.008

198. Crittenden, R G, Playne, M J. Production, properties and application of food grade oligosaccharides. Trends in Food Science & Technology. 1996;71:353-361.

199. Sarabia-Sainz, H M, Armenta-Ruiz, C, Sarabia-Sainz, J A, Guzmán-Partida, A M, Ledesma-Osuna, A I, Vázquez-Moreno, L, Montfort, G R C. Adhesion of enterotoxigenic *Escherichia coli* strains to neoglycans synthesised with prebiotic galactooligosaccharides. Food Chemistry. 2013;141:2727-2734. DOI:10.1016/j.foodchem.2013.05.040

200. Puccio, G, Cajozzo, C, Meli, F, Rochat, F, Grathwohl, D, Steenhout, P. Clinical evaluation of a new starter formula for infants containing live*Bifidobacterium longum* BL999 and prebiotics. Nutrition. 2007;23:1-8. DOI:10.1016/j.nut.2006.09.007

201. Zhong, Y, Cai, D, Cai, W, Geng, S, Chen, L, Han, T. Protective effect of galactooligosaccharide-supplemented enteral nutrition on intestinal barrier function in rats with severe acute pancreatitis. Clinical Nutrition. 2009;28:575-580. DOI:10.1016/j.clnu.2009.04.026

202. Sangwan, V, Tomar, S K, Ali, B, Singh, R R B, Singh, A K. Galactooligosaccharides reduce infection caused by *Listeria monocytogenes* and modulate IgG and IgA levels in mice. International Dairy Journal. 2015;41:58-63. DOI:10.1016/j.idairyj.2014.09.010

203. Hernandez, O, Ruiz-Matute, A I, Olano, A, Moreno, F J, Sanz, M L. Comparison of fractionation techniques to obtain prebiotic galactooligosaccharides. International Dairy Journal. 2009;19:531-536. DOI:10.1016/j.idairyj.2009.03.002

204. Frenzel, M, Zerge, K, Clawin-Radecker, I, Lorenzen, P C. Comparison of the galacto-oligosaccharide forming activity of different β-galactosidases. LWT - Food Science and Technology. 2015;60:1068-1071. DOI:10.1016/j.lwt.2014.10.064

205. Tymczyszyn, E E, Sosa, N, Gerbino, E, Hugo, A, Gómez-Zavaglia, A, Schebor, C. Effect of physical properties on the stability of *Lactobacillus bulgaricus* in a freeze-dried galacto-oligosaccharides matrix. International Journal of Food Microbiology. 2012;155:217-221. DOI:10.1016/j.ijfoodmicro.2012.02.008

206. Bruno-Barcena, J M, Azcarate-Peril, M A. Galacto-oligosaccharides and colorectal cancer: Feeding our intestinal probiome. Journal of Functional Foods. 2015;12:92-108. DOI:10.1016/j.jff.2014.10.029

207. Torres, D P M, Bastos, M, Gonçalves, M P F, Teixeira, J A, Rodrigues, L R. Water sorption and plasticization of an amorphous galacto-

oligosaccharide mixture. Carbohydrate Polymers. 2011;83:831-835. DOI: 10.1111/j.1541-4337.2010.00119.x

208. Adebola, O, Corcoran, O, Morgan, W,A. Protective effects of prebiotics inulin and lactulose from cytotoxicity and genotoxicity in human colon adenocarcinoma cells. Food Research International. 2013;52:269-274. DOI:10.1016/j.foodres.2013.03.024

209. Venema, K. Intestinal fermentation of lactose and prebiotic lactose derivatives, including human milk oligosaccharides. International Dairy Journal. 2012;22:123-140. DOI:10.1016/j.idairyj.2011.10.011

210. Mayer, J., Kranz, B., Fischer, L. Continuous production of lactulose by immobilized thermostable -glycosidase from *Pyrococcus furiosus*. Journal of Biotechnology. 2010;145:387-393. DOI:10.1016/j. jbiotec.2009.12.017

211. Schuster-Wolff-Bühring, R, Fischer, L, Hinrichs, J. Production and physiological action of the disaccharide lactulose. International Dairy Journal. 2010;20:731-741. DOI:10.1016/j.idairyj.2010.05.004

212. Seki, N, Saito, H. Lactose as a source for lactulose and other functional lactose derivatives. International Dairy Journal. 2012;22:110-115. DOI:10.1016/j.idairyj.2011.09.016.

213. Muzzarelli, R A A, Boudrant, J, Meyer, D, Manno, N, DeMarchis, M, Paoletti, M G. Current views on fungal chitin/chitosan, human chitinases, food preservation, glucans, pectins and inulin: A tribute to Henri Braconnot, precursor of the carbohydrate polymers science, on the chitin bicentennial. Carbohydrate Polymers. 2012;87:995-1012. DOI:10.1016/j.carbpol.2011.09.063

214. Prashanth, K V H, Tharanathan, R N. Chitin/chitosan: modifications and their unlimited application potential: an overview. Trends in Food Science & Technology. 2007;18:117-131. DOI:10.1016/j.tifs.2006.10.022

215. Chung, Y C, Hsub, C K, Koa, C Y, Chana, Y C. Dietary intake of xylooligosaccharides improves the intestinal microbiota, fecal moisture, and pH value in the elderly. Nutrition Research. 2007;27:756-761. DOI:10.1016/j.nutres.2007.09.014

216. Carvalho, A F A, Oliva Neto, P, Silva, D,F, Pastore, G,M. Xylo-oligosaccharides from lignocellulosic materials: Chemical structure, health benefits and production by chemical and enzymatic hydrolysis. Food Research International. 2013;51:75-85. DOI:10.1016/j. foodres.2012.11.021

217. Veenashri, B R, Muralikrishna, G. *In vitro* anti-oxidant activity of xylo-oligosaccharides derived from cereal and millet brans – A comparative

study. Food Chemistry. 2011;126:1475-1481. DOI:10.1016/j.foodchem.2010.11.163

218. Bharti, S,K, Kumar, A, Krishnan, S, Gupta, A K, Kumar A. Mechanism-based antidiabetic activity of Fructo- and isomalto-oligosaccharides: Validation by *in vivo, in silico* and *in vitro* interaction potential. Process Biochemistry. 2015;50:317-327. DOI:10.1016/j.procbio.2014.10.014

219. Candela, M, Maccaferri, S, Turroni, S, Carnevali, P, Brigidi, P. Functional intestinal microbiome, new frontiers in prebiotic design. International Journal of Food Microbiology. 2010;140:93-101. DOI:10.1016/j.ijfoodmicro.2010.04.017

220. Nguyen, T H, Haltrich, D. Microbial production of prebiotic oligosaccharides. In: McNeil B, Archer D, Giavasis I, Harvey L. Microbial Production of Food Ingredients, Enzymes and Nutraceuticals. Woodhead Publishing; 2013. p. 494-530. DOI: 10.1533/9780857093547.2.494. ch18.

221. Chen, H L, Lu, Y H, Lin, J, Ko, L I. Effects of fructooligosaccharide on bowel function and indicators of nutritional status in constipated elderly men. Nutrition Research. 2000;20:1725-1733.

222. Cheng, W T, Lin, S Y. Processes of dehydration and rehydration of raffinose pentahydrate investigated by thermal analysis and FT-IR/DSC microscopic system. Carbohydrate Polymers. 2006;64:212-217. DOI:10.1016/j.carbpol.2005.11.024

223. Huebner, J., Wehling, R L, Hutkins, R W. Functional activity of commercial prebiotics. International Dairy Journal. 2007;17:770-775. DOI:10.1016/j.idairyj.2006.10.006

224. Anthony, J C, Merriman, T N, Heimbach, J T. 90-Day oral (gavage) study in rats with galactooligosaccharides syrup. Food and Chemical Toxicology. 2006;44:819-826. DOI:10.1016/j.fct.2005.10.012

225. Osman, A, Tzortzis, G, Rastall, R A, Charalampopoulos, D. A comprehensive investigation of the synthesis of prebiotic galactooligosaccharides by whole cells of *Bifidobacterium bifidum* NCIMB 41171. Journal of Biotechnology. 2010;150:140-148. DOI:10.1016/j.jbiotec.2010.08.008

226. Davis, L M G, Martínez, I, Walter, J, Hutkins R. A dose dependent impact of prebiotic galactooligosaccharides on the intestinal microbiota of healthy adults. International Journal of Food Microbiology. 2010;144:285-292. DOI:10.1111/jam.12415

227. Goulas, A, Tzortzis, G, Gibson, G R. Development of a process for the production and purification of α- and β-galactooligosaccharides from *Bifidobacterium bifidum* NCIMB 41171. International Dairy Journal.

2007;17:648-656. DOI:10.1016/j.idairyj.2006.08.010

228. Searle, L E J, Jones, G, Tzortzis, G, Woodward, M J, Rastall, R A, Gibson, G R, La Ragione, R M. Low molecular weight fractions of BiMuno exert immunostimulatory properties in murine macrophages. Journal of Functional Foods. 2012;4:941-953. DOI:10.1016/j.jff.2012.07.002

229. Förster-Fromme, K, Schuster-Wolff-Bühring, R, Hartwig, A, Holder, A, Schwiertz, A, Bischoff, S C, Hinrichs J. A new enzymatically produced 1-lactulose: A pilot study to test the bifidogenic effects. International Dairy Journal. 2011;21:940-948. DOI:10.1016/j.idairyj.2011.07.002

230. Shen, Q, Yang, R, Hua, X, Ye, F, Wang, H, Zhao, W, Wang K. Enzymatic synthesis and identification of oligosaccharides obtained by transgalactosylation of lactose in the presence of fructose using b-galactosidase from *Kluyveromyces lactis*. Food Chemistry. 2012;135:1547-1554. DOI:10.1016/j.foodchem.2012.05.115

231. Santos, M I, Gerbino, E, Araujo-Andrade, C, Tymczyszyn, E E, Gómez-Zavaglia, A. Stability of freeze-dried *Lactobacillus delbrueckii* subsp. *bulgaricus* in the presence of galacto-oligosaccharides and lactulose as determined by near infrared spectroscopy. Food Research International. 2014;59:53-60. DOI:10.1016/j.foodres.2014.01.054

232. Dilokpimol, A, Nakai, H, Gotfredsen, C H, Appeldoorn, M, Baumann, M J, Nakai, N, Schols H A, Hachem, M A, Svensson B. Enzymatic synthesis of b-xylosyl-oligosaccharides by transxylosylation using two b-xylosidases of glycoside hydrolase family 3 from *Aspergillus nidulans* FGSC A4. Carbohydrate Research. 2011;346:421-429. DOI: 10.1016/j.carres.2010.12.010

233. Aam, B B, Heggset, E B, Norberg, A L, Sørlie, M, Vårum, K M, Eijsink, V G H. Production of Chitooligosaccharides and Their Potential Applications in Medicine. Marine Drugs. 2010;8:1482-1517. DOI:10.3390/md8051482

234. Jeon, Y J, Park, P J, Kim, S K. Antimicrobial effect of chitooligosaccharides produced by bioreactor. Carbohydrate Polymers. 2001;44:71-76.

235. Jeon, Y J, Kim, S K. Production of chitooligosaccharides using an ultrafiltration membrane reactor and their antibacterial activity. Carbohydrate Polymers. 2000;41:133-141.

236. Kim, S K. Chitin, Chitosan, Oligosaccharides and Their Derivatives - Biological Activities and Applications. CRC Press; 2011.

237. Kim, Y M, Kang, H K, Moon, Y H, Nguyen, T T H, Day, D F, Kim, D. Production and Bioactivity of Glucooligosaccharides and Glucosides Synthesized using Glucansucrases. In: Moreno, F J, Sanz, M L, editors.

Food Oligosaccharides - Production, Analysis and Bioactivity. Wiley Blackwell, IFT Press; 2014. ch10.

238. Vardakou, M, Palop, C N, Christakopoulos, P, Faulds, C B, Gasson, M A, Narbad A. Evaluation of the prebiotic properties of wheat arabinoxylan fractions and induction of hydrolase activity in gut microflora. International Journal of Food Microbiology. 2008;123:166-170. DOI:10.1016/j.ijfoodmicro.2007.11.007

239. Wicker, L, Kim, Y, Kim, M J, Thirkield, B, Lin, Z, Jung, J. Pectin as a bioactive polysaccharide e Extracting tailored function from less. Trends in Food Science & Technology. 2014;42:251-259. DOI:10.1016/j.foodhyd.2014.01.002

240. Holck, J, Hjernø, K, Lorentzen, A, Vigsnæs, L K, Hemmingsen, L, Licht, T R, Mikkelsena, J D, Meyer, A S. Tailored enzymatic production of oligosaccharides from sugar beet pectin and evidence of differential effects of a single DP chain length difference on human faecal microbiota composition after *in vitro* fermentation. Process Biochemistry. 2011;46:1039-1049. DOI:10.1016/j.procbio.2011.01.013

241. Rastall, R A, Gibson, G R. Recent developments in prebiotics to selectively impact beneficial microbes and promote intestinal health. Current Opinion in Biotechnology. 2015,32:42-46. DOI:10.1016/j.copbio.2014.11.002

Chapter 7

APPLICATION OF CHROMATOGRAPHIC AND INFRA-RED SPECTROSCOPIC TECHNIQUES FOR DETECTION OF ADULTERATION IN FOOD LIPIDS: A REVIEW

J. M. N. Marikkar, M.E.S Mirghani and I. Jaswir

International Institute for Halal Research and Training, International Islamic University Malaysia, Malaysia

ABSTRACT

Adulteration of oils and fats is an important commercial issue, which needs intervention from regulatory agencies. Tremendous amount of research has been carried out during the past several decades to address this, starting from classical methods to more sophisticated instrumental techniques. Instrumental techniques based on chromatography and infrared spectroscopy have received particular attention from researchers worldwide since they are fast and efficient. Majority of the past studies suggested the use of assays based on fatty acids, triacylglycerol components, minor constituents, and spectral characteristics as they are really useful to determine the adulteration of food lipids. A discussion on the specificity and sensitivity of these assays in solving adulteration issues of oils and fats is timely. Hence, the purpose of this review is to present an update of the current literature in this topic and provide some directions for future research.

INTRODUCTION

Food authentication has become an important aspect of food quality control. As there are numerous practices of fraudulent nature in the food sector [1], it has become highly essential to establish procedures to monitor food quality at various stages of production, processing and distribution. Fat and oil processing industry is no exception to fraudulent practices. There are many reported cases of adulteration practices for highly priced vegetable oils and fats such as virgin olive oil, cocoa butter etc. [1-4]. As extra virgin olive oil is a premium-product, which is short in supply and high in demand, there has always been temptations for its adulteration with cheaper oils [5] or other sub-branded olive products such refined olive oil and olive pomace oil [6]. This tendency has ultimately prompted many European olive-growing countries to impose high tariffs on cottonseed oil imports and adopt a common legislation to protect olive producers and consumers [7]. In recent times, coconut oil industry has shown lack of competitiveness with other major vegetable oils due to its ever-increasing cost of production. Palm olein on the other hand, is a much cheaper product to be imported for blending purposes and hence, coconut oil adulterated with palm olein are sold as 'genuine' product in many Asian countries.

Animal body fats such as lard, beef and mutton tallow could find some applications in the food sector. They are cheap to be used as substitutes since voluminous amounts of animal fats are generated by the carcass industry [8]. Lard, for instance, was a major source of shortening in North America and other Western European countries for a long time. Apart from its use as a component in food applications, deliberate mixing of lard in vegetable oils, fats and dairy products could be possible due to economic reasons [9-12]. Since lard and palm oil have some similarities with respect to chemical composition and physical characteristics [13], it could be mixed easily with palm oil. Seriburi & Akoh [14] demonstrated that mixing of lard with sunflower oil in different ratios could produce a variety of plastic shortenings. Lard has often been a potential substitute for dairy products like ghee [15] and butter [16] since it showed good compatibility to these products. However, mixing of animal fats with plant oils may not be desirable due to religious restrictions and negative nutritional perception regarding the consumption of animal fats. Hence, a great deal of research has gone into development of methods to authenticate food against the adulteration of animal fats.

Studies on food authentication issues have been conducted for the past several decades. As there has been a huge influx of information on food authenticity, many attempts were made time to time to present reviews on this area. As early as nineteen eighties, Rossell and co-workers [4] summarized different classical approaches for detection of adulterations in oils and fats. Rossell [17] also discussed the criteria for determining the purity of selected edible vegetable oils using both classical and modern instrumental approaches. As adulteration of virgin olive oil has always been an issue of increasing importance, there were several attempts to update the developments in the detection of adulterations in virgin olive oil [18, 19]. In the recent past, Reid and co-workers [20] presented a general overview on the applications of spectroscopic (mid infrared, near infrared, Raman spect), chromatographic (gas liquid chromatography and high performance liquid chromatography), and thermo analytical techniques (differential scanning calorimetry) on determining the authenticity of various kinds of foods. As the technological advancement in food authentication takes place at a faster rate, there is still a need for an updated review of the current literature with a specific focus on the quality assurance of oils and fats. Hence, this article is intended to present a review of the studies carried out on the applications of chromatographic and infra-red spectroscopic techniques on detection of adulteration in vegetable oils and fats.

DIFFERENT ANALYTICAL APPROACHES

Fatty Acid Analyses

Several scientific investigations were carried out in the past to develop analytical methodologies based on fatty acid (FA) compositional data to detect and quantify adulterations in fats and oils [Table 1]. Fatty acid analysis by gas liquid chromatography (GLC) system equipped with a polar capillary column and flame ionization detector (FID) was useful to establish the purity of oils and fats. More recent work on oil authenticity has concentrated on the compilation of FA composition for various oils and fats. The Codex Committee on Fats and Oils which was established by the joint FAO/ WHO Codex Alimentarious Commission published FA composition ranges for typical commercial samples of bona fide fats and oils. However, it was recognized that these ranges were not definitive and hence, it had to undergo revision from time to time. This was due to the fact that FA compositional changes were possible based on varietal differences, differences in geographical origins, and the influence of seed maturity [21]. For the detection of adulteration, the relative abundance of individual FA in a given sample is needed to be crosschecked with a reference FA data base. When the determined FA values of a sample deviates significantly from the range found in the reference, it could be suspected to have undergone adulteration. This approach has been useful to detect groundnut oil contamination with soybean oil [17], detection of either vegetable oils or animal body fats in ghee [22], as well as animal fat adulterations in palm kernel oil [8]. Selection of a marker fatty acid is an important step to detect adulterations in oils and fats. For instance, lauric acid was the marker to detect coconut oil adulteration in soybean oil, cottonseed oil and tallow as it did not occur in significant proportions in these oils. Occurrence of castor oil contamination in some vegetable oils was ascertained by means of ricinoleic or hydroxystearic acid content [23]. In another instance, 11, 14-eicosadienoic acid (C20:2) was the marker to detect lard adulteration in beef and mutton fat [12] though later investigations proved that the use of 11, 14-eicosadienoic acid (C20:2) as sole indicator to detect lard would not be reliable [24].

Generally, getting a positive indication of adulteration would become difficult, if the FA compositions of the contaminant and the original oil were closely similar. Most of the time, this remained as a challenge in several adulteration cases. In such instances, alternative strategies such as fractional crystallization of lipids, regio-specific analysis of FA using pancreatic lipolysis, FA ratio calculations and principle component analysis (PCA) of FA data have been adopted by investigators [7, 25]. When extra-virgin olive oil was adulterated either by refined olive oil or olive pomace oil, detection of adulteration especially at lower levels (< 5%) became quite difficult by mere comparison of the overall fatty acid data [26]. Detection of adulteration of olive oil with other seed oils (cottonseed, sesame, corn, and soybean) [23] and detection of lard in butter fat (ghee) were resolved using fractional crystallization [27]. Gamazo-Vazquez and co-workers [28] proposed to take oleic and linoleic acid ratio as the diagnosis parameter since detection of olive oil adulteration by other seed oils below 5 (%) became practically difficult. Interestingly, this approach enabled the detection of contamination of olive oil with sunflower oil at the lowest of 1% level. Similarly, Seo and co-workers [29] suggested to the use of stearic acid content in combination with the ratio between linoleic and oleic acid to detect sesame oil adulteration by corn oil. This method, however, could detect sesame oil adulteration only above 5% level. For detection of milk fat adulteration with foreign fats such as lard, tallow and palm oil, FA ratios C14:0/C18:2 and C18:2/C18:0 were used as parameters. For successful discrimination between pure milk fat and admixtures containing more than 3% lard could be achieved by the application of linear-discrimination analysis to FA data [22]. In another case, Dourtoglou and co-workers [7] applied PCA to the total and region-specific FA data to discriminate pure olive oil from those adulterated with corn, soybean, sunflower and cottonseed oils. According to this study, even samples adulterated at 5% level could be discriminated along with the possibility of knowing the type of the adulterant.

As lard is reported to have some unique pattern of FA substitution in its TAG structure, looking into positional distribution of FA within the glycerol backbone of the suspected lipid sample would be a more reliable approach to detect lard in food systems. This was tested to detect vegetable oil contamination with lard [25], to trace products, which were deep-fried in lard [30], as well as meat products contaminated with pork [11]. As lard in comparison to any other animal fat or plant oil would possess excessive amounts of palmitic acid in its sn-2 position [31], the percentage of palmitic acid content at the sn-2 position of lipids extracted from product contaminated with lard would also be high. The validity of this approach was even tested to differentiate genuine olive oil from those which were modified. According to Firestone and co-workers [6], determination of the percent palmitic and stearic acids at the sn-2 position of TAG molecules was helpful to distinguish genuine olive oil from esterified olive oil samples. However, this approach is laborious and timeconsuming since it involves several steps to achieve the results.

Investigations were also carried out find biomarkers from isomeric forms of unsaturated fatty acids to detect adulterations. On several occasions, conventional GC systems coupled with FID did not have the capacity to resolve isomeric forms of unsaturated fatty acids. Hence, efforts were diverted to employ advanced version of GLC systems. While GLC hyphenated with mass spectrometer was found to give greater details of FA composition [32], comprehensive two-dimensional gas chromatography (GCxGC) was able to unravel the entire spectrum of individual components, including those which occur in lower abundance or in different isomeric forms [33]. According to some recent reports, these advanced forms of GC systems were able to distinguish lard from other animal fats [34, 35]. As these preliminary studies have been mainly focused on distinguishing lard from other animal fats, further investigation would be necessary to show their potential applications in detecting lard in complex lipid mixtures.

Triacylglycerol Compositional Analyses

TAG compositional analyses have been extensively explored as means of detection of adulterations in oils and fats [Table 1] [36]. The use of packed column GC to determine TAG compositions of lipid products helped to detect adulterations in milk fat [37] and cocoa butter [38]. In both of these, high temperature short columns were employed to develop equations incorporating the major TAG peaks to discriminate pure samples (milk fat and cocoa butter) from the adulterated ones. As these equations were meant to define pure milk fat, their use helped to detect non-milk fat adulterants present in milk products at 5% level with 99% confidence [37]. As genuine CB is comprised of only three major TAG species namely, POP, POS and SOS, a straight line relationship was found between C_{50} and C_{54} contents based on the natural variability of POP, POS and SOS [38]. According to this method, the minimum detection limit of cocoa butter equivalents in chocolate fat was 15% at 95% confidence limit [38]. Quantitatively, this method also helped to estimate the amount of cocoa butter equivalents in suspected samples of chocolate fat. TAG compositional analyses using GC fitted with high temperature columns were also employed to detect lard in food systems [39]. As lard was found to have six dominant TAG (C_{46}, C_{48}, C_{50}, C_{52}, C_{54}, and C_{56}) with C_{52} being the major TAG [39, 40], C_{52}/C_{38} ratio was found to be a sensitive index for detection of lard in butter fat [38]. After analyzing TAG composition of four different Canadian butter fats (ranging in iodine value from 34.8 to 39.1), Parodi [40] found that the lower limit of detection for lard in butterfat was 5 to 10%.

With modernization, packed column GC system was largely replaced by high-temperature capillary column GC system due to rapidity in providing TAG profiles. The improved system was successful in detecting non-milk fat component in goat milk fat [41] as well as lard in ewe's milk fat [42]. Fontecha and co-workers [41] employed a fairly high number of goat milk samples from five different herds of goats to develop a formula to define pure goat milk fat in terms of TAG composition. The formula was found to be useful to indicate deviations of goat milk due to 3 to 5% adulterations either by palm oil or tallow. According to Goudjil and coworkers [42], use of capillary column GC system would be able to detect lard adulterations exceeding 5% in ewe's milk fat based on the multiple linear regression equations obtained using the TAG composition of pure ewe's milk fat and its adulterated blends [Table 1] [42]. In another study, Simoneau and co-workers [43] employed capillary GC column to detect plant based cocoa butter equivalents (CBE) in CB. As TAG compositions of CBEs and CB were not identical, the differences between them was useful to detect the presence of the CBEs in confectionery products [43, 44]. In this method, plots of percentages of specific TAG of CB would help to identify deviations due to adulteration practices. As various plant fats exhibited different deviating values from those of CB, several plots were needed to provide optimal detection and quantification level [43].

Analysis of TAG composition by reversed-phase (RP) HPLC for detection of adulteration in oils and fats has seen another phase of development. The use of RP-HPLC in the detection of oil adulteration gained much attention due to ease of sample preparation as well as the natural variations in fatty acid composition do not affect the characteristic TAG profile of several oils and fats [45]. This approach was particularly useful for detection of adulteration of virgin olive oil with various seed oils rich in linoleic acid [45] as well as vegetable oils mixed with lard [33]. In majority of the cases, vegetable oil adulterations with animal fats are generally found to cause some deviations mainly on the existing TAG molecular species of the TAG elusion profiles [36]. Adulteration of palm kernel oil with animal fats such as lard and beef tallow, however, was tend to cause additional TAG peaks visible even at adulteration level as low as 5% [36]. Kapoulas and Andrikopoulos [45] introduced a HPLC-method which enabled the detection of low levels of seed oils such as sunflower, soybean, cottonseed and corn oils as adulterants in olive oil. According to this method, olive oil adulteration was detected using trilinoleoylglycerol (LLL) TAG peak as marker since it was a TAG almost absent in several olive oil samples. The validity of this approach was later tested by Antoniosi and co-workers [46] who used soybean oil as the potential adulterant in olive oil. The detection limit of this method was as low as 4%.

Table 1: A summary overview of chromatographic methods used for detection of adulterations in food lipids

Product	Objective to detect	Analytical technique	Main results	Ref. no.
Meat lipids of fresh meat species and canned meat products	Lard in meat lipids	GLC coupled with FID detector to determine overall fatty acid data	Presence of 11, 14-eicosadienoic acid (C20:2) indicated pork in meat products. Minimum detection was 1% (w/w)	[12]
Milk lipids of cow and buffalo ghees	Lard in milk lipids	GLC coupled with FID detector to determine overall fatty acid data	Deviations in behenic (C22:0), oleic (C18:1), stearic (C18:0) and palmitic (C16:0) helped for detection. Minimum detection was 5% (w/w)	[27]
Milk lipids of cow and buffalo ghees	Lard in milk lipids	GLC coupled with FID detector to determine component unsaponifiable matter	Deviation in n-nonacosane concentration helped for detection. Minimum detection was 5% (w/w)	[15]
Lipids of fried peanut, tempeh, chicken and beef	Fried oils contaminated with lard	GLC coupled with FID detector to determine fatty acid data of the sn-2 position	High value for palmitic acid enrichment factor was indicator of lard	[30]
Milk fat of cow and ewe	Non milk fat mixed with milk fat	GLC coupled with FID detector to determine overall fatty acid data	C14:0/C18:2 and C18:2/C18:0 were sensitive parameters. Minimum detection 10% (w/w)	[22]
Milk fat of cow and ewe	Lard and palm oil in milk fats of cow and ewe	GC fitted with high-temperature capillary column to determine	The deviation in M value of the model $\sum ai\,Ci = M + c$ using values of C42, C44, C48, C50, and C52 indicated adulteration. Minimum detection was 5% (w/w)	[42]
Milk fat of goat	Palm oil and tallow mixed in goat milk fat	GC fitted with short capillary column	The deviation in M value of the model $\sum ai\,Ci = M + c$ using values of C42, C44, C50, and C52 indicated adulteration. Minimum detection was 5% (w/w)	[41]
Cocoa butter	Plant fats in cocoa butter	High resolution GC fitted with high-temperature capillary column	Quantification of added cocoa butter equivalence could be achieved down to a 5% (w/w). The identification of the nature of the foreign fat added was also possible.	[43]
Milk lipids of butter	Non-milk fat adulterants in milk	Packed column GC to determine C_{α} content	Minimum detection limit of 5% (w/w) was achieved	[39, 40]

Milk lipids	Non-milk fat mixed with milk fat	Packed column GC to determine TAG compositions	Minimum detection limit of 5% (w/w) was achieved	[37]
Cocoa butter	Cocoa butter equivalents in chocolate fat	Packed column GC to determine TAG compositions	Minimum detection limit of 15% (w/w) was achieved	[38]
Vegetable oils of palm, Palm kernel, and canola	Vegetable oils adulterated with lard	GLC coupled with FID detector to determine fatty acid data of the sn-2 position	CANDISC analysis of fatty acid data of the sn-2 position helped to discriminate oils contaminated with lard	[25]
Meat lipids of fresh meat species and canned meat products	Lard in meat lipids	HPLC coupled with RI detector to TAG profiling of samples	Significant increase in SSU/SUS ratio when pork was added 1% (w/w) into beef and 3% (w/w) into mutton.	[11]
Vegetable oils of palm, Palm kernel, and canola	Vegetable oils adulterated with lard	TAG profiling of the lipids of different vegetable oils using HPLC couples with RI detector	CANDISC analysis of TAG compositional data helped to discriminate oils contaminated with lard	[36]
Virgin olive oil	Linoleic rich seed oils in virgin olive oil	HPLC coupled with RI detector to TAG profiling of samples	Tri-linoleoylglycerol (LLL) TAG peak was identified as a good biomarker to detect olive oil adulteration	[45, 46]
Lipids of fried peanut, tempeh, chicken and beef	Fried oils mixed with lard	HPLC coupled with RI detector to TAG profiling	The characteristic TAG profiling provided direct evidence for lard contamination in these products	[30]
Olive oil	Hazel nut oil in olive oil	Reversed-phase liquid chromatography coupled to gas chromatography	(E)-5-methylhept-2-en-4-one was found to be a potential biomarker for the rapid recognition of olive oils adulterated with hazelnut oils	[53]
Olive oil and butter	Milk fat mixed with palm oil / olive oil mixed with sunflower oil	GLC coupled with FID detector to determine sterol compositional data	Possible to detect the presence of sunflower oil in olive oil or of palm oil in milk fat, down to as low as 5% (w/w)	[54]

Olive oil	Olive oil adulteration by other seed oils	GLC analysis of sterols using polar capillary column	An olive oil authenticity factor based on the summation of campesterol and stigmasterol percentages was established as an indicator of olive oil adulteration with vegetable oils. The minimum detection limit was 5% (w/w).	[55]
Extra virgin olive oil	Adulteration of extra virgin olive oil with other seed oils	RP-HPLC with florescence detector	The ratio of a-/(b+c)-tocopherol concentrations as a first screening marker of the authenticity of extra vergin olive oil was established.	[65]
Dark chocolate	Palm mid-fraction in dark chocolate	HPLC system with florescence detector	a-tocotrienol was a potential biomarker to detect palm mid fraction in cocoa butter. Minimum detection limit was 5% (w/w)	[68, 69]

On several occasions, HPLC detection of adulteration in olive oils became more difficult when the TAG composition of the adulterant was almost similar to that of main oil [47]. As canola oil is also characterized by high content of monounsaturated TAG, which was within the limits of olive oil, detection of canola oil in olive oil would become more difficult [47]. A similar situation was encountered in the case of olive oil adulteration with high-oleic sunflower oil. In such cases, alternative strategies were adopted to resolve the adulteration issues. Use of multivariate data analysis techniques such as PCA became inevitable to find solutions.

Application of PCA to TAG compositional data obtained by reversed-phase HPLC offer was found to be an effective way to discriminate pure olive oil from those adulterated with maize, rapeseed, cottonseed, sunflower, and soybean oils. It was possible to discriminate authentic olive oil samples from oil samples adulterated at 10% level and above [48, 49].

Use of multivariate statistical approaches was also investigated for detection of lard in common vegetable oil such as palm oil, palm kernel oil, and canola. When discriminant analysis was performed to liquid chromatographic data of the adulterated samples, it was possible to discriminate vegetable oil samples adulterated with lard [36]. On another instance, plotting a graph using the linear relationship between LOO/ LOP and OOO/POO ratios as variables was found to be one of the effective methods of authentication of virgin olive oil from other sub-branded olive oils [49]. This became possible using a database of TAG compositions of various grades of commercial olive oils coming from major olive producing countries [49]. In this approach, deviations in genuine products could be pinpointed easily as lack of adherence to this line would mean that a sample was defective in some manner.

Minor Component Analyses

Most vegetable oils and fats are generally found to contain some minor components as unsaponifiable matter. These minor components could be sterols, triterpene alcohols, or hydrocarbons such as n-eicosane, n-docosane, squalene, carrotinoids, etc [50]. Cholesterol is present as the minor component in majority of the animal body fats as well as milk fats [51]. Although researchers in the past have shown much interest to study minor components of oils and fat due to nutritional significance [52], their determination has helped for detection of adulterations. Sterol analysis, for instance, has been used for detection of vegetable fats in milk fat, margarine in butter, various seed oils in olive oil [53], and animal fats in vegetable fats [54]. Recently, Azadmard-Damirchi [18] presented a review on the use of phytosterols in the detection of olive oil adulteration with hazelnut oil, or other sub-branded olive products.

According to conventional methods of analysis, the unsaponifiable matter of lipids is extracted and analyzed directly using GC systems as total sterols. This approach has greatly helped to handle adulteration of olive oil with other seed oils such as corn, sunflower, soybean and cottonseed oils. The usefulness of sterols analysis by GLC has been demonstrated to detect olive oil adulteration by other seed oils using thermo stable polar columns [55]. In majority of olive oils, campesterol, stigmasterol, β-sitosterol and Δ 5-avenasterol were found as major sterols that account for more than 95% of total sterols present. Interestingly, the relative proportions of these four sterols were found to be fitting into an equation as given below (Eq 1). The Af value of the majority of olive oil samples were in the range 19.7-25.45 with an average of 21.99 ± 1.65, while those of corn, sunflower, soybean and cottonseed oils were in the range 2.04-2.9. Hence, the addition of 5% of corn, sunflower, soybean and cottonseed oils tended to decrease the Af value of olive oil to 9.9, 13.5, 12.5 and 13.7. This kind of deviations in test samples may indicate possible adulteration in olive oil. However, this method might not be applicable to detect hazelnut and lampante oils since the relatively proportions of campesterol and stigmasterol in these two oils were extremely lower.

$$Af = [100 - (Campesterol\% + stigmasterol\%)] / (Campesterol\% + stigmasterol\%) \qquad (1)$$

Alternatively, sterol fraction may be isolated into individual component by using either thin layer chromatography or chromatography on a silica gel column and subsequently analyzed using GC either as free sterols or trimethylsilyl derivatives. For better accuracy, determination of free and esterified sterols in oils and fats can be done using either capillary GC after methanolysis or on-line coupled liquid chromatography-gas chromatography (LC-GC) [56]. The method of capillary GC followed by methanolysis could help reduce the total analysis time as it eliminates the need for saponification, extraction, and derivatization steps [54]. On-line coupling of LC-GC might offer rapid separation of free sterols in edible oils and fats since sample preparation can be integrated into the chromatographic procedure [57]. A detailed account of the developments in the analysis of sterols using on-line coupled LC-GC systems could be found elsewhere in the literature [58].

The possibility of detecting lard in ghee samples using cholesterol content as a marker was investigated [15]. As both cow and buffalo ghee samples had higher concentration of cholesterol in comparison to lard, decreasing trend in cholesterol content of admixtures was noticed with the increasing level of adulteration. This method enabled detection of lard as low as 5% in admixtures of both cow and buffalo gees. However, Alonso and co-workers [54] proposed an alternative procedure for direct analysis of sterols after methanolysis, which effectively eliminated the need for extraction of the saponifiable matter. This has been first tested to detect milk fat adulterations with palm oil, which is generally known to possess campesterol, stigmasterol, and β-sitosterol in higher proportions. As milk fat is usually found to contain higher concentration of cholesterol, a quantitative comparison between the chromatograms of the authentic sample and its admixtures would indicate either a reducing tendency in cholesterol or additional peaks appearing due to the presence of plant sterols originating from palm oil. Apart from showing a good repeatability, this method was claimed to be more rapid in comparison to other methods.

Analysis of hydrocarbons in plant and animal lipids has been the interest of researchers for many reasons. Determination of squalene, for instance, could be particularly useful for the detection of adulteration in olive oil, as it occurs in abundance in olive oil compared to any other vegetable oil [59]. The conventional methods of squalene analysis involve multi-steps, namely the isolation of the unsaponifiable matter, the fractionation of it into several sub-classes, and their subsequent analysis by GC. According to the European Union official methods [60], squalene can be determined simultaneously with waxes using a short capillary column coated with a low-polarity phase. Analysis of hydrocarbons in the unsaponifiable matter has also been useful to detect either lard or margarine in cow and buffalo ghees [15]. As lard was found to have a very high concentration of n-nonacosane in comparison to ghees of cow and buffalo, it could be used as parameter to detect adulterations in both cow and buffalo ghees at the minimum of 5% (w/w). On the other hand, n-dotriacontane could be used as a marker to detect the presence of margarine in ghee since it did not occur as a minor constituent in ghee originated from either cow or buffalo milk. Inclusion of a set of hydrocarbons and sterols as standards as well as a check on the detection limit using proper statistical techniques were merits of this study.

Analysis of carotenoids could also be made use for authentication of oil and fats. Being a group of pigments in certain oils and fats, (all-E)- α- and (all-E)–β-carotene, (allE)-lutein, and (all-E)-zeaxanthin are reported as the most dominant constituents of this group. Analysis of carotenoids in lipids is usually carried out either by spectrophotometry or HPLC system using UV detector. Recently, Franke and coworkers [61] made a comparison between photometric and liquid chromatographic determinations of total carotenoid contents in selected seeds, their oils as well as the press cakes of them. In certain cases, carotenoid contents by photometric method were significantly higher than those obtained by HPLC method. Unlike the HPLC method, the calculation of carotenoid content by using photometric method is done according to absorbance maximum at 446 nm for the whole group of carotenoids though some of them might show slight differences in absorbance maxima. Hence, there could be a possibility for an over-estimation of total carotenoid content by photometric method. Apart from their significance as precursor in the biosynthesis of Vitamin A as well as protective agents against carcinogenesis, they are also useful as biomarkers to detect adulteration in some food lipids. Among the lipids originating from plants, palm oil is well-known to possess high concentration of carotenoids while coconut oil does not possess any of the carotenoids in significant amounts. Hence, the determination of β-carotene was found to be helpful to detect coconut oil adulteration with palm olein [62].

Tocopherols are yet another class of minor components used for quality assurance of oils and fats. Soybean, wheat germ, canola, peanut, and cottonseed oils are some of the richest sources of tocopherols, of which α, β, γ, and δ-tocopherols are the important constituents. Apart from being an oil with high amount of tocopherols, palm oil is used to have still another group of minor compounds known as tocotrienols. Analysis of tocopherols present in oils and fats could be easily done using HPLC system equipped with normal-phase column and a florescence detector [63]. However, HPLC systems equipped with reversed-phase column (RP-HPLC) and UV detection were also found to give good results [64]. According to some recent reports, RP-HPLC also has the advantage of faster chromatographic runs, faster equilibration time and better reproducibility of retention times [65]. In certain cases, better efficiency in separation was achieved through purification of oils using gel permeation chromatography as well as detection by evaporative light scattering detector (ELSD) [66]. The usefulness of tocopherol analysis has been demonstrated for detection of extra virgin olive oil (EVOO) with other seed oil [65], milk fat adulteration with vegetable fats [54, 67], sunflower oil with groundnut oil [17], and butter with margarine. For authentication of EVOO, calculation of the ratio of a-/(b+c)-tocopherol concentration by measuring tocopherols (a-, (b+c)- and d-tocopherols) in EVOO and adulterated samples with RP-HPLC was found to be useful. According to some recent reports, the presence of palm mid fraction in CB as low as 5% (w/w) could be monitored by using HPLC analysis of α-tocotrienols [68, 69]. Authentication methods based on the composition of minor constituents of oils has to be applied cautiously as changes in the proportions of tocopherol and carotenoid contents could be possible due to bleaching and deodorization activities. While bleaching earth could adsorb part of the plant pigments, deodorization might cause thermal deterioration of them [61]. Apart from this, natural decomposition of tocopherols might also be possible during the storage due to the influence of light, oxygen, and temperature [63]. Hence, these factors should be taken into account while interpreting analytical results in authentication studies.

Mid-Infrared Spectroscopic Analyses

In recent years, FTIR spectroscopy has emerged as a major analytical technique for a wide range of food analyses. It has become an attractive option because of its high speed in analysis and ease of operation. Considerable efforts have been made so far to use this technique to detect adulterations in oils and fats. In the early nineteen nineties, FTIR spectroscopy has been employed to characterize of pure vegetable oils [70, 71], butters fat, and margarines [72] to distinguish between plant oils and lard [71] as well as lard and other animal fats [73]. All these studies concluded that the mid-IR spectra of most oils and fats were apparently similar despite some dissimilarities in spectral features of certain regions. The observed differences in spectral characteristics were due to molecular compositional and structural differences such as degree of unsaturation and chain length, monounsaturated to polyunsaturated acyl group ratio, variations in trans fatty acid content among the oils [70- 72]. These differences were subsequently exploited to check adulteration practices in oils and fats [74]. A summary of the several studies reported on the detection of adulteration in vegetable oils, namely canola oil, corn oil, cod-liver oil, extra virgin olive oil, soybean oil, and sunflower oil has been given in Table 2.

Depending on the nature of the adulteration, various approaches have been used by researchers [Table 2]. For detection of lard adulteration in beef fat, goat fat, and chicken fat, spectral changes in four frequency regions have been considered: (A): 3009-3000, (B): 1418-1417, (C): 1116-1098 and (D): 968-966 cm^{-1}. In this case, the variations in the spectral properties have been exploited to develop predictive models for quantification of lard by using simple regression [73]. As the mid-IR spectra of majority of oils and fats are apparently similar, several researchers employed chemo metric techniques such as discriminant analysis to exploit subtle differences in the spectra. This approach has been successfully applied to helped differentiate extra virgin olive oil from other seed oils [5], pork from other meat species [10], cocoa butter from cocoa butter mixed with other vegetable fats [75], codliver oil from cod-liver oil mixed with lard [76], and lard from lard mixed with beef fat, goat fat, and chicken fat [77, 78].

Table 2: A summary overview of infrared spectrometric methods used for detection of adulterations in food lipids

Product	Objective	Method	Wavenumber	Main results	Ref. no.
Meat lipids of fresh meat species and canned meat products	To differentiate meat lipids of different animal species	FITR spectroscopy with ATR element and PCA analysis	1800-1000 cm⁻¹	PCA application for FTIR data helped to distinguish pork from chicken and turkey	[10]
Fresh Animal body fats (Lamb, cow, and chicken)	To differentiate meat lipids of different animal species	FITR spectroscopy with simple regression analysis	3008-3000, 1418-1417, 1385-1370, 1126-1085, 968-965 cm⁻¹	Able to differentiate lard from meat lipids of different animal species. Regression models were useful to quantify lard contamination in these lipids	[73]
Fresh Animal body fats (Lamb, cow, and chicken)	To differentiate meat lipids of different animal species	FITR spectroscopy with deuterated triglycine sulfate detector and PLS regression analysis	3010-2000, 1220-1095, 968-965 cm⁻¹	PLS models were useful to quantify lard contamination in these lipids	[78]
Fresh Animal body fats (Lamb, cow, and chicken)	To quantify the proportion of lard contamination in meat lipids	FITR spectroscopy with PLS regression and discriminant analysis	3300-700, 1, 500-900 cm⁻¹	PLS models were useful to quantify lard contamination in these lipids	[77]
Cake lipids	To quantify the proportion of lard contamination in cake lipids	FITR spectroscopy with ATR element and PLS regression analysis	1117-1097, 990-950 cm⁻¹	PLS models for quantification of lard with a minimum detection of 4 % (w/w)	[83]
Chocolate and its products	To quantify the proportion of lard contamination in chocolate lipids	FITR spectroscopy with ATR element and PLS regression analysis	4000-650 cm⁻¹	PLS models for quantification of lard with a minimum detection of 3 % (w/w)	[84]
Cod liver oil	To quantify the proportion of lard contamination in cod liver oil	FITR spectroscopy with ATR and PLS regression analysis	1035-1030 cm⁻¹	PLS models for quantification of lard with a minimum detection of 1% (v/v)	[76]
Some vegetable oils (Canola oil, corn oil, extra virgin olive oil)	To quantify the proportion of lard contamination in the vegetable oils	FITR spectroscopy with ATR and PLS, PCR, and DA	1500-1000 cm⁻¹	PLS models for quantification of lard with a minimum detection of 1% (v/v)	[79]
Extra virgin olive oil	To quantify the proportion of refined olive oil and walnut oil in extra virgin olive oil	FITR spectroscopy with ATR and PLS regression	3100-2800, 1800-1000 cm⁻¹	PLS models for quantification of walnut oil in extra virgin olive oil with a minimum detection of 5% (v/v)	[82]
Biscuit lipids	To quantify the proportion of lard contamination	FITR spectroscopy with ATR element and PLS-DA regression analysis	3050-2800, 1800-1600, 1500-650 cm⁻¹	Cooman plot showed that vegetable fats/oils and animal fats were clustered into distinct groups	[85]
Cocoa butter	To differentiate cocoa butter and cocoa butter equivalents	FITR spectroscopy with principle component analysis	4000-600 cm⁻¹	Application of principle component analysis was able to distinguish between cocoa butter and cocoa butter equivalents	[75]
Red fruit oil	To quantify the proportion of either corn oil or soybean oil contamination	FITR spectroscopy with ATR element and PLS regression analysis	4000-600 cm⁻¹	PLS models for quantification of either corn oil or soybean oil in red fruit oil	[81]
Extra virgin olive oil	To differentiate extra virgin olive oil from adulterated samples and quantify the proportion of adulterant	FITR spectroscopy with ATR element and PCR and PLS regression analysis	3007, 2922, 2853, 1754, 1160, 1117 cm⁻¹	Able to differentiate between pure EVOO and EVOO adulterated with 5% V/V	[74]
Fish packing oils	To differentiate extra virgin olive oil from high-oleic sunflower oil	FITR spectroscopy with ATR element and PLS-DA regression analysis	3030-2800 cm⁻¹	Able to discriminate olive oil from high oleic-sunflower oil	[80]

According to the overall findings of these studies, the finger print region (1500–1000 cm-1) would be most appropriate to be used in discriminant analysis methodology to differentiate pure oils from their admixtures containing lard or any other plant oil [79]. However, some other studies differed from this view point as they made use of the dissimilarities in the spectral bands of other regions of the spectrum. Vasconcelos and coworkers [74] showed that 3007, 2922, 2853, 1754, 1160, and 1117 cm-1 were best suited to differentiate extra virgin olive oil from those adulterated peanut oil while Dominguez-vidal and co-workers [80] indicated that the strongest contribution to differentiate extra virgin olive oil from high-oleic sunflower oil came from 3010, 2920, and 2852 cm⁻¹. Rohman and coworkers [81] used the entire frequency region (4000–600 cm-1) to deal with adulteration of red fruit oil with corn and soybean oil mixture.

For estimation of the proportion of adulterant quantitatively, application of partial least square (PLS) methodology was considered in majority of the cases to select the most suitable spectral regions showing best correlations between FTIR spectra and the concentration of the analyte(s) of interest [82]. However, some other researchers pointed out that application of PLS regression was required for the entire finger print region (1500–900 cm-1) for the quantification of the adulterant [77, 78, 81]. They insisted that same approach was still required to determine lard adulteration level in lipids extracted from product such as cake [83], chocolate [84], biscuits [85], etc. At times, semi-quantitative approaches were proposed to estimate lard content in the lipids extracted from these products since steady changes in absorbance values were noticed in different regions of the spectra with the increasing proportion of lard. For instance, increasing lard content in cake formulation caused a decreased in absorbance values of spectral peaks in the frequency region of 990–950 cm-1. This could be probably due to the fact that shortening used in the cake formulation had high trans fatty acid content [83]. However, this particular frequency region has not been found to be useful for lard detection in chocolate as genuine cocoa butter was not found to possess any trans fatty acid content [84]. Because of this reason, the PLS calibration model to quantify lard content in chocolate was developed on the basis of spectral data in the entire frequency region 4000–650 cm^{-1}. These findings clearly showed a common region could not be used to run the PLS regression for quantification of lard content in lipids extracted from different food products. Instead, spectral regions suitable for quantification of adulterant needs to be selected based on the differences in sample matrix.

FUTURE PROSPECTS

In the recent literature, there are several reports on the use of modern techniques such as FT-NIR spectroscopy [86, 87], ^{1}H and ^{13}C nuclear magnetic resonance (NMR) spectroscopy [88], FT-Raman spectroscopy [89], isotope ratio mass spectrometry [90] for the detection of various food adulteration practices. FT-NIR spectroscopy, for instance, has already been recognized for its uses in the measurement of adulterations in milk fat [86], olive oil [87], etc. Likewise, the potential applications of FT-Raman spectroscopy in detection of virgin olive oil adulterations by pomace, soybean, and corn oils have also been highlighted [89]. As 13C NMR has been recognized as a valuable tool for analysis of the most abundant fatty acids of various oils, there has been a growing interest among researchers to use it to detect adulterations of virgin olive oil [88]. However, the literature on the use of these modern analytical techniques for detection of animal fats in food systems, plant fats in cocoa butter or milk fat is limited. Hence, there is much scope to expand the investigations on lipid adulterations in new directions.

CONCLUDING REMARKS

This review highlighted the developments on the use of analytical techniques such as GLC, HPLC, and FTIR spectroscopy to control adulteration practices in food lipids. As the nature of the FA and TAG compositions of lipids and their adulterants vary drastically, devising generalized criteria for purity determination is a challenging task. However, establishment of comprehensive data bases giving acceptable ranges of FA and TAGs of individual lipids prone to adulteration would be essential for regulatory purposes. There is an increasing trend to use chemo metrics techniques to analyze the FA and TAG compositional data either for differentiating genuine products from adulterated ones or for estimation of the level of adulteration. Chromatographic techniques have also given several successes to detect adulterations in food lipids by monitoring cholesterol and phytosterols as biomarkers. When compared to chromatographic techniques, Mid-IR spectroscopy is an attractive option for detection of adulterations in oils and fats due to the speed of analysis, and minimal sample preparation. Since adulteration in oils and fats could bring about deviations in different spectral regions of vegetable oils, discriminant analysis methodology could be able to differentiate adulterated ones from real samples. For this purpose, selecting the finger print region of the spectra has yielded considerable success in majority of the cases. For quantitative estimation of the adulteration, PLS based calibration models have been considered in majority of the cases. However, a PLS model developed for a particular oil might not be applicable to some other oils. Hence, there is a need to develop PLS models for individual food system using different spectral regions.

REFERENCES

1. Manning L, Soon JM. 2014. Developing systems to control food adulteration. Food Policy 49(1): 23-32. doi: 10.1016/j. foodpol.2014.06.005

2. Kamal M, Karoui R. 2015. Analytical methods coupled with chemometric tools for determining the authenticity and detecting the adulteration of dairy products: A review. Trend Food Sci Technol 46(1): 27-48. doi: 10.1016/j.tifs.2015.07.007

3. Rossell JB, King B, Downes MJ. 1985. Composition of oils. J Am Oil Chem Soc 62(2): 221-230. doi: 10.1007/BF02541382

4. Rossell JB, King B, Downes MJ. 1983. Detection of adulteration. J Am

Oil Chem Soc 60(2): 333-339. doi: 10.1007/BF02543513

5. Lai YW, Kemsley EK, Wilson RH. 1994. Potential of fourier transform infrared spectroscopy for the authentication of vegetable oils. J Agric Food Chem 42(5): 1154 -1159. doi: 10.1021/jf00041a020

6. Firestone D, Summers JL, Reina RJ, Adams WS. 1985. Detection of adulterated and misbranded olive oil products. J Am Oil Chem Soc 62(11): 1558-1562. doi: 10.1007/BF02541684

7. Dourtoglou VG, Dourtoglou T, Antonopoulos A, Stefanou E, Lalas S, et al. 2003. Detection of olive oil adulteration using principal component analysis applied on total and regio FA content. J Am Oil Chem Soc 80(3): 203-208. doi: 10.1007/s11746-003-0677-1

8. Marikkar JMN, Lai OM. 2004. Detection of adulteration in lauric oils: use of different analytical techniques. In: Peiris TSG and Ranasinghe CS (eds) Proceedings II, 270-284: International Conference of the Coconut Research Institute of Sri Lanka, Lunuwila, Sri Lanka.

9. Sosa JFM, Pesini ER, Montoya J, Roncales P, Perez MJL, et al. 2000. Direct and highly species-specific detection of pork meat and fat in meat products by PCR amplification of mitochondrial DNA. J Agric Food Chem 48(7): 2829-2832. doi: 10.1021/jf9907438

10. Jowder OA, Kemsley EK, Wilson RH. 1997. Mid-infrared spectroscopy and authenticity problems in selected meats: a feasibility study. Food Chem 59(2): 195-201. doi: 10.1016/S0308-8146(96)00289-0

11. Saeed T, Ali SG, Rahman HA, Saway WN. 1989. Detection of pork and lard as adulterants in processed meat: liquid chromatographic analysis of derivatized triglycerides. J Ass Offi Anal Chem 72(6): 921-925.

12. Saeed T, Abu-Dagga F, Rahman HA. 1986. Detection of pork and lard as adulterants in beef and mutton mixtures. J Ass Offi Anal Chem 69: 999-1002.

13. Marikkar JMN, Lai OM, Ghazali HM, Che Man YB. 2001. Detection of lard and randomized lard as adulterants in RBD palm oil by differential scanning calorimetry. J Am Oil Chem Soc 78(11): 1113-1119. doi: 10.1007/s11746-001-0398-5

14. Seriburi V, Akoh CC. 1998. Enzymatic interesterification of lard and higholeic sunflower oil with candida anatartica lipase to produce plastic fats. J Am Oil Chem Soc 75(10): 1339-1345. doi: 10.1007/s11746-998-0181-x

15. Farag RS, Ahmed FA, Shihata AA, Aboraya SH, Abdalla AF. 1982. Use of unsaponifiable matter for detection of ghee adulteration with other

fats. J Am Oil Chem Soc 59(12): 557-560. doi: 10.1007/BF02636323

16. Lambelet P, Ganguli NC. 1983. Detection of pig and buffalo body fat in cow and buffalo ghees by differential scanning calorimetry. J Am Oil Chem Soc 60(5): 1005-1008. doi: 10.1007/BF02660216

17. Rossell JB. 1998. Development of purity criteria for edible vegetable oils. In: Hamilton RJ (ed) Blackie Academic and Professional, London, UK, pp 265-289.

18. Azadmard-Damirchi S. 2010. Review of the use of phytosterols as a detection tool for adulteration of olive oils with hazelnut oil. Food Add and Contam 27(1): 1-10. doi: 10.1080/02652030903225773

19. Li-Chan E. 1994. Developments in the detection of adulteration of olive oil. Trends Food Sci Technol 5(1): 3-11. doi: 10.1016/0924-2244(94)90042-6

20. Reid LM, O'Donnell CP, Downey G. 2006. Recent technological advances for the determination of food authenticity. Trends Food Sci Technol 17(7): 344-353. doi: 10.1016/j.tifs.2006.01.006

21. Raihana AR, Marikkar JMN, Ismail A, Musthafa S. 2015. A review on food values of selected fruits' seeds. Int J Food Prop 18(11): 2380-2392. doi: 10.1080/10942912.2014.980946

22. Ulberth F. 1994. Detection of milk fat adulteration by linear discriminant analysis of fatty acid data. J AOAC Int 77(5): 1326-1334.

23. Norris FA. 1982. Analytical methods. In: Swern D (ed) Bailey's Industrial Oil and Fat Products, John Wiley and Sons, Inc., New York, USA, pp 407-525.

24. Firestone D. 1988. General referee reports: oils and fats. J Ass Offi Anal Chem 71: 76-78.

25. Marikkar JMN, Ghazali HM, Che Man YB, Peiris TSG, Lai OM. 2005. Use of gas chromatography in combination with pancreatic lipolysis and multivariate data analysis techniques for identification of lard contamination in some vegetable oils. Food Chem 90(1-2): 23-30. doi: 10.1016/j.foodchem.2004.03.021

26. Firestone D, Carson KL, Reina RJ. 1988. Update on control of olive oil adulteration and misbranding in the United States. J Am Oil Chem Soc 65(5): 788-792. doi: 10.1007/BF02542533

27. Farag RS, Aboraya SH, Ahmed FA, Hewedi FM, Khalifa HH. 1983. Fractional crystallization and gas chromatographic analysis of fatty acids as a means of detecting butter fat adulteration. J Am Oil Chem Soc 60(9): 1665-1669. doi: 10.1007/BF02662429

28. Gamazo-Vazquez J, Garcia-Falco MS, Simal-Gandara J. 2003. Control of olive oil contamination by sunflower seed oil by GC-MS of fatty acid methyl esters. Food Control 14(7): 463-467. doi: 10.1016/S0956-7135(02)00102-0

29. Seo HY, Ha J, Shin DB, Shim SL, No KM, et al. 2010. Detection of corn oil in adulterated sesame oil by chromatography and carbon isotope analysis. J Am Oil Chem Soc 87(6): 621-626. doi: 10.1007/ s11746-010-1545-6

30. Marikkar JMN, Ghazali HM, Long K, Lai OM. 2003. Lard uptake and its detection in selected food products deep-fried in lard. Food Res Int 36(9-10): 1047-1060. doi: 10.1016/j.foodres.2003.08.003

31. Chacko GK, Perkins EG. 1965. Anatomical variation in fatty acid composition and triglyceride distribution in animal depot fats. J Am Oil Chem Soc 42(12): 1121-1124. doi: 10.1007/BF02636926

32. Huang Z, Wang B, Crenshaw AA. 2006. A simple method for the analysis of trans fatty acid with GC–MS and ATe-Silar-90 capillary column. Food Chem 98(4): 593-598. doi: 10.1016/j.foodchem.2005.05.013

33. Philips JB, Beens J. 1999. Comprehensive two-dimensional gas chromatography: a hyphenated method with strong coupling between the two dimensions. J Chrom A 856(1-2): 331-347. doi: 10.1016/S0021-9673(99)00815-8

34. Indrasti D, Che Man YB, Mustafa S, Hashim DM. 2010. Lard detection based on fatty acids profiles using comprehensive gas chromatography hyphenated with time-of-flight mass spectrometry. Food Chem 122(4): 1273-1277. doi: 10.1016/j.foodchem.2010.03.082

35. Chin ST, Che Man YB, Tan CP, Hashim DM. 2009. Rapid profiling of animal-derived fatty acids sing fats GC X GC coupled to time-of-flight mass spectrometry. J Am Oil Chem Soc 86(10): 949-958. doi: 10.1007/ s11746-009-1427-y

36. Marikkar JMN, Ghazali HM, Che Man YB, Peiris TSG, Lai OM. 2005. Distinguishing lard from other animal fats in admixtures of some vegetable oils using liquid chromatographic data coupled with multivariate data analysis. Food Chem 91(1): 5-14. doi: 10.1016/j. foodchem.2004.01.080

37. Timms RE. 1980. Detection and quantification of non-milk fat in mixtures of milk and non-milk fats. J Dairy Res 47(3): 295-303. doi: 10.1017/ S002202990002118X

38. Padley FB, Timme RE. 1980. The determination of cocoa butter equivalents in chocolate. J Am Oil Chem Soc 57(9): 286-293. doi:

10.1007/BF02662209

39. Rashood KA, Shaaban RRA, Moety EMA, Rauf A. 1996. Compositional and thermal characterization of genuine and randomized lard: A comparative study. J Am Oil Chem Soc 73(3): 303-309. doi: 10.1007/BF02523423

40. Parodi PW. 1973. Detection of synthetic and adulterated butterfat 4. GLC triglyceride analysis. Aus J Dairy Technol 28(1): 38-42.

41. Fontecha J, Diaz V, Fraga MJ, Juarez M. 1998. Triglyceride analysis by gas chromatography in assessment of authenticity of goat milk fat. J Am Oil Chem Soc 75(12): 1893-1896. doi: 10.1007/s11746-998-0347-6

42. Goudjil H, Fontecha J, Fraga MJ, Juarez M. 2003. TAG composition of Eve's milk fat. Detection of foreign fats. J Am Oil Chem Soc 80(3): 219-222. doi: 10.1007/s11746-003-0680-6

43. Simoneau C, Hannaert P, Anklam E. 1999. Detection and quantification of cocoa butter equivalents in chocolate model system: analysis of triglyceride profiles by high resolution GC. Food Chem 65(1): 111-116. doi: 10.1016/S0308-8146(98)00106-X

44. Lipp M, Anklam E. 1998. Review of cocoa butter and alternative fats for use in chocolate - Part B. analytical approaches for identification and determination. Food Chem 62(1): 99-108. doi: 10.1016/S0308-8146(97)00161-1

45. Kapoulas VM, Andrikopoulos NK. 1986. Detection of olive oil adulteration with linoleic acid-rich oils by reversed-phase highperformance liquid chromatography J Chromatogra A 366: 311-320. doi: 10.1016/S0021-9673(01)93478-8

46. Antoniosi NR, Carrilho E, Lancas FM. 1993. Fast quantitative analysis of soybean oil in olive oil by high-temperature capillary gas chromatography. J Am Oil Chem Soc 70(10): 1051-1053. doi: 10.1007/ BF02543037

47. Salivaras E, McCurdy AR. 1992. Detection of olive oil adulteration with canola oil from triacylglycerol analysis by reversed-phase highperformance liquid chromatography. J Am Oil Chem Soc 69(9): 935- 938. doi: 10.1007/BF02636347

48. Tsimidou M, Macrae R, Wilson I. 1987. Authentication of virgin olive oils using principal component analysis of triglyceride and fatty acid profiles: part 2-detection of adulteration with other vegetable oils. Food Chem 25(4): 251-258. doi: 10.1016/0308-8146(87)90011-2

49. Flor RV, Hecking LT, Martin BD. 1993. Development of highperformance liquid chromatography criteria for determination of grades of commercial

olive oils. Part I. the normal ranges for the triacylglycerols. J Am Oil Chem Soc 70(2): 199-203. doi: 10.1007/BF02542626

50. Cert W, Moreda MC, Pe´rez-Camino A. 2000. Chromatographic analysis of minor constituents in vegetable oils. J Chromatogr A 881(1- 2): 131-148. doi: 10.1016/S0021-9673(00)00389-7

51. Bragagnolo N, Rodriguez-Amaya, DB. 2002. Simultaneous determination of total lipid, cholesterol and fatty acids in meat and back fat of suckling and adult pigs. Food Chem 79(2): 255 -260. doi: 10.1016/ S0308-8146(02)00136-X

52. Boskou D. 2008. Phenolic compounds in olives and olive oil. In: Boskou D (ed) Olive Oil: Minor Constituents and Health. CRC Press, Florida, USA, pp 11-45.

53. Castillo MLRD, Caja MDM, Herraiz M, Gracia P. 1998. Blanch. Rapid recognition of olive oil adulterated with hazelnut oil by direct analysis of the enantiomeric composition of filbertone. J Agric Food Chem 46(12): 5128-5131. doi: 10.1021/jf9807014

54. Alonso L, Fontecha J, Lozada L, Juarez M. 1997. Determination of mixtures in vegetable oils and milk fat by analysis of sterol fraction by gas chromatography. J Am Oil Chem Soc 74(2): 131-135. doi: 10.1007/ s11746-997-0157-2

55. Al-Ismail KM, Alsaed AK, Ahmad R, Al-Dabbas M. 2010. Detection of olive oil adulteration with some plant oils by GLC analysis of sterols using polar column. Food Chem 121(4): 1255-1259. doi: 10.1016/j. foodchem.2010.01.016

56. Plank C, Lorbeer E. 1993. Analysis of free and esterified sterols in vegetable oil methyl esters by capillary GC. J High Resol Chromatogr 16(8): 483-487. doi: 10.1002/jhrc.1240160808

57. Senorans FJ, Tabera J, Herraiz M. 1996. Rapid separation of free sterols in edible oils by on-line coupled reversed phase liquid chromatographygas chromatography. J Agric Food Chem 44(10): 3189-3192. doi: 10.1021/ jf960071a

58. Villen J, Blanch GP, Castillo MLR, Herraiz M. 1998. Rapid and simultaneous analysis of free sterols, tocopherols, and squalene in edible oils by coupled reversed-phase liquid chromatography-gas chromatography. J Agri Food Chem 46(4): 1419-1422. doi: 10.1021/ jf9707061

59. Nenadis N, Tsimidou M. 2002. Determination of squalene in olive oil

using fractional crystallization for sample preparation. J Am Oil Chem Soc 79(3): 257-259. doi: 10.1007/s11746-002-0470-1

60. European Union Commission 1993. Regulation EEC/183/ 93, Off J Eu Comm, L248.

61. Franke S, Frohlich K, Werner S, Bohm V, Schone F. 2010. Analysis of carotinoids and vitamin E in selected oil seeds, press cakes and oils. Eu J Lipid Sci Technol 112(10): 1122-1129. doi: 10.1002/ejlt.200900251

62. Viver MJ. 1999. A simple method for detection of palm olein in coconut oil. Indian Coco J 11: 6-7.

63. Coors U, Montag A. 1988. Untersuchungen zur Stabilität des Tocopherolgehaltes pflanzlicher Öle. Fett/Lipid 90(4):   129-136. doi: 10.1002/lipi.19880900402.

64. Andrikopoulos NK, Brueschweilet H, Felber H, Taeschler CH. 1991. HPLC analysis of phenolic antioxidants, tocopherols, and triglycerides. J Am Oil Chem Soc 68(6): 359-364. doi: 10.1007/BF02663750

65. Huilun C, Marco A, Carlo F, Elpidio T, Giuseppe S, et al. 2011. Tocopherol speciation as first screening for the assessment of extra virgin olive oil quality by reversed-phase high-performance liquid chromatography/ fluorescence detector. Food Chem 125(4): 1423-1429. doi: 10.1016/j. foodchem.2010.10.026

66. Chase GW Jr., Akoh CC, Eitenmiller RR. 1994. Analysis of tocopherols in vegetable oils by high-performance liquid chromatography: comparison of fluorescence and evaporative light-scattering detection. J Am Oil Chem Soc 71(8): 877-880. doi: 10.1007/BF02540466

67. Keeney M, Bachman KC, Tikriti HH, King RL. 1971. Rapid vitamin E method for detecting adulteration of dairy products with non-coconut vegetable oils. J Dairy Sci 54(11): 1702-1703. doi: 10.3168/jds. S0022-0302(71)86092-7

68. Elham MF, Jinap S, Che Man YB, Noraini I. 2008. Application of alpha-tocotrienol for detection of palm mid-fraction in dark chocolate formulation. Eu Food Res Technol 228(2): 163-168. doi: 10.1007/ s00217-008-0919-6

69. Elham MF, Jinap S, Che Man YB, Noraini I. 2008. Detection and quantification of palm mid fraction in chocolate model system. Int J Food Sci Technol 43(6): 1083-1087. doi: 10.1111/j.1365-2621.2007. 01570.x

70. Guillen MD, Ruiz A, Cabo N, Chirinos R, Pascual G. 2003. Characterization of sacha inchi (Plukenetia volubilis L.) oil by FTIR

spectroscopy and 1 H NMR. Comparison with line seed oil. J Am Oil Chem Soc 80(8): 755-762. doi: 10.1007/s11746-003-0768-z

71. Guillen MD, Cabo N. 1997. Characterization of edible oils and lard by fourier transform infrared spectroscopy. Relationships between composition and frequency of concrete bands in the fingerprint region. J Am Oil Chem Soc 74(10): 1281-1286. doi: 10.1007/s11746-997-0058-4

72. Safar M, Bertrand D, Robert P, Devaux, MF, Genot C. 1994. Characterization of edible oils, butters and margarines by fourier transform infrared spectroscopy with attenuated total reflectance. J Am Oil Chem Soc 71(4): 371-377. doi: 10.1007/BF02540516

73. Che Man YB, Mirgani MES. 2001. Detection of lard mixed with body fats of chicken, lamb, and cow by FTIR spectroscopy. J Am Oil Chem Soc 78(7): 753–761. doi: 10.1007/s11746-001-0338-4

74. Vasconcelos M, Coelho L, Barros A, Martins de Almeida JMM. 2015. Study of adulteration of extra virgin olive oil with peanut oil using FTIR spectroscopy and chemometrics. Cogent Food Agric 1: 1-13. doi: 10.1080/23311932.2015.1018695

75. Goodacre R, Anklam E. 2001. Fourier transform infrared spectroscopy and chemometrics as a tool for rapid detection of other vegetable fats mixed in cocoa butter. J Am Oil Chem Soc 78(10): 993-1000. doi: 10.1007/s11746-001-0377-x

76. Rohman A, Che Man YB. 2009. Analysis of cod-liver oil adulteration using fourier transform infrared (FTIR) spectroscopy. J Am Oil Chem Soc 86: 1149-1153. doi: 10.1007/s11746-009-1453-9

77. Rohman A, Che Man YB. 2010. FTIR spectroscopy combined with chemometrics for analysis of lard in the mixtures with body fats of lamb, cow, and chicken. Int Food Res J 17: 519-526.

78. Jaswir I, Mirghani MES, Hassan TH, Said MZM. 2003. Determination of lard in mixture of body fats of mutton and cow by fourier transform infrared spectroscopy. J Oleo Sci 52(12): 633-638. doi: 10.5650/jos.52.633

79. Rohman A, Che Man YB, Ismail A, Puziah H. 2011. FTIR spectroscopy combined with chemometrics for analysis of lard in some vegetable oils. Cyta-J Food 9(2): 96-101. doi: 10.1080/19476331003774639

80. Dominguez-vidal A, Rosa JP, Cuadros-Rodriguez LC, Canda MJA. 2016. Authentication of canned fish packing oils by means of Fourier transform infrared spectroscopy. Food Chem 190: 122-127. doi: 10.1016/j.foodchem.2015.05.064

81. Rohman A, Setyaningrum DL, Riyanto S 2014. FTIR spectroscopy combined with partial least square for analysis of red fruit oil in ternary mixture system. Intl J Spectr 2014: 785914. doi: 10.1155/2014/785914

82. Lai YW, Kemsley EK, Wilson RH. 1995. Quantitative Analysis of potential adulterants of extra virgin olive oil using infrared spectroscopy. Food Chem 53(1): 95-98. doi: 10.1016/0308-8146(95)95793-6

83. Che Man YB, Syariza ZA, Mirghani MES, Jinap S, Bakar J. 2005. Detection of lard adulteration in cake formulation by Fourier transform infrared (FTIR) spectroscopy. Food Chem 92(2): 365-371. doi: 10.1016/j.foodchem.2004.10.039

84. Che Man YB, Syariza ZA, Mirghani MES, Jinap S, Bakar J. 2005. Analysis of potential lard adulteration in chocolate and chocolate products using fourier transform infrared spectroscopy. Food Chem 90(4): 815-819. doi: 10.1016/j.foodchem.2004.05.029

85. Che Man YB, Syahariza ZA, Rohman A. 2011. Discriminant analysis of selected edible fats and oils and those in biscuit formulation using (FTIR) spectroscopy. Food Anal Meth 4(3): 404-409. doi: 10.1007/ s12161-010-9184-y

86. Sato T, Kawano S, Iwamoto M. 1990. Detection of foreign fat adulteration of milk by near infrared spectroscopic method. J Dairy Sci 73(12): 3408-3413. doi: 10.3168/jds. S0022-0302(90)79037-6

87. Wesley IJ, Barnes RJ, McGill AEJ. 1995. Measurement of adulteration of olive oils by near-infrared spectroscopy. J Am Oil Chem Soc 72(3): 289-292. doi: 10.1007/BF02541084

88. Mavromoustakos T, Zervou M, Bonas G, Kolocouris A, Petrakis P. 2000. A novel analytical method to detect adulteration of virgin olive oil by other oils. J Am Oil Chem Soc 77(4): 405-411. doi: 10.1007/s11746-000-0065-x

89. Baeten V, Meurens M, Morales MT, Aparicio R. 1996. Detection of virgin olive oil adulteration by fourier transform Raman spectroscopy. J Agric Food Chem 44(8): 2225-2230. doi: 10.1021/jf9600115

90. Nina Naqiyah AN, Marikkar JMN, Dzulkifly MH. 2013. Differentiation of lard, chicken fat, beef fat, and mutton fat by GCMS and EA-IRMS techniques. J Oleo Sci 63(7): 459-464. doi: 10.5650/jos.62.459

Chapter 8

VIRGIN OLIVE OILS: ENVIRONMENTAL CONDITIONS, AGRONOMICAL FACTORS AND PROCESSING TECHNOLOGY AFFECTING THE CHEMISTRY OF FLAVOR PROFILE

Nicola Caporaso[1,2]

[1]University of Naples Federico II, Department of Agriculture, Via Università 100, Portici 80055 (NA), Italy

[2]The University of Nottingham, Division of Food Sciences, Sutton Bonington, Loughborough, LE12 5RD, United Kingdom

ABSTRACT

Olive oil, and especially virgin olive oil (VOO) and extra VOO, has great interest worldwide because of its unique fatty acid profile, phenolic compounds with positive health properties for the human health and appreciated aroma profile. However, the composition of VOO can change dramatically depending on the olive variety, field management, olive fruit maturity degree, harvesting and processing, which can affect its final characteristic flavour. At industrial level, the extraction conditions applied can affect the concentration and composition of phenolic and volatile compounds, with possible negative consequence and the formation of off-flavours. Olive crushing, malaxation, centrifugation, filtration and storage were reported to be all possible factors affecting VOO flavour. The present review paper describes the factors affecting the virgin olive oils composition, with a focus on the aroma compounds, biophenols and the resulting sensory profile. A particular focus was put on the volatile and phenolic compounds and their relation with the sensory description and impact of the final flavour. In conclusion, many factors still need to be studied to fully understand the complex composition and interactions of this appreciated product. This information is also relevant and useful both for scientific knowledge and for its industrial application, as producers can modulate the final characteristics of the VOO.

INTRODUCTION

Olive oil has a limited production with respect to other vegetable oils, but it has a paramount importance in terms of economic significance in the producing countries and interest for consumers worldwide, which is driven by its health properties and peculiar sensory properties. The major olive oil production is localised in the Mediterranean area, being the European Union (EU) the first producer with almost 2.5 million tons in 2013/2014. After non-European countries, Syria, Turkey and Tunisia are the other important producing countries, with 180,000, 135,000 and 70,000 tons in 2013/2014, whereas a great year-byyear variability is observed because of the olive plant physiology. The EU accounts to 80-90% of the world olive production, and specifically Spain, Italy and Greece produce 45, 25 and 20% of the world olive oil production, respectively [1].

Olive oil is the product obtained from olive fruit and it is mainly composed by triacylglycerols, which accounts to 98% of the total composition. The remaining fraction comprises free fatty acids, phenols, tocopherols, sterols, phospholipids, waxes, squalene and other hydrocarbons. For "olive oil", many commercial categories exist: olive oil itself has been defined as a mixture of oil obtained from olives and that undergo refining process, with virgin olive oil (VOO). This latter is a product that is extracted from olives without any chemical process. Among VOOs, extra virgin olive oil has particularly high standards both in terms of composition, namely acidity, peroxide values and oxidation indices, as well as sensory characteristics assessed by recognised panels [2].

Several studies have reported on the quality of extra VOOs from the retail market, as this is relevant for the consumers in terms of the oil composition, health benefits and organoleptic properties [3]. However, it is important to understand the compositional changes due to other factors that influence VOO composition, including the agronomical practices and extraction technology. The olive variety, or "cultivar", the climatic conditions and geographical location of the olive orchard, as well as the agronomic practices and olive ripening degrees influence the final composition of the oil and therefore its flavour [4, 5]. All these factors are described in the present review paper.

Virgin olive oil (VOO) contains an unsaponifiable fraction including waxes, phospholipids, phenolics, pigments and carotenoids. The most abundant ones are represented by phenolic compounds (oleuropein aglycone, hydroxytyrosol, lignani, ecc.), tocopherols, squalene and β-sitosterol [6].

The known antioxidant activity of olive oils is mainly due to the presence of phenolic and ortho-diphenolic compounds, and secondarily to the fatty acid composition [7]. Olive fruit - contrary to other vegetable fats which usually derive from seeds - is properly a fruit. The major olive components are water (40-70%) and fat (6-25%), mainly present in the mesocarp. The fruit also contains simple sugars, organic acids, nitrogen compounds and bio phenols. Particularly important for olive composition is the phenolic fraction, mainly constituted by oleuropein, demethyl oleuropein and verbascoside. These compounds are present in concentration comprised between 0.5 and 2.5% of the fresh weight. The oil fraction is present as oil droplets in the pulp (16.5-23.5% fresh weight), while little amounts are also found in the seed (1-1.5%).

An eco-physiological role of the phenolic compounds in many fruits and vegetables seems to be related to their protection against insects or other plants "enemies". Indeed, the reaction of polyphenols after their esterification by esterase's during fruit drupe maturation exerts a plant defence mechanism against biotic attacks [8]. Once oleuropein and ligstroside are accumulated in the fruit, the enzymatic hydrolysis produces hydroxytyrosol, tyrosol, demethyloleuropein and secoiridoide glycoside, arising from the degradation of complex polyphenols.

OLIVE OIL AROMA BIOGENESIS

Virgin olive oil has unique characteristics which results from its fatty acid composition and the presence of minor compounds including volatile and phenolic compounds. Some of these are biosynthesized during the olive fruit ripening, especially during the climacteric period. Some volatile compounds are still present in the fruit belonging from the lipids or amino acids metabolism [9, 10]. However, the majority of the olive aroma derives from enzymatic oxidation of the fatty acids, linoleic and linolenic [11]. Saturated aldehydes having carbon number comprised between 7 and 12 are the major responsible of VOO aroma. This class of volatile compounds is usually present at concentrations of 50 and 75% of the total volatiles in unripe and fully mature olive drupes, respectively [12] [Table 1].

Table 1: Volatile compounds and odour descriptors attributed to VOO (from Kiritsakis [12])

Compound	Sniffing	MDS (multidimensional scaling)
Methyl acetate		Green (nuts)
Octene	Solvent-like	Green (grass)
Ethyl acetate	Sweet, aromatic	Slightly bitter/pungent
2-Butanone	Fragrant, pleasant	Tomato, apple
3-Methylbutanal	Sweet, fruity	Apple
1,3-Hexadien-5-yne		Green (green olives)
An alcohol	Sweet, apple	Other ripe fruit
Ethylfuran	Sweet	Sweet
Ethyl propanoate	Sweet, strawberry, apple	Sweet
An alcohol hydrocarbon	Pungent, acid fruit	
3-Pentanone	Sweet	Sweet

4-Methylpentan-2-one	Sweet	Green
Pent-1-en-3-one	Sweet, strawberry	Sweet
2-Methylbut-2-enal	Solvent -like	Ripe fruit (olives, dry wood)
2-Methylbut-3-enol		Slightly bitter
A hydrocarbon	Sweet, apple	Sweet
Methylbenzene	Glue, solvent-like	Overripe fruit
Butyl acetate	Green, pungent, sweet	Sweet
Hexanal	Green, apple	Sweet
A hydrocarbon	Sweet, aromatic	Sweet
2-Methylbutyl propanoate	Aromatic, ketone	Olive, apple
2-Methyl-1-propanol	Ethyl acetate-like	Green

(E)-2-Pentenal	Green, apple	Ripe fruit (soft fruit)
An alcohol	Greasy	Undesirable (rancid)
(Z)-2-Pentenal	Green, pleasant	Overripe fruit
Ethylbenzene	Strong	Bitter taste (dried green herbs)
(E)-3-Hexenal	Artichoke, green	Artichoke
(Z)-3-Hexenal	Green, green leaves	Green
1-Penten-3-ol	Wet earth	Undesirable
3-Methylbutyl acetate	Banana	Slightly fruity
Heptan-2-one	Fruity	Ripe fruit
(E)-2-Hexenal	Bitter almonds	Bitter
(Z)-2-Hexenal	Fruity, almonds	Almond odor, bitter taste
2-Methylbutan-1-ol	Fish oil	
3-Methy1-2-butenyl acetate	Putty-like unpleasant	Ripe fruit

Dodecene		Slightly bitter -taste
Pentan-1-ol	Pungent	Ripe fruit
Ethenylbenzene		Fruity
3-Methyl-butanol		Undesirable (yeast)
Hexyl acetate	Sweet, fruity	Green (grassy)
A ketone	Fruity, mushroom -like	Green
Octan-2-one	Moldy	Undesirable
3-(4-rnethyl-3-pentertyl)Furan	Paint-like strong	Overripe
3-Hexenyl acetate	Green banana, green leaves	Green
(Z)-2-Penten-l-ol	Banana	Green (grass)
6-Methyl-5-heptert-2-one	Fruity	Bitter taste (dried green herbs)
Nonan-2-one	Fruity	Apple

Hexan-1-ol	Fruity, aromatic, soft	Rough mouthfeel and rancid odor
(E)-3-Hexen-1-ol		Green leaf, nuts
2,4-Hexadienal		Ripe fruit
(E)-2-Hexen-1-ol	Green, grassy	Green (cut green grassy)
Acetic acid		Undesirable
Methyl decanoate	Fresh	Green leaf, nuts
Tridecene		Bitter (almond)
(Z)-3-Hexen-1-ol	Banana	Green banana

Aliphatic and aromatic hydrocarbons, aliphatic and triterpenic alcohols, aldehydes, ketones, ethers, esters and furan derivatives are present in the VOO aroma. The most abundant individual compounds are hexanal, trans-2-hexenal, hexanol and 3-methylbutanol [11, 12]. Among the volatile compounds, those attributed to the sensory note described as "green" are C6 aliphatic compounds, hexanal, cis-3-hexenal, trans-2-hexenal, hexanol, cis-3-hexenol, trans-2-hexen-1-ol and the corresponding esters [13].

The following volatile compounds account to circa 80% of the total headspace composition:

• C6 aldehydes (hexanal, cis-3-hexanal, trans-2-hexanal);

• Alcohols (hexanol, cis-3-hexenol, trans-2-hexenol);

• Esters (hexyl acetate, cis-3-hexenyl acetate [9, 11, 14].

The Lipoxygenase (LOX) pathway is a prominent enzymatic route for olive oil aroma generation, and involves a variety of enzymes acting on unsaturated fatty acids. LOX is activated when the olive drupe is crushed, which is the first step of olive oil extraction process. During olive crushing and the next malaxation phase, the typical olive oil aroma is produced [14]. The LOX pathway involves the activity of four enzymes, namely lipoxygenase, hydroperoxide lyase, alcohol dehydrogenase and alcohol acyltransferase [9, 13, 14].

When the integrity of olive drupe is compromised, 9- and 13-hydroperoxides of linolenic and linoleic acids are produced [4, 13].

The hydroperoxide lyase produces aldehydes, subsequently reduced in alcohols by the alcohol dehydrogenase. These compounds are then esterified by the alcohol acetyltransferase to produce esters [9, 14].

The 13-hydroperoxide undergoes an enzymatic scission by the hydroperoxide lyase, with the subsequent production of hexanal and cis-3-hexenal. The hydroperoxide lyase found in the olive pulp is highly specific for 13-hydroperoxides [9, 15]. The 13-hydroperoxide formed from linolenic acid is rapidly reduced to cis-3-hexenol, from which cis-3-hexenyl acetate is produced. Alternatively, it is isomerised to form trans-2- hexenal, which is then reduced to trans-2-hexenal by following a different biosynthetic route [4].

INFLUENCE OF BOTANICAL AND AGRONOMICAL FACTORS

Variety

Olive cultivar has been reported as one of the most crucial factors responsible for olive oil flavour variability [16]. Virgin olive oil extracted in the same conditions can show dramatic differences in their composition and sensory characteristics because they are extracted from different varieties [17] [Figure 1]. Volatile compounds having six carbon atoms, especially trans-2-hexenal, have been regarded as good indicators for differentiation of single-variety olive oils [18].

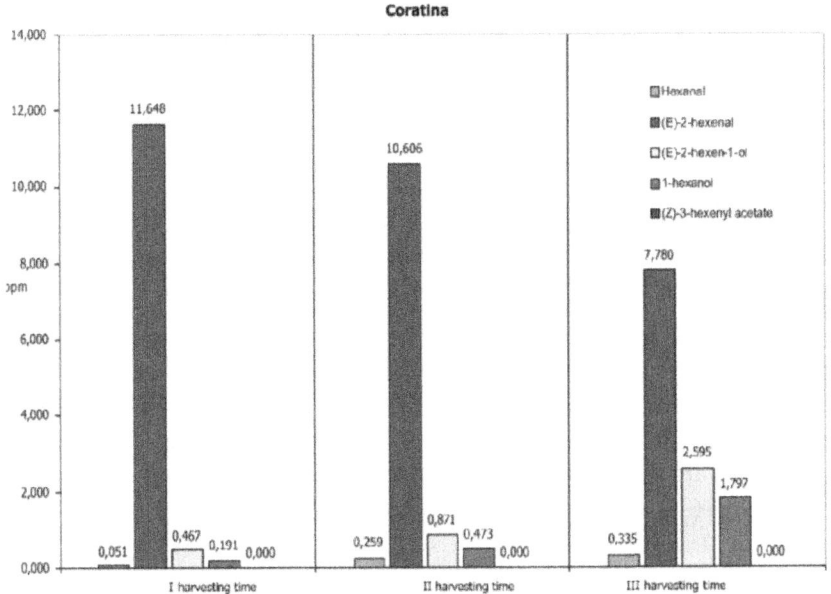

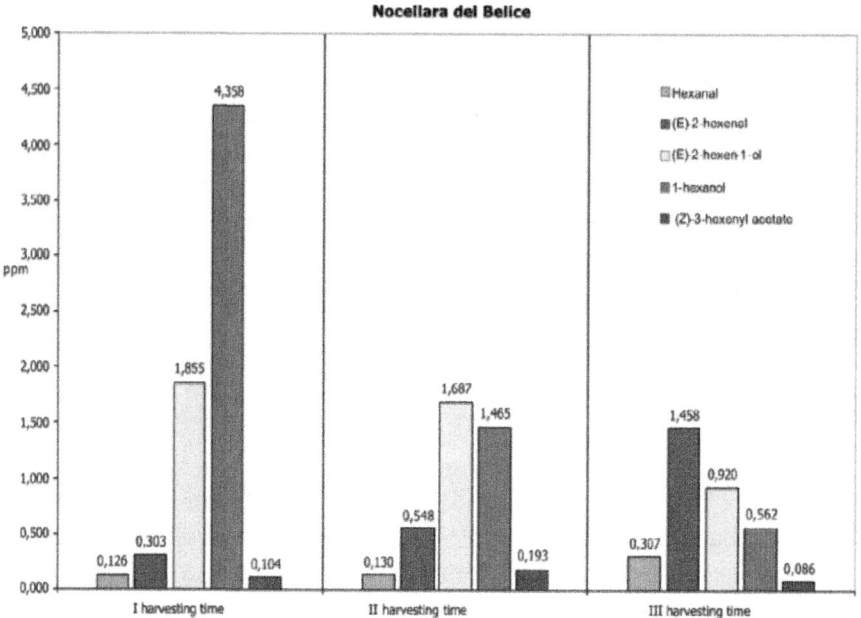

Figure 1: Variation of some volatile compounds in VOOs obtained from single olive varieties, from olives harvested at longer maturation degrees (adapted from Benincasa et al. [5]).

The olive variety also influences the concentration of phenolic compounds, whereas the major influence was attributed to the olive ripening degrees and environmental conditions [19]. Dhifi et al. [20] reported the concentration of aroma compounds in VOO extracted from four Tunisian olive cultivars, i.e. Chetoui, Chemlali, Chemchali, Oueslati. A wide range was described depending on the variety, with cv. Chemchali showing 59.7 µg/mL, while cv. Oueslati had the lowest concentration, i.e. 15.7 µg/ml.

Other researchers have compared Tunisian cultivars with Italian ones, reporting that cv. Chetoui is characterized by a low content of C6 aldehydes, responsible of the sensory notes "green" and "fruity" [21]. Ranalli et al. [22] reported concentrations of 230.4, 381.1 and 401.7 ppm total volatiles for the Italian cultivars Cipressino, Cassanese e Leccino, respectively. According to Tura et al. [16], the major volatile compounds affected by the olive cultivars are the following ones: ethanol, 2-methyl propanol, pentanol, cis-2-penten-1-ol, cis-3-hexenol and octanol. They have been associated to the following sensory notes, respectively: floral, banana, apple, walnut, hay, butter, sweet and fruity [16].

Ripening Degree

The oil accumulation in the olive drupe starts about 5 weeks after anthesis, and its major increase is between 60 and 120 days after flowering [9]. During the olive maturation phase, the fruit undergoes several modifications: exocarp color change from green to violet-black, lower fruit consistency, higher free acidity, lower pigment concentration, phenolic compounds concentration, which concentration is maximum when the darkening phase starts and a continuous decrease follows [23].

Fatty acid composition is quantitatively affected by two main factors, i.e. the olive variety used and the ripening stage at which the olives were harvested. The differences in the fatty acid profiles in different varieties were partially related to the biosynthesis of fatty acids during ripening. Specifically, stearoyl-acyl carrier protein (Stearoyl-ACP) is obtained from Acetyl-CoA by the action of an enzymatic complex called fatty acid synthase I and III (FAS I/III) followed by FAS II. Stearoyl-ACP is desaturated to oleoyl-ACP (C18:1-ACP) by the stearoyl-ACP 9-desaturase, which is highly active in the plastids. As growth proceeds, desaturase transcripts accumulate at higher levels, and the high transcription rate remains up to 28 weeks after flowering. This transcription pattern is observed during fruit development parallels with the synthesis of oleic acid in olives [24].

The fatty acid composition of VOOs is known to depend mainly on genetic factors, i.e. olive cultivar, and therefore several investigations have tried to study whether it is possible to discriminate the VOO variety based on the fatty acid profile, e.g. in Spanish cultivars Arbequina, Hojiblanca and Picual, which are among the most important ones worldwide. The authors reported a very good discrimination power by Partial Least Square, whereas they also add the phenolic composition and not just fatty acid profile [25].

The oil quality is strictly linked to the fruit physiological conditions, as the progress of the ripening phase cause higher production of free acidity, with a contemporary decrease of phenolic compounds responsible for the bitter-pungent notes [26] [Figure 2].

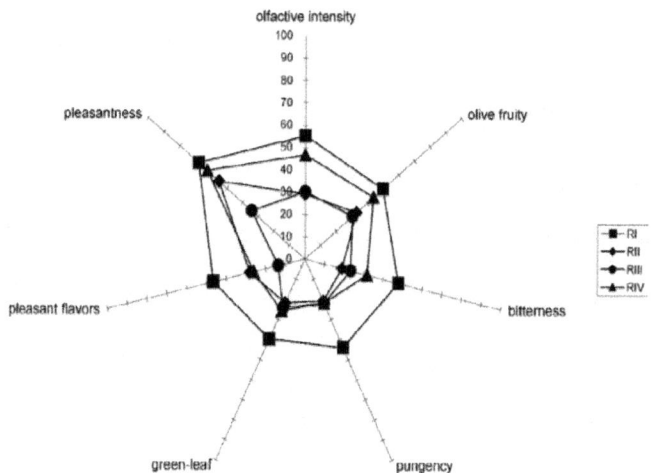

Figure 2: Sensory profile of VOOs obtained from cv. Nostrana di Brisighella at increasing ripening degrees (adapted from Rotondi et al. [30]).

To define the olive drupe maturity degree, the most common method applied rely on the observation of the fruit skin and pulp colors, according to the method reported by Uceda and Frias [27]. It describes seven maturity states of the fruit (0, bright green skin; 1, green-yellowish skin; 2, green skin with reddish spots; 3, reddish-brown skin; 4, black skin with white flesh; 5, black skin with < 50% purple flesh; 6, black skin with ≥ 50 and >100% purple flesh; and 7, black skin and purple flesh), and the agronomist or farmer randomly takes 100 fruits from the field, then segregates them into groups and calculates the number of each fruit per group color, by using a color chart.

Olive maturity degree also influences the generation of aroma compounds, being the maximum concentration observed when the fruit starts the color changing phase, and followed by a continuous decrease [12, 17, 28].

It is known that at higher ripening stages, the aromatic note describes as "fruity-grassy" is significantly lower than oils obtained from greener olives. The sensory notes of bitterpungent and other positive aromatic notes decrease with increasing ripening degree [23]. The increase in olive maturity leads to the increase in 1-penten-3-ol. The compounds hexanal, trans-3-hexenol, cis-3-hexen-1-ol and cis-2-hexenol were reported to have high selectivity for unripe olives. The intermediate maturation phase was characterised using hexyl acetate, while over-ripe drupes can be discriminated using trans-3-hexenal, cis-2-hexenal, trans-3-hexenol and cis-3- hexenol [28].

Some triterpenic alcohols were be proposed as markers of fruit maturity. Sakouhi et al. [29] demonstrated that triterpenic alcohols and 4-monomethyl sterols concentrations in the olive fruit increase from 18 to 30% during ripening of the cv. Picholine.

A negative correlation (r^2 =-0.88) between the drupe ripening degree and phenolic content in VOO was also reported by Rotondi et al. [30]. Gomez-Rico et al. [31] studied the combined effect of maturity and irrigation on the phenolic and volatile compounds of VOOs obtained from cv. Cornicabra. They reported that complex phenolics undergo the major changes, with a stronger loss depending on the irrigation. This was linked to the higher activity of L-phenylalanine ammonia-lyase in condition of plant stress.

During drupe maturation, an increase of the triterpenic acid 24-methylene-cycloartenol has been reported, with a contemporary decrease in cycloartenol. Also, sterol compounds undergo significant changes, e.g. increasing the sterol Δ5- avenasterol and decrease of β-sitosterol [32].

A recent paper reported that trans-2-hexenal concentration decreased as maturity progressed, for some Turkish varieties [33]. The authors also described that the concentration of all aroma compounds, except trans-2-hexenol and hexyl acetate, decreased with the degree of ripeness, and that altitude plays a significant but less intense effect on the volatile composition [33].

The olive fly (Bactrocera olea) is the most important biotic stress affecting the drupe quality and the resulting olive oil. VOO produced from fruits attacked by the olive fruits has higher free acidity value and peroxide number, different fatty acid composition and concentration of phenolic compounds [34]. Significant differences from the sensory panel scores of these oils were also reported [35]. The sensory profile of such VOOs had lower scores when the olives are harvested at full maturity stage, with the appearance or increase of negative sensory notes such as fusty, mouldy and winey–vinegary. Olive fly attacks cause a significant decrease of the attribute "pungent-bitter", due to the hydrolysis of complex phenolic compounds caused by the insect damages [36].

Environmental and Climatic Conditions

Olive is a xerophyte plant, and it needs a mild climate, with no sudden temperature changes below -8 °C, and with its full potential between 22 and 32 °C. At higher latitudes, an increase in oleic acid content and a higher ratio of unsaturated to saturated fatty acids has been reported. The environmental conditions, slowing down the ripening of fruit, can diminish the activity of certain hydrolytic enzymes which act on phenols [37].

Several studies have been conducted on the effect of soil salinity and irrigation water on the production and VOO fatty acid profile, but still scarce and contradictory results have been published so far in relation to the effects on their organoleptic characteristics. For example, Ahmed et al. [38] reported an increase in the concentration of total phenolic compounds, especially tyrosol, hydroxytyrosol and vanillic acid, and a reduction in the ratio unsaturated to saturated fatty acids, with increasing salinity of the water.

The influence of the organic agronomical practice on the VOO composition and flavour characteristics has been scarcely investigated. In particular, Ninfali et al. [39] reported higher concentrations of phenolic compounds from organic than non-organic VOOs in the first year of the experiment, while this difference was not significant over the second and third year. From the panel test, organic VOOs have shown a stronger note of "hay" and "artichoke". Non-organic oils presented more markedly floral notes of "fresh grass" and "fruity". The results are not consistent, however, in all the years, and it is strongly influenced by climate change over the years [39].

Water deficit represents one of the main environmental factors limiting its production potential, caused by the decrease of the fruit endogenous esterase's. Rainfall has prominent influence with respect to the environmental temperature, and some volatile compounds such as hexanal and isobutyl acetate were negatively correlated to the rainfall [18].

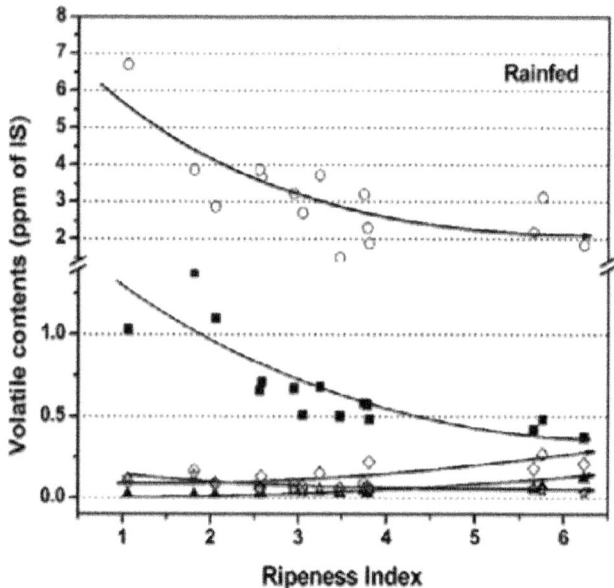

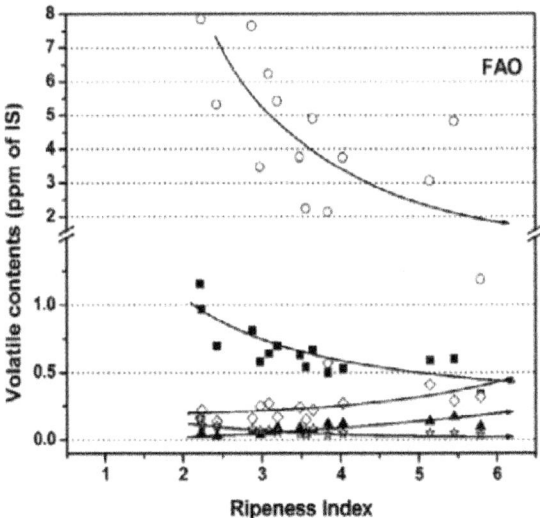

Figure 3: Evolution of main volatile compounds in VOOs under rain-fed conditions and FAO irrigation strategy during fruit ripening (crop season 2003/2004). White circle: trans-2-hexenal; square: hexanal; white rhombus: cis- 3-hexenl-1-ol; triangle: hexan-1-ol; black rhombus: trans-2-hexen-1ol (adapted from Gomez-Rico et al. [31]).

The amount of water applied to the orchard can modify the fatty acids ratio, with a higher concentration of monounsaturated and polyunsaturated fatty acids at higher water volumes applied. Salas et al. [40] reported a clear inverse correlation between the amount of irrigation water applied and the intensity of the bitter note in the VOO. Also Gomez-Rico et al. [31] confirmed previous findings piloting three water management systems: no added water, fully water addition, controlled deficit irrigation using by the method proposed by FAO [Figure 3].

Greater water availability seems to also affect VOO volatile compounds concentration, especially trans-2-hexenal, cis-3- hexen-1-ol and hexanol, in the sense of their decrease with increasing irrigation volume. Some other volatile compounds were not reported to be subject to variations, such as 1-penten-3-one and 1-penten-3-ol, as they were linked more to seasonal factors and ripening degree than the irrigation practices [31]. In subsequent studies, the same authors [41] verified that the hexanal also shows an inverse variation of its concentration with increasing water availability [Figure 4].

Opposite conclusions were reported by Stefanoudaki et al. [42], where lower values of 1-penten-3-ol and 1-penten-3-one found as a result of the irrigation, for cv. Leccino but not for cv. Cornicabra, in which there were no significant modifications. It was reported a decrease in concentration of the total volatile compounds in olive oils irrigated. The C5 compounds undergo a drastic decrease but, in contrast to other authors, no increases were reported for C6 compounds.

Cornicabra variety is of great importance in Spain because comprises about 10-12% of its production, and is one of the most important after the Spanish Picual variety. Cornicabra variety shows late maturation than other olive varieties with the pulp and the skin being greener also at high maturity stage. This suggests that the ripening has a different effect on olive oil composition and quality, compared to other olive oil varieties. This aspect is very interesting and needs further research to evaluate the specific effect of ripening on different varieties.

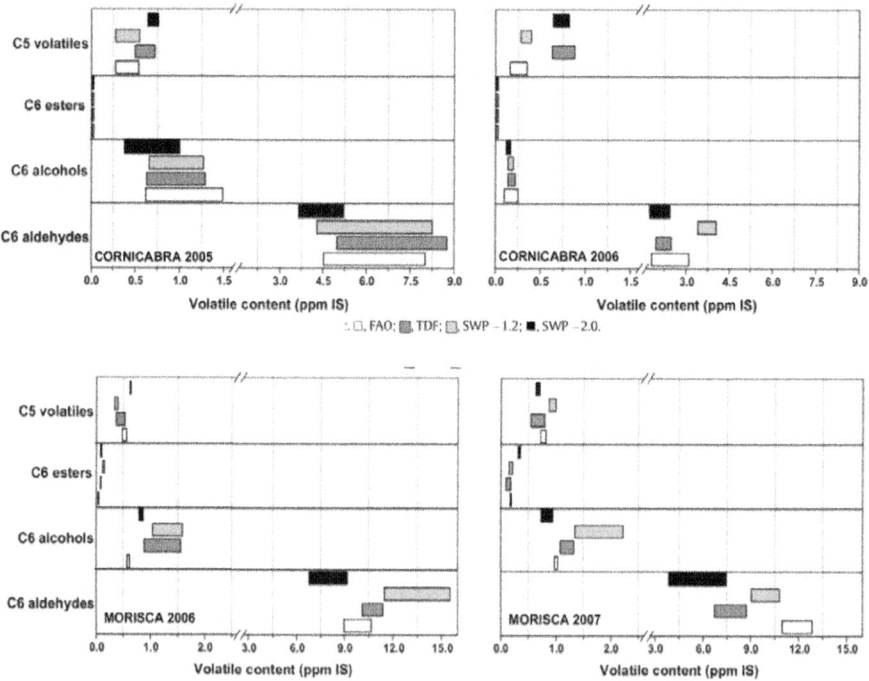

Figure 4: Concentration of volatile compounds C5 and C6 derived from the lipoxygenase pathway in VOOs from cv. Cornicabra and Morisca, depending on the water strategy applied on the orchard (ripening degree: 3-3.5) (adapted from Gomez-Rico et al. [41]).

The irrigation strategy has been reported to be more effective than the total amount of rainfall during the year, as the concentration of phenolic compounds depends on the hydric stress [31, 42]. Also fatty acids undergo modifications, with an increase in the monounsaturated, i.e. oleic acid, to polyunsaturated fatty acids [43]. The sensory evaluation of VOOs from irrigated orchards seem to show lower intensity of the attributes fruity, bitter and pungent [42].

A recent research paper reported on the influence of climate, soil composition, agricultural practices (fertilization and irrigation) and variety on the compounds responsible for the flavour of extra VOO (volatiles and phenols) and how these compounds can explain the differences in chemical profiles by geographical origin, cultivar and fruit ripeness stage [44]. The authors reported that the soil and the climate of the Chilean regions have much more influence than cultivars on the concentration of sensory quality compounds. The difference in latitude between the orchards increases the importance of the geographical origin on the virgin olive oil chemical composition, while at the same time, full irrigation decreases the influence of the olive variety, which seems to be an opposite trend with respect to the Mediterranean cultivation area [44].

GEOGRAPHIC LOCATION

Studies by Montedoro et al. [45] reported the possibility to discriminate VOOs from different Italian regions through the analysis of volatile compounds. VOO from Italy were reported as richer in C6 aldehydes and lower in fruity esters then Moroccan oils [46]. Vichi et al. [47] did not observe any difference in the concentration of C5 compounds in different varieties, thus suggested that these volatiles might be influenced mainly by the geographical location. VOO from Chetoui varieties from Tunisia had different profiles according to different cultivation sites, particularly for the fatty acid profile and phenolic compounds concentration [48].

Many authors have tried to discriminate VOO according to their geographical origin. Issaoui et al. [49] studied the effect of the difference in altitude, latitude and climatic conditions (north and south of Tunisia) in the cv. Chemlali and Chetoui.,VOOs from the northern regions of Tunisia contain mainly the following volatile compounds: cis-3- hexenyl acetate, cis-β-ocimene, hexyl acetate and cis-cis-α- farnesene.

There seems to be a general agreement about some volatile compounds which are dependent on the geographical area of origin, i.e. hexanal, hexanol, trans-2-hexenal, cis-3-hexenal, trans-3-hexenol, cis-3-hexenol and trans-2-hexenol [47].

The application of cluster analysis based on "green" volatile compounds allowed to group 39 varieties with the aim to find markers for volatile compounds that could discriminate oils coming from different areas [50]. The following compounds were described as the most important ones for Italian and Spanish VOOs: 2-methylbutyl acetate, 2-methyl-4-pentenal and hexanol. For the distinction of Italian oils by the Greek ones, pentane-3-one, cis-3-hexenal and a hydrocarbon were identified. Regression analysis allowed to classify oils derived from macro-areas of origin (Spain, Italy and Greece) by the following volatile compounds: hexyl acetate, ethyl benzene, 2-methyl-4-pentenal, ethyl furan, trans-2-pentenal, 1,2,3-trimethylbenzene, 3-methyl butanol and a hydrocarbon [50].

HARVESTING SYSTEMS

The olive harvesting system exerts a dramatic influence on the final VOO flavour. Oils derived from olives picked up from the soil are qualitatively poorer, and both the chemical parameters (acidity, peroxides, UV) and the organoleptic characteristics might be affected. With the protracted contact times of the olives with the ground, trans-2-hexenal and cis-3- hexenyl acetate ("green notes") concentration decrease, and the concentration of hydrocarbons, acetic acid and other carbonyl compounds increase, with consequent negative organoleptic notes [18].

The mechanical harvesting of the olives seems not to be of significant influence on the chemical composition of VOOs. The olive harvesting needs to be performed trying not to damage the drupe exocarp and mesocarp, to prevent spontaneous fermentation and mold growth that would result in quality deterioration and occurrence of defects oil.

Tura et al. [16] reported the importance of the degree of maturity on the content in phenolic compounds and volatile, but also fruit storage can result in a change of the flavour profile, due to the reduction of esters and aldehydes responsible for the positive notes [46].

In single-variety oils, cv. Correggiolo obtained from olives with maturation index (expressed as an index of Jaén) of less than 2.5, with the increase of time of storage of olives there is a general decrease of the content of volatile substances, in particular of those responsible for the sensations of fruity, almond, green notes (trans-2-hexenal, hexanal, cis-3- hexenol) and sweet notes (1-penten-3-one, 3-pentanone), simultaneously with an increase of compounds which are associated notes odorous (3-methyl-1-butanol, 2-methyl-1- propanol).

INFLUENCE OF THE OLIVE OIL EXTRACTION SYSTEM

Olive Crushing

The majority of VOO aroma compounds are produced during the crushing phase, when the enzymatic oxidation of linoleic and linolenic acid takes place [21]. The type of milling system thus influences the final oil quality, by changing the quali-quantitative composition of some minor constituents, with consequent influence on the organoleptic properties and product stability.

There are two major types or olive crushers: granite stone mill and metallic crusher. This letter can be divided into two types, i.e. disc and hammer crushers. The hammer crushers have stronger effects and cause a strong emulsification on the olive paste and an increase of its temperature, due to friction phenomena, with disadvantages on the yield and on the oil quality. It should be also noted that the stone mill is a discontinuous system, while the metallic crusher system is used in continuous and therefore their performance in terms of processed olive is much higher. VOO obtained using stone crusher is more aromatic and harmonic than oils obtained using the other two types of crushers. The disc crusher, however, leads to the formation of oil richer in phenolic compounds, more bitter and stable to lipid oxidation over storage [51]. Olias et al. [13], on the contrary, reported that stone mill leads to higher concentration of volatile compounds than metallic crushers, e.g. trans-2-hexenal, hexanal and cis-3-hexenol. For cv. Coratina and Oliarola, Servili et al. [52] reported that disc crusher cause higher concentrations of C6 aldehydes and some esters such as hexyl acetate, 3-hexenyl acetate and cis-4-hexenyl acetate, than hammer crusher. Therefore, the olive milling technology is the first parameter to be considered for the setting of the best extracting conditions.

Malaxation

The malaxation is a fundamental stage in VOO extraction, which involves the continuous slowly mixing of the olive paste to aggregate the oil/water emulsion. Malaxation time and temperature are parameters that are usually controlled by the industry during VOO extraction process, which can potentially affect the sensory profile and composition of the final product [46] [Figure 5]. The factor "malaxation time" was reported to be positively correlated to the total content of volatile compounds, but negatively correlated with the concentration of total phenolics [22]. Longer malaxation times promote the accumulation of alcohols and C5 other compounds, particularly hexanal. The temperature increase speeds up the activity of oxidative enzymes such as polyphenol oxidase, lipoxygenase and peroxidase [11].

(a)   Hexanal response surface

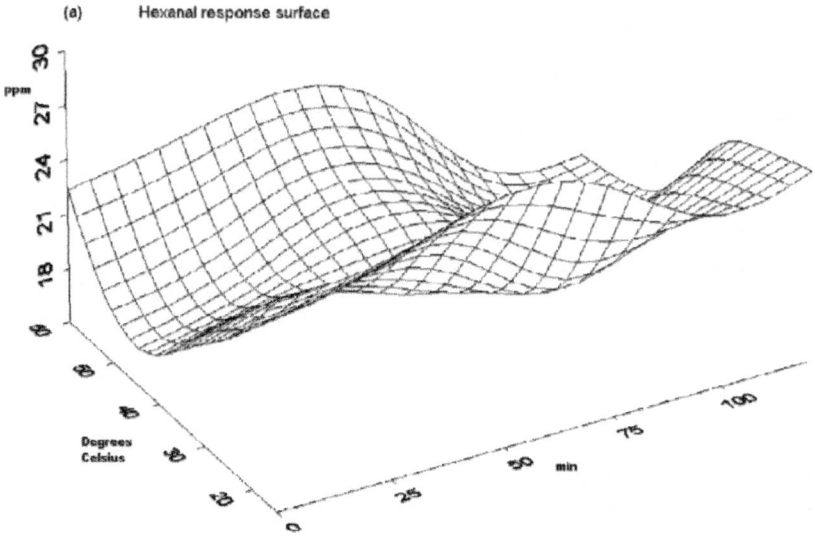

Figure 5: Malaxation time-temperature response surface for hexanal generation (adapted from Kalua et al. [10]).

However, there is a loss of volatile compounds when high malaxation temperatures are applied, attributed to the enzymes deactivation, especially hydroperoxide lyase [14]. Indeed, the increase of malaxation temperature causes a considerable decrease of C6 compounds, cis-3-hexenol and C5 metabolites, as well as an increase in hexanol and trans-2-hexen-1-ol and a loss of the bitter-pungent notes [53].

Angerosa et al. [53] studied the combined effect of malaxation temperature and time on the VOO aroma from cv. Coratina and Frantoio, and highlighting that malaxation negatively affects the total amount of secoiridoids. Some phenolic compounds, such as tyrosol, seem not undergo major changes as affected by the malaxation time, but they are only affected by the temperature adopted. The opposite trend was reported for the phenolic compound 3,4-DHPEADEDA, for which the maximum concentration was obtained at short malaxation time. The loss of phenolic compounds during malaxation is due to the enzymatic oxidation of the secoiridoid derivatives, catalysed by polyphenol oxidase and peroxidase [54]. The hydrolysis of complex phenols (oleuropein, dimethyl oleuropein, ligstroside, verbascoside, rutin and luteolin- 7-glucoside) by the endogenous glucosidase results in the production of aglycones and simple phenols, which dissolve appreciably in the oil phase, while the repartition of unhydrolysed glycosidic phenols is more toward the aqueous phase [55]. The following compounds were proposed as markers for malaxation temperature:

1-penten-3-ol, cis-3-hexenal and octane, which were proposed as markers to discriminate malaxation temperatures of 15, 30 and 35 °C, respectively [10] [Table 2]. The concentration of oxygen in the paste headspace during the malaxation process can be manipulated to achieve significant modification of the VOO aroma, in relation to the desired characteristics and industrial needs [54].

Extraction Phase

The separation of the liquid phase and solid particles from the olive paste is usually performed using two major systems, i.e. pressure or centrifugation. The subsequent step involves the separation of the oil from the oil-water mixture, which is performed by centrifugation.

Table 2: Discrimination of individual malaxation times and temperatures based on VOO volatile and phenolic compounds (from Kalua et al. [10])

Discriminating Variables	
Time	
30 min	Hexanal, 3,4-DHPEA-DEDA[a], and FFA[b]
60 min	Z-2-Penten-1-ol, Hexanal, Acetoxypinoresional, and FFA[b]
90 min	Z-2-Penten-1-ol, Acetoxypinoresinol, and yield
120 min	Z-2-Penten-1-ol, Acetoxypinoresinol, and yield
Temperature	
15 °C	1-Penten-3-ol, E-2 Hexenal, and Vanillic acid
30 °C	E-2-Hexenal
45 °C	Tyrosol and FFA[b]
60 °C	Hexanal, Octane, and 3,4-DHPEA-DEDA[a]

[a] 3,4-Dihydroyphenyl ethyl alcohol-decarboxymethyl elenolic acid dialdehyde.

[b] Free fatty acid expressed as percent of oleic acid.

When separation systems based on pressure area applied, the VOOs tend to be higher in the fruity sensory score and with a high concentration of volatile alcohol, but possible fermentation and/or degradation phenomena can take place, with consequent sensory defects [11].

Differences in phenolic content in the oils was also observed between the conventional centrifugal ("three-phase") and the "two-phase". The latter does not involve the addition of water (or very little amounts), which results in a lower dilution of hydrophilic phenolic compounds such as orthodiphenols and hydroxytyrosol.

The use of two phases centrifuge, compared to the threephase one, allows the production of VOOs with higher concentrations of trans-2-hexenal and greater total aromatic content, but with lower concentration of pigments, aliphatic and triterpene alcohols, sterols and waxes [56]. The use of the three-phases centrifuge causes decrease in the content of C6 aldehydes, hexanol and trans-2-hexenol compared with the pressure extraction, probably because of the addition of hot water [18]. Incorrect or improper management of the pressure extraction system may lead to olive paste fermentation that and the subsequent off-flavours [57].

Olive Oil Filtration and Storage

VOO filtration may have important effects on its sensory properties and shelf life. Unfiltered oils contain a certain amount of water (from 2 g/kg to 4 g/kg) dispersed in microdroplets present in dispersion as part of the colloidal system of the fruit. The filtration might influence the bitter-pungent sensory note, caused by the hydrolysis of phenolic compounds. It has been recently reported that the industrial-scale filtration of highly bitter-pungent extra VOO has influence on the release of key aroma compounds in the product after storage, and therefore the filtration process should be also regarded as one of the possible parameter influencing VOO flavour [58]. VOO quality, similarly to other vegetable oils, is affected by the storage conditions. Oxygen and light exposure, high temperatures and presence of metal ions as trace can accelerate the lipid oxidation and then shorten the shelf life of the product. The stability of VOO depends on its fatty acid composition, in particular by the ratio of the oleic to linoleic acid, and presence of minor compounds such as α-tocopherol, carotenoids, squalene and phenolics [19].

Among the saturated aldehydes, nonanal and hexanal suffer a sharp increase as oxidation rate increases. Some authors suggested to use the relationship hexanal/nonanal headspace concentration as an indicator of the oxidation status, while others reported that trans-2-heptenal is associated with the perception of rancid defect [46, 47]. Trans-2-Hexenal is the most abundant volatile compound present in VOOs, and it undergoes continuous decrease over storage. The profile of phenolic compounds in VOO can continuously change over storage, as hydrolytic processes take place on complex forms of oleuropein derivatives, as well as oxidation of ortho-diphenols [18]. It is known, in fact, that prolonged storage cause the increase in simple phenolic compounds such as tyrosol and hydroxytyrosol with a contemporary decrease in complex forms and an effect also on the sensory aspect.

The type of container where the oil is stored was also reported to influence the lipid oxidation status and the chemical composition of VOOs under normal retail storage conditions [59], and therefore this parameter should be carefully considered to avoid quality loss over storage.

Olive Oil Aromatization

A niche market for olive oil involves the flavouring by the addition of natural aromatic agents, particularly spices and herbs. This is usually done to give extra flavour and new or unexpected aromas, and it is typical for some countries of the Mediterranean area. For example, the aromatisation of olive oil using red dried chili pepper by the traditional infusion method was recently studied by Caporaso et al. [60]. The authors reported that the addition of dried chili pepper caused a significant increase in hexanal, related to oxidation processes. 2-methylbutanal, 3-methylbutanal and 6-methyl- 5-hepten-2-one were also detected in flavoured VOO and derived from chili as degradation products of the drying process. Apart of the volatile compounds, also a considerable amount of capsaicinoids were released in the flavoured oils, with consequent spicy and pungent taste [60].

Sensory Description of VOO Aroma Compounds

The most abundant volatile compounds in VOO are aldehydes and C6 alcohols, which were related to the sensory note of sweetness. Aldehydes and C5 alcohols contribute to other positive attributes of VOOs. The odour threshold of VOO volatiles is dramatically different depending on their chemical structure and physio-chemical characteristics. For example, cis-3-hexenal, which gives the typical odour of freshly cut grass and despite its low concentration, greatly contributes to the aroma as it has a very low odour threshold. On the contrary, trans-2-hexenal has much higher perception threshold and it contributes minimally to the final VOO aroma, despite of being the most abundant volatile in the oil [46].

Trans-2-Hexenal and trans-2-hexenol have been associated to VOOs obtained from olives in good conditions. The following compounds contribute mostly to the "green note": cis-3-hexenal, cis-3-hexenol (grass and banana) and cis-3-hexenyl acetate (fruity and green leaves). Important in defining the complex flavour are also trans-2-hexenal, hexanol and trans-3-hexenol. Hexyl acetate contributes to perceptions of fruity and sweet, while hexanal is responsible for the "green" and apple note. This latter compound is not only produced during the lipoxygenase pathway, but is also generated from lipid auto-oxidation, together with other oxidation products [22]. cis-3-Hexenol has been associated to the bitter note, as well as the attributes of "apple", "tomato", "vegetable bitter", "grass" and "fruity olive oil", along with some C5 alcohols (cis-2-penten-1-ol and 1-penten-3-ol) and trans-2-hexenal [61]. Other compounds such as toluene, octane, octene and 3-methyl butanol arise from different routes from the LOX pathway, and are not related to positive sensory attributes of VOO.

The "fruity" note of VOO was positively correlated with the cis-3-hexenol and negatively correlated with 3-pentanone. The "ripe fruit" sensory description is correlated with 3-pentanone, while "leaf" is correlates positively with hexyl acetate, 1-penten-3-ol, cis-2-penten-1-ol and negatively correlated with hexanol, associated with the "grass" note. The note of "almond" correlates positively with the hexyl acetate, 1-penten-3-ol and cis-2-penten-1-ol and negatively with the hexanol; cis-3-hexenol correlates positively with the scent of "tomato" and "bitter vegetable"; 3-pentanone correlates negatively with the attributes of "bitter", "spicy", "tomato" and "vegetable bitter" [11].

The attributes of bitter and spicy oil are due to the presence of phenolic compounds. The bitter taste is attributed to compounds aglycone form the dialdehydic form of the decarboxymethyl oleuropein, and other forms of the oleuropein aglycone; the "pungent" note has been attributed to the aglycone form the dialdehydic of the decarboxymethyl ligstroside [62].

The sensory defect of "winey-vinegary" is associated with fermentative processes especially from Lactobacillus, when olives are left on the ground for prolonged times. The consequent production of acetic acid is a marker of olives collected from the ground. A high concentration of acetic acid and octane was correlated to the defect of "reheating", a consequence of the activity of Enterobacteriaceae the genus Aerobacter and Escherichia during the first days the permanence of the olives to the ground. The genera Pseudomonas, Clostridium and Serratia appear if the olives are left in bags for a long time after harvesting. The activity of these microorganisms results in the presence of some volatile compounds at high concentrations. The defect of "mouldhumidity" has been correlated to the presence of 3-methyl-

1-butanol. Other compounds that are important in defining the overall aroma oil, were reported in low concentrations in the olive paste: the hexyl acetate, characterized by the smell fruity and marker of good quality oils, it does not contribute to the aroma of the paste due to its high perception threshold (~ 1:04 mg/kg) and low concentration in the paste (maximum at 1.00 mg/kg) [63].

CONCLUSIONS

Virgin olive oil has unique characteristics among all other vegetable oils, and its minor compounds are of paramount importance for their health properties and peculiar sensory notes, in terms of aroma and taste. However, a sole VOO does not exist, as the olive variety, field management, olive fruit maturity degree, harvesting and further processing affect dramatically the final VOO characteristics.

At industrial level, the extraction conditions applied can affect the concentration and composition of phenolic and volatile compounds, with possible negative consequence and the formation of offflavours. Olive crushing, malaxation, centrifugation, filtration and storage were reported to be all possible factors affecting VOO flavour. The factors affecting the presence and amount of positive aroma compounds, as well as the generation of volatile related to the off-flavour should be known to produce a VOO with improved sensory and nutritional characteristics, by the modulation of agronomic and technological factors. These factors include the harvesting, crushing, malaxation, centrifugation, filtration and storage conditions which can all dramatically influence the composition and final flavour of the product. All these major variables have been reviewed in detail in this paper to have a wide vision of the complexity of this product, with particular emphasis on the volatile compounds related to VOO aroma.

REFERENCES

1.   International Olive Council (IOC). 2015. World olive oil production statistics.

2.   EC Regulation No. 2568/91/EEC, 1991. Official Journal of the European Communities L248, pp 1-83.

3.   Caporaso N, Savarese M, Paduano A, Guidone G, De Marco E, et al. 2015. Nutritional quality assessment of extra virgin olive oil from the Italian retail market: Do natural antioxidants satisfy EFSA health claims?. J Food Compost Anal 40: 154-162.

4.   Angerosa F, Basti C, Vito R. 1999. Virgin olive oil volatile compounds from lipoxygenase pathway and characterization of some Italian cultivars.

J Agric Food Chem 47(3): 836-839. doi: 10.1021/jf980911g

5. Benincasa MMP. 2003. Plant Growth Analysis: Basic Knowledge. 2nd ed, Funep, Jaboticabal, pp 41.

6. Boskou D. 2008. Olive Oil: Minor Constituents and Health. CRC Press.

7. Visioli F, Poli A, Gall C. 2002. Antioxidant and other biological activities of phenols from olives and olive oil. Med Res Rev 22(1): 65- 75. doi: 10.1002/med.1028

8. Piperno A, Toscano M, Uccella NA. 2004. The Cannizzaro-like metabolites of secoiridoid glucosides in some olive cultivars. J Sci Food Agric 84(4): 341-349. doi: 10.1002/jsfa.1640

9. Conde C, Delrot S, Geros H. 2008. Physiological, biochemical and molecular changes occurring during olive development and ripening. J Plant Physiol 165(15): 1545-1562. doi: 10.1016/j.jplph.2008.04.018

10. Kalua CM, Bedgood DR, Bishop AG, Prenzler PD. 2006. Changes in volatile and phenolic compounds with malaxation time and temperature during virgin olive oil production J Agric Food Chem 54(20): 7641-7651. doi: 10.1021/jf061122z

11. Angerosa F. 2002. Influence of volatile compounds on virgin olive oil quality evaluated by analytical approaches and sensor panels. Eur J Lipid Sci Technol 104(9-10): 639-660. doi: 10.1002/1438-9312(200210)104:9/103.0.CO;2-U

12. Kiritsakis AK. 1998. Flavour components of olive oil - A review. J Am Oil Chem Soc 75(6): 673-681. doi: 10.1007/s11746-998-0205-6

13. Olias JM, Perez AG, Rios JJ, Sanz LC. 1993. Aroma of virgin olive oil: biogenesis of the "green" odor notes. J Agric Food Chem 41(12): 2368-2373. doi: 10.1021/jf00036a029

14. Sanchez J, Harwood JL. 2002. Biosynthesis of triacylglycerols and volatiles in olives. Eur J Lipid Sci Technol 104(9-10): 564-573. doi: 10.1002/1438-9312(200210)104:9/103.0.CO;2-5

15. Salas JJ, Sanchez J. 1999. The decrease of virgin olive oil flavor produced by high malaxation temperature is due to inactivation of hydroperoxide lyase. J Agric Food Chem 47(3): 809-812. doi: 10.1021/jf981261j

16. Tura D, Failla O, Bassi D, Pedo S, Serraiocco A. 2008. Cultivar influence on virgin olive (Olea europea L.) oil flavor based on aromatic compounds and sensorial profile. Sci Hortic 118(2): 139-148. doi: 10.1016/j.scienta.2008.05.030

17. Montedoro G, Bertuccioli M, Anichini F. 1987. Aroma analysis of virgine olive oil by head space volatiles extraction techniques, In: Flavor

of Foods and Beverages, Academic Press, New York, USA, pp 247-281. doi:10.1016/B978-0-12-169060-1.50023-0

18. Angerosa F, Servili M, Selvaggini R, Taticchi A, Esposito S, et al. 2004. Volatile compounds in virgin olive oil: occurrence and their relationship with the quality. J Chromatogr A 1054(1-2): 17-31. doi: 10.1016/j. chroma.2004.07.093

19. Skevin D, Rade D, Strucelj D, Mokrovcak Z, Nederal S, et al. 2003. The influence of variety and harvest time on the bitterness and phenolic compounds of olive oil. Eur J Lipid Sci Technol 105(9): 536-541. doi: 10.1002/ejlt.200300782

20. Dhifi W, Hamrouni I, Ayachi S, Chahed T, Saidani M, et al. 2004. Biochemical characterization of some tunisian olive oils. J Food Lipids 11(4): 287-296. doi: 10.1111/j.1745-4522.2004. 01148.x

21. Baccouri O, Bendini A, Cerretani L, Guerfel M, Baccouri B, et al. 2008. Comparative study on volatile compounds from Tunisian and Sicilian monovarietal virgin olive oils. Food Chem 111(2): 322-328. doi: 10.1016/j.foodchem.2008.03.066

22. Ranalli A, Malfatti A, Cabras P. 2001. Composition and quality of pressed virgin olive oils extracted with new enzyme processing aid. J Food Sci 66(4): 592-603. doi: 10.1111/j.1365-2621. 2001.tb04607.x

23. Salvador MD, Aranda F, Fregapane G. 2001. Influence of fruit ripening on 'Cornicabra'virgin olive oil quality A study of four successive crop seasons. Food Chemistry 73(1): 45-53. doi:10.1016/S0308-8146(00)00276-4a

24. Haralampidis K, Milioni D, Sanchez J, Baltrusch M, Heinz E, et al. 1998. Temporal and transient expression of stearoyl-ACP carrier protein desaturase gene during olive fruit development. J Exp Bot 49(327): 1661-1669. doi: 10.1093/jxb/49.327.1661s

25. Lerma-García MJ, Ramis-Ramos G, Herrero-Martínez JM, Simó-Alfonso EF. 2008. Classification of vegetable oils according to their botanical origin using sterol profiles established by direct infusion mass spectrometry. Rapid Commun Mass Spectrom 22(7): 973–978. doi: 10.1002/rcm.3459

26. García JM, Seller S, Pérez-Camino MC. 1996. Influence of fruit ripening on olive oil quality. J Agric Food Chem 44(11): 3516-3520. doi: 10.1021/ jf950585u

27. Uceda M, Frias L. 1975. Época de recolección. Evolución del contenido graso y de la composición y la calidad del aceite. In: Proceeding II Seminario Olcícola Internacional; Córdoba, Spain.

28. Aparicio R, Morales MT. 1998. Characterization of olive ripeness by green aroma compounds of virgin olive oil. J Agric Food Chem 46(3): 1116-1122. doi: 10.1021/jf970540o

29. Sakouhi F, Absalon C, Sebei K, Fouquet E, Boukhchina S, et al. 2009. Gas chromatographic–mass spectrometric characterization of triterpene alcohols and monomethyl sterols in developing Olea europaea L. fruits. Food chem 116(1): 345-350. doi: 10.1016/j.foodchem.2009.01.094

30. Rotondi A, Bendini A, Cerretani L, Mari M, Lercker G. 2004. Effect of olive ripening degree on the oxidative stability and organoleptic properties of cv. Nostrana di Brisighella extra virgin olive oil. J Agric Food chem 52(11): 3649-3654. doi: 10.1021/jf049845a

31. Gomez-Rico A, Salvador MD, Greca ML, Fregapane G. 2006. Phenolic and volatile compounds od extra virgin olive oil (Olea europaea L. cv. Cornicabra) with regard to fruit ripening and irrigation management. J Agric Food Chem 54(19): 7130-7136. doi: 10.1021/jf060798r

32. Luna G, Aparicio R. 2002. Characterization of monovarietal virgin olive oils. Eur J Lipid Sci Technol 104(9-10): 614-627. doi: 10.1002/1438-9312(200210)104:9

33. Toker C, Aksoy U, Ertaş H. 2015. The effect of fruit ripening, altitude and harvest year on volatile compounds of virgin olive oil obtained from the Ayvalık variety. Flavour Fragr J. doi: 10.1002/ffj.3300

34. Gomez-Caravaca AMC, Cerretani L, Bendini A, Carretero AS, Gutiérrez AR, et al. 2008. Effects of fly attack (Bactrocera oleae) on the phenolic profile and selected chemical parameters of olive oil. J Agr Food Chem 56(12): 4577-4583. doi: 10.1021/jf800118t

35. Angerosa F, Di Giacinto L, Solinas M. 1992. Influence of Dacus Oleae infestation on flavor of oils, extracted from attacked olive fruits, by HPLC and HRGC analyses of volatile compounds. Grasas y Aceites 43(3): 134-142. doi: 10.3989/gya. 1992.v43. i3.1165

36. Tamendjari A, Angerosa F, Mettouchi S, Bellal MM. 2009. The effect of fly attack (Bactrocera oleae) on the quality and phenolic content of Chemlal olive oil. Grasas y aceites 60(5): 509-515. doi:10.3989/gya.032209

37. Patumi M, d'Andria R, Fontanazza G, Morelli G, Giorgio P, et al. 1999. Yield and oil quality of intensively trained trees of three cultivars of olive (Olea europaea L.) under different irrigation regimes. Journal of Horticultural Science and Biotechnology 74(6): 729-737.

38. Ahmed CB, Rouina BB, Sensoy S, Boukhriss M. 2009. Saline water effects on fruit development, quality and phenolic composition of virgin

olive oils, cv. Chemlali. J Agric Food Chem 57(7): 2803-2811. doi: 10.1021/jf8034379

39. Ninfali P, Bacchiocca M, Biagiotti E, Esposto S, Servili M, et al. 2008. A 3-year study on quality, nutritional and organoleptic evaluation of organic and conventional extra-virgin olive oils. J Am Oil Chem Soc 85(2): 151-158. doi: 10.1007/s11746-007-1171-0

40. Salas JJ, Sanchez J. 1997. Biogenesis of alcohols present in the aroma of virgin olive oil. In: Williams JP, Khan MD, Lem NW (eds) Physiology, Biochemistry and Molecular Biology of Plant Lipids. Springer, The Netherlands, pp 328-330. doi: 10.1007/978-94-017-2662-7_104

41. Gomez-Rico A, Salvador MD, Fregapane G. 2009. Virgin olive oil and olive fruit minor consistuents as affected by irrigation management based on SWP and TDF as compared to ETc in medium-density young olive orchards (Olea europaea L. cv. Cornicabra and Morisca). Food Res Int 42(8): 1067-1076. doi: 10.1016/j.foodres.2009.05.003

42. Stefanoudaki E, Williams M, Chartzoulakis K, Harwood J. 2009. Effect of irrigation on quality attributes of olive oil. J Agric Food Chem 57(15): 7048-7055. doi: 10.1021/jf900862w

43. Tognetti R, D'Andria R, Sacchi R, Lavini A, Morelli G, et al. 2007. Deficit irrigation affects seasonal changes in leaf physiology and oil quality of Olea europea (cultivars Frantobio and Leccino). Annals of Applied Biology 150(2): 169-186. doi: 10.1111/j.1744-7348.2007.00117.x

44. Romero N, Saavedra J, Tapia F, Sepúlveda B, Aparicio R. 2016. Influence of agroclimatic parameters on phenolic and volatile compounds of Chilean virgin olive oils and characterization based on geographical origin, cultivar and ripening stage. J Sci Food Agric 96(2): 583-592. doi: 10.1002/jsfa.7127

45. Montedoro G, Servili M, Baldiol M, Miniati E. 1992. Simple and hydrolyzable phenolic compounds in virgin olive oil. 1. Their extraction, separation and quantitative and semiquantitative evaluation by HPLC. J Agric Food Chem 40(9): 1571-1576. doi: 10.1021/jf00021a019

46. Kalua CM, Allen MS, Bedgood DR Jr., Bishop AG, Prenzler PD, et al. 2007. Olive oil volatile compounds, flavour development and quality: A critical review. Food Chem 100(1): 273-286. doi: 10.1016/j.foodchem.2005.09.059

47. Vichi S, Pizzale L, Conte LS, Buxaderas S, Lopez-Tamames E. 2003. Solid-Phase microextraction in the analysis of virgin olive oil volatile fraction: modifications induced by oxidation and suitable markers of oxidative status. J Agric Food Chem 51(22): 6564-6571. doi: 10.1021/

jf030268k

48. Temime SB, Wael T, Bechir B, Leila A, Douja A, et al. 2006. Changes in olive oil quality of chetoui variety according to origin of plantation. Journal of Food Lipids 13: 88-99. doi: 10.1111/j.1745-4522.2006. 00036.x

49. Issaoui M, Jlamini G, Brahmi F, Dabbou S, Hassine KB, et al. 2009. Effect of the growing area conditions on differentiation between Chemlali and Chetoui olive oils. Food Chem 119(1): 220-225. doi:10.1016/j. foodchem.2009.06.012

50. Luna G, Morales MT, Aparicio R. 2006. Characterization of 39 varietal virgin olive oils by their volatile compositions. Food Chem 98(2): 243-252. doi:10.1016/j.foodchem.2005.05.069

51. Angerosa F, Di Giacinto L. 1995. Caratteristiche di qualita dellolio vergine in relazione ai metodi di frangitura. Nota II (Features of virgin oil quality in relation to the pressing methods. The Note). Riv Ital Sostanze Grasse 72: 1-4.

52. Servili M, Montedoro G. 2002. Contribution of phenolic compounds to virgin olive oil quality. Eur J Lipid Sci Technol 104(9-10): 602- 613. doi: 10.1002/1438-9312(200210)104:9/103.0.CO;2-X

53. Angerosa F, Mostallino R, Basti C, Vito R. 2001. Influence of malaxation temperature and time on the quality of virgin olive oils. Food Chem 72(1): 19-28. doi: 10.1016/S0308-8146(00)00194-1

54. Servili M, Taticchi A, Esposto S, Urbani S, Selvaggini R, et al. 2008. Influence of the decrease in oxygen during malaxation of olive paste on the composition of volatiles and phenolic compounds in virgin olive oil. J Agric Food Chem 56(21): 10048-10055. doi: 10.1021/jf800694h

55. Ranalli A, Contento S, Lucera L, Pavone G, Di Giacomo G, et al. 2004. Characterization of carrot root oil arising from supercritical fluid carbon dioxide extraction. J Agric Food Chem 52(15): 4795-4801. doi: 10.1021/jf049713h

56. Aparicio R, Romero M, Khouri N, Rojas LB, Usubillaga A. 2002. Volatile constituents from the leaves of three Coespeletia species from the Venezuelan Andes. Journal of Essential Oil Research 14(1): 37-39. doi: 10.1080/10412905.2002.9699755

57. Di Giovacchino L, Solinas Mz, Miccoli M. 1994. Aspetti qualitativi e quantitativi delle produzioni olearie ottenute dalla lavorazione delle olive con i differenti sistemi di estrazione. Riv Ital Sostanze Grasse 12: 587-594.

58. Sacchi R, Caporaso N, Paduano A, Genovese A. 2015. Industrial-scale filtration affects volatile compounds in extra virgin olive oil cv. Ravece. Eur J Lipid Sci Technol 117(12): 2007-2014. doi: 10.1002/ejlt.201400456

59. Savarese M, Caporaso N, De Marco E, Sacchi R. 2013. Extra virgin olive oil overall quality assessment during prolonged storage in PET containers. Proceedings in GV –Global Virtual Conference 1(1): 674679.

60. Caporaso N, Paduano A, Nicoletti G, Sacchi R. 2013. Capsaicinoids, antioxidant activity, and volatile compounds in olive oil flavored with dried chili pepper (Capsicum annuum). Eur J Lipid Sci Technol 115(12): 1434-1442. doi: 10.1002/ejlt.201300158

61. Caporale G, Policastro, S, Monteleone E. 2004. Bitterness enhancement induced by cut grass odorant (cis-3-hexen-1-ol) in a model olive oil. Food Quality and Preference 15(3): 219-227. doi:10.1016/S0950-3293(03)00061-2

62. Servili M, Esposto S, Fabiani R, Urbani S, Taticchi A, et al. 2009. Phenolic compounds in olive oil: antioxidant, health and organoleptic activities according to their chemical structure. Inflammopharmacology 17(2): 76-84. doi: 10.1007/s10787-008-8014-y

63. Garcia-Gonzales DL, Tena N, Aparicio R. 2007. Characterization of olive paste volatiles to predict the sensory quality of virgin olive oil. Eur J Lipid Sci Technol 109(7): 663-672. doi: 10.1002/ejlt.200700056

CITATION

CHAPTER 1

Suzymeire Baroni, Izabel Aparecida Soares, Rodrigo Patera Barcelos, Alexandre Carvalho de Moura, Fabiana Gisele da Silva Pinto and Carmem Lucia de Mello Sartori Cardoso da Rocha (2013). Microbiological Contamination of Homemade Food, Food Industry, Dr. Innocenzo Muzzalupo (Ed.), ISBN: 978-953-51-0911-2, InTech, DOI: 10.5772/53170.

CHAPTER 2

Ana Rodríguez-Bernaldo de Quirós, Noelia Viqueira Varela and Raquel Sendón, Study of the Migration of Three Model Substances from Low Density Polyethylene into Food Simulant and Fruit Juices, doi:10.3390/beverages1030159

CHAPTER 3

Makun Hussaini Anthony, Dutton Michael Francis, Njobeh Patrick Berka, Gbodi Timothy Ayinla and Ogbadu Godwin Haruna (2012). Aflatoxin Contamination in Foods and Feeds: A Special Focus on Africa, Trends in Vital Food and Control Engineering, Prof. Ayman Amer Eissa (Ed.), ISBN: 978-953-51-0449-0, InTech, DOI: 10.5772/24919.

CHAPTER 4

Makoto Kanauchi, Sakiko Hatanaka and Makoto Shimoyamada (2015). New Cheese-Like Food Production from Soy Milk — Utility of Soy Milk Curdling Yeast, Food Production and Industry, Prof. Ayman Amer Eissa (Ed.), ISBN: 978-953-51-2191-6, InTech, DOI: 10.5772/60848.

CHAPTER 5

Seyed Fazel Nabavi, Arianna Di Lorenzo, Morteza Izadi, Eduardo Sobarzo-Sánchez, Maria Daglia,and Seyed Mohammad Nabavi, Antibacterial Effects of Cinnamon: From Farm to Food, Cosmetic and Pharmaceutical Industries, doi:10.3390/nu7095359

CHAPTER 6

Tathiana Souza Martins Meyer, Ângelo Samir Melim Miguel, Daniel Ernesto Rodríguez Fernández and Gisela Maria Dellamora Ortiz (2015). Biotechnological Production of Oligosaccharides — Applications in the Food Industry, Food Production and Industry, Prof. Ayman Amer Eissa (Ed.), ISBN: 978-953-51-2191-6, InTech, DOI: 10.5772/60934.

CHAPTER 7

Marikkar JNM, Mirghani MES, Jaswir I. 2016. Application of Chromatographic and Infra-Red Spectroscopic Techniques for Detection of Adulteration in Food Lipids: A Review. J Food Chem Nanotechnol 2(1): 32-41.

CHAPTER 8

Caporaso N. 2016. Virgin Olive Oils: Environmental Conditions, Agronomical Factors and Processing Technology Affecting the Chemistry of Flavor Profile. J Food Chem Nanotechnol 2(1): 21-31.

INDEX